D1645748

# Disc
# M

Nu     ers        eyon

he date last stamped
d.
220 or via the

*0096388*

51 BAR
96388

**International Mathematics Series**

*Consulting Editor: A Jeffrey, University of Newcastle upon Tyne*

Forthcoming titles in the Series

*Introduction to Numerical Analysis*
A Wood

*Complex Variables and their Applications*
A Osbourne

# Discrete Mathematics
## Numbers and Beyond

**STEPHEN BARNETT**

*Department of Applied Mathematical Studies*
*University of Leeds*

**ADDISON WESLEY LONGMAN**

HARLOW, ENGLAND ■ READING, MASSACHUSETTS ■ NEW YORK

MENLO PARK, CALIFORNIA ■ DON MILLS, ONTARIO ■ AMSTERDAM

BONN ■ SYDNEY ■ SINGAPORE ■ TOKYO ■ MADRID ■ SAN JUAN

MILAN ■ MEXICO CITY ■ SEOUL ■ TAIPEI

© Addison Wesley Longman Limited 1998

Addison Wesley Longman Limited
Edinburgh Gate
Harlow
Essex CM20 2JE
England

and Associated Companies throughout the world.

The right of Stephen Barnett to be identified as author of this Work has been asserted by him in accordance with the Copyright, Designs and Patents Act 1988.

All rights reserved. No part of this publication may be reproduced, stored in a retrieval system, or transmitted in any form or by any means, electronic, mechanical, photocopying, recording, or otherwise, without either the prior written permission of the publisher or a license permitting restricted copying in the United Kingdom issued by the Copyright Licensing Agency Ltd, 90 Tottenham Court Road, London W1P 9HE.

Many designations used by manufacturers and sellers to distinguish their products are claimed as trademarks. Addison Wesley Longman has made every attempt to supply trademark information about manufacturers and their products mentioned in this work. Readers are advised that the following company and product names may be trademarks of their respective owners: Abbot Ale; Federal Express; Kellogg Company (Corn Flakes, Rice Crispies, Special K); Maille Provencal mustard; Sacla pesto sauce; Safeway; Tesco; United Parcel Services; Volvic bottled water.

We are grateful to Kellogg Company for permission to reproduce the barcode on page 204.

Typeset by 32
Produced by Addison Wesley Longman Singapore (Pte) Ltd., Printed in Singapore

First printed 1998

ISBN 0-201-34292-8

**British Library Cataloguing-in-Publication Data**
A catalogue record for this book is available from the British Library

# Contents

# Preface

Just as steam was the motive force in the nineteenth century, so the engine which now powers the developed world is the digital computer. This is nothing more than a machine which processes numbers, specifically the binary digits, or 'bits', 0 and 1. The only 'words' a computer understands, even when voice input is used, are strings of zeros and ones. In the early days of computing, half a century ago, so-called analogue computers were quite popular. They used some variable physical quantity, such as voltage, to represent numbers and were thought by some to have a great potential. However, the invention of microchip electronics killed such ideas. Indeed, other analogue systems used in radio, television and telecommunications, either have been, or probably soon will be, superseded by digital versions. Think of digital audio (compact discs), digital television and VCRs, digital telephones, digital barcodes for product identification, personal identity numbers (PINs) – the list goes on and on. For the foreseeable future, numbers seem set to rule many aspects of our lives. In his book *Being Digital* (Hodder and Stoughton, 1995) Nicholas Negroponte argues that 'The change from atoms to bits is irrevocable and unstoppable'. By 'atoms' he means the goods and objects which have in the past been the constituents of world trade; whereas the future will be dominated by electronic transfers of information, expressed in terms of bits.

Broadly speaking, the principles and applications of manipulating and processing numbers constitute the area of **discrete mathematics**. As there are many textbooks already available in the field, it is important to state why I feel this book has something distinctive to offer.

■ The coverage is based on numbers, counting and numerical processes – abstract topics such as Boolean algebra and symbolic logic are not included, these being mainly of concern to computer science specialists.

■ In order to arouse interest and set the scene, an introductory chapter explores some fascinating elementary results which are seldom contained in a book of this type.

■ The book is aimed mainly at students taking a first-year, or possibly a second-year, module in discrete mathematics, as part of an undergraduate degree programme in areas such as finance, economics, management, business, life sciences or engineering. I have taken into account the likelihood that many contemporary students may have lower levels of formal mathematical skills than used to be the case, for whatever reasons.

■ The style of the book is relatively informal, with the emphasis on trying to make topics, and their applications, interesting and understandable. Many existing textbooks, although admirable, tend to prefer a rigorous presentation which does not go down well with many of today's students. There is no reason why being able to cope with the rather forbidding traditional approach which so many authors adopt should be a prerequisite for learning mathematics.

■ Large numbers of exercises are provided throughout. Each chapter ends with somewhat harder problems, together with student projects which are more challenging and may require a team approach for their solution. Nothing more than a pocket calculator with mathematical functions is needed, but readers are encouraged to refer to the further reading which is listed for each chapter.

To summarize: my objective has been to make the material accessible to a wide audience whose mathematical background is limited. In the words of a reviewer of my initial proposal, I hope to have 'found a niche between the superficiality of school/college texts and the rigour of an introduction for students who are primarily mathematicians'. Discrete mathematics can be readily understood by many students who may well feel that traditional calculus courses are difficult, and not very appropriate to their needs. The book reveals that discrete mathematics has a wide and growing range of applications in the contemporary world, and therein lies one of its main appeals. There is also something intrinsically intriguing about the properties of numbers – if computers could express opinions, they would surely agree!

I wish to thank Robert Aykroyd, Mike Gover and Alan Slomson for their helpful comments on first drafts of chapters. I am also very grateful to Valerie Hunter who made a superb job of producing the typescript, as she has done for several of my previous books.

**Stephen Barnett**
*Leeds, January 1998*

# 1 Introduction: the Fascination of Numbers

Mathematics began many thousands of years ago with numbers and counting. Addition, subtraction, multiplication and division were all developed for practical reasons, such as keeping track of the number of sheep in a flock, measuring the area of fields, following the rhythm of the seasons. Even today, everyone begins mathematics by learning to count and to manipulate numbers. The word 'arithmetic' comes from the Greek words 'arithmein', meaning 'to count', and 'arithmos', meaning 'number'; and the word 'digit', meaning any single numerical character, comes from the Latin 'digitus', meaning 'finger', because people first counted on their fingers.

Mathematics remains the most useful of subjects, indispensable for almost every aspect of contemporary life. Indeed, everyday things now taken for granted – such as portable phones, CD players, product barcodes, cashcard machines, personal computers – all rely on digital technology, so 'whole numbers' are more important than ever.

However, despite this practicality, numbers are artificially created abstractions – have you ever seen a number floating around in the air, or ever put a number in your pocket? It therefore remains a source of wonderment that numbers seem to have a sort of 'inner life' of their own, with endlessly intriguing and unexpected features. The objective of this chapter is to introduce you to a few items from the fascinating world of numbers. If these capture your imagination, then you will be well prepared to tackle the remainder of the book with enthusiasm.

## 1.1 Some peculiarities

### 1.1.1 The number 6174

The number 6174 doesn't appear to be anything special, but let's play around with it in the following way. Write its digits in *decreasing* order, that is from largest to smallest, giving 7641. Now write the digits in *increasing* order from smallest to largest, giving 1467. Finally, subtract the smaller number from the larger one:

$$7641 - 1467 = 6174$$

which is the original number back again. 'So what?' you may well be thinking, 'there are probably lots of other numbers like this'. Well, let's try another one, say the year of my birth 1938. Apply the same operations to it, as follows:

9831  digits in decreasing order
1389  digits in increasing order
$9831 - 1389 = 8442$  difference between them

Repeat the procedure with this new number, whose digits are fortunately already in decreasing order:

$$8442 - 2448 = 5994$$

Continue in the same way with 5994:

$$9954 - 4599 = 5355$$

Keep on doing this:

$5553 - 3555 = 1998$
$9981 - 1899 = 8082$
$8820 - 0288 = 8532$  (*)
$8532 - 2358 = 6174$

Bingo! But can this be a coincidence? Try it yourself using the year of your own birth, or any other four-digit numbers you fancy, and sure enough you will end up with 6174. Notice that if at some stage you get a number with zero as one of the digits, then the smaller number starts with zero, as illustrated in the line marked with an asterisk in the example above.

In fact it can be proved that whatever four-digit number you start off with you will *always* end up with 6174. The only proviso is that the original number mustn't have all its digits the same; this is obvious, for example if you started with 3333 then the first step would give $3333 - 3333 = 0000$, which is not interesting! Moreover, the procedure never needs more than a total of seven subtractions (as in the example of my birth year) to produce 6174.

Of course, trying lots of examples which all end up with 6174 only provides verification, not a proof, of the stated facts. A proof occupies several pages, and you are invited to tackle it as a project (Problem 1.22). This unexpected property of four-digit numbers in general and 6174 in particular is certainly intriguing. An underlying reason for what happens is that we looked at numbers in **decimal** form (that is, using 'base' 10). Number bases will be studied in Chapter 2.

**Exercise**    **1.1**  Apply the same procedure to three-digit numbers to see if you can find a number which behaves in a corresponding way to 6174.

**Exercise**    **1.2**  Apply the following procedure to a four-digit number, again with not all its digits the same. Write the digits in *decreasing* order to obtain *abcd*. Reverse the first pair of digits, and the second pair of digits, to get *badc*. Next, compute the difference *badc − cdab*, where the second number is obtained from the first by reversing the order of the digits. For example, 1938 gives *abcd* = 9831 and the required difference is

$$8913 - 3198 = 5715$$
$$\phantom{8913}\ badc \quad\ \ cdab$$

Repeat this procedure, so the next two steps are

$$7551 \longrightarrow 5715 - 5175 = 540$$
$$5400 \longrightarrow 4500 - 0054 = 4446$$

Continue this process. What do you end up with? Try some more examples.

### 1.1.2    The number 142857

With your calculator, check the following table of multiplications of 142857 by the numbers 2, 3, 4, 5, 6:

$$142857 \times 2 = 285714$$
$$\times 3 = 428571$$
$$\times 4 = 571428$$
$$\times 5 = 714285$$
$$\times 6 = 857142$$

In each case the product on the right hand side consists of the same six digits as in the original number. What's more, these digits are in the same **cyclic order**. To explain what this means, suppose that the digits of the original number are placed around a circle in clockwise order, as shown in Figure 1.1.

Then each of the products listed is obtained by starting at one of the digits and going round the circle in the clockwise direction (indicated by the arrows in the figure). For example, the product 142857 × 4 begins with the digit 5, and going round the circle (or 'cycle') produces 571428.

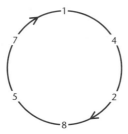

*Figure 1.1*  Cyclic arrangement

This interesting property of 142857 is not the whole story by any means. Here are some further curious facts.

(a) Write down again the above multiplication table, this time starting with $142857 \times 1$:

|  |  |  |  |  |  | Row sum |
|---|---|---|---|---|---|---|
| 1 | 4 | 2 | 8 | 5 | 7 | 27 |
| 2 | 8 | 5 | 7 | 1 | 4 | 27 |
| 4 | 2 | 8 | 5 | 7 | 1 | 27 |
| 5 | 7 | 1 | 4 | 2 | 8 | 27 |
| 7 | 1 | 4 | 2 | 8 | 5 | 27 |
| 8 | 5 | 7 | 1 | 4 | 2 | 27 |

Column sum    27    27    27    27    27    27

The sum of the numbers in each row (that is, added horizontally) is 27, and the sum of the numbers in each column (that is, added vertically) is also 27.

(b) Construct the following table for the multiplication of 7 by increasing powers of 2, and add up the products as shown:

$$
\begin{array}{llll}
7 \times 2 & = 7 \times 2 & = 1\ 4 \\
7 \times 2^2 & = 7 \times 4 & = \ .\ .\ 2\ 8 \\
7 \times 2^3 & = 7 \times 8 & = \ .\ .\ .\ .\ 5\ 6 \\
7 \times 2^4 & = 7 \times 16 & = \ .\ .\ .\ .\ .\ 1\ 1\ 2 \\
7 \times 2^5 & = 7 \times 32 & = \ .\ .\ .\ .\ .\ .\ .\ 2\ 2\ 4 \\
7 \times 2^6 & = 7 \times 64 & = \ .\ .\ .\ .\ .\ .\ .\ .\ .\ 4\ 4\ 8 \\
7 \times 2^7 & = 7 \times 128 & = \ .\ .\ .\ .\ .\ .\ .\ .\ .\ .\ .\ 8\ 9\ 6 \\
7 \times 2^8 & = 7 \times 256 & = \ .\ .\ .\ .\ .\ .\ .\ .\ .\ .\ .\ .\ 1\ 7\ 9\ \quad 2 \\
7 \times 2^9 & = 7 \times 512 & = \ .\ .\ .\ .\ .\ .\ .\ .\ .\ .\ .\ .\ .\ .\ .\ 3\ \quad 5\ 8\ 4 \\
\end{array}
$$

$$1\ 4\ 2\ 8\ 5\ 7\ 1\ 4\ 2\ 8\ 5\ 7\ 1\ 4\ 2\ (\ 7\ 8\ 4\ )$$

The sequence 142857 has come up again, with repetitions, apart from the last three digits in brackets. However, you can continue the calculation with $2^{10}$, $2^{11}$ and so on, and as the sum builds up more repetitions of 142857 will appear.

(c) Continue the multiplication table of 142857 with the factors 8 to 13 (the multiplication by 7 will be deferred for a little while). You can work out:

$$
\begin{array}{rl}
142857 \times\ 8 & = 1142856 \\
\times\ 9 & = 1285713 \\
\times 10 & = 1428570 \\
\times 11 & = 1571427 \\
\times 12 & = 1714284 \\
\times 13 & = 1857141 \\
\end{array}
$$

Notice that each of these products is a seven-digit number, and that in each case five of the six original digits appear. The remaining two digits in each product, when added together, produce the 'missing' sixth original digit. For example.

$$142857 \times 12 = 17 \quad 1 \quad 4 \quad 2 \quad 8 \quad 4$$
$$\qquad\qquad\quad \uparrow \quad \uparrow \quad \uparrow \quad \uparrow \quad \uparrow$$

The five digits belonging to the original number are indicated by arrows; the two digits left give $1 + 4 = 5$, which completes the original set of six. You should verify that this works for the other products (the product by 10 is an exception to this rule).

(d) If 142857 is multiplied by a very large number then another strange thing happens. For example:

$$142857 \times 52284662476 = 7469{,}230027{,}333932$$

Break the product up into groups of six digits, starting from the right, as shown by the commas, and add up these numbers:

```
333932
230027
  7469
------
571428
```

Once again the original six digits appear, as if by magic! If the sum has more than six digits the 'breaking up and addition' procedure has to be done again, but the six original digits will be recovered in the correct cyclic order. For example:

$$142857 \times 45213646 = 6{,}459085{,}826622$$

$$826622 + 459085 + 6 = 1{,}285713$$

$$285713 + 1 = 285714$$

(e) It's now time to look at the neglected product

$$142857 \times 7 = 999999$$

which can be rearranged as

$$\frac{999999}{7} = 142857$$

The number 999999 is close enough to 1000000 to make you think of investigating what happens when 1000000 is divided by 7. Your calculator isn't much help here – even with a ten-digit display you'll only get 142857.1429 which is not an exact result; don't be misled by the fact that if you multiply your result by 7 you'll get back to 1000000. You must realize that a calculator doesn't do exact arithmetic. It can only work to a *finite number* (possibly very large in the case of a computer) of decimal places. Here's a situation where 'old-fashioned' division by hand is superior, and provides the full picture. Look at the division of 1000000 by 7 in the following 'long-division' layout:

$$
\begin{array}{r}
142857 \\
7 \overline{)1000000} \\
\underline{7} \\
30 \\
\underline{28} \\
20 \\
\underline{14} \\
60 \\
\underline{56} \\
40 \\
\underline{35} \\
50 \\
\underline{49} \\
1
\end{array}
$$

There is a remainder of 1. The division can therefore be continued after the decimal point, giving exactly the same six digits in the quotient, with a remainder of 1 yet again, so the process continues *without end*. In other words, you get

$$
\frac{1000000}{7} = 142857.142857142857142857\ldots \qquad (+)
$$

where the dots indicate that our old friends the six digits 142857 repeat for ever. Notice that the calculator produced its result by rounding off to four decimal places, so the interesting structure of the repeating digits was completely lost. You may prefer to change the expression for 1000000/7 into the neater fraction 1/7. To do this, simply divide both sides of the expression (+) by 1000000. This has the effect of moving the decimal point on the right-hand side of the expression (+) six places to the left, so you'll end up with

$$
\frac{1}{7} = 0.142857142857142857\ldots
$$

The decimal expansion of 1/7 is an example of a **recurring decimal**, so-called because the set of six digits is repeated endlessly, in the same order. You've probably come across simple recurring decimals like

$$
\frac{1}{3} = 0.33333\ldots
$$

where just a single digit is repeated.

You might think that recurring decimals are an exceptional kind of number, but consider another example, say 0.151515... Call this recurring decimal

$$x = 0.15151515...$$

Multiplying both sides by 100 gives

$$100x = 15.151515...$$

which has the same recurring decimal part. Subtracting one expression from the other produces

$$100x - x = 15.151515... - 0.151515...$$

so the recurring parts nicely cancel out, leaving

$$99x = 15$$

Finally,

$$x = \frac{15}{99} = \frac{5}{33}$$

which is the representation of $x$ as a 'fraction', or to use the mathematical term, as a **rational number**, which is the **ratio** of one integer divided by another.

There will be more to say about rational numbers later in the chapter (Section 1.4), but you should be aware that the method used above for converting a recurring decimal into a rational number does have pitfalls. For example, if

$$x = 0.999...$$

then

$$10x - x = 9.999... - 0.999...$$

so

$$9x = 9$$

that is, $x = 1$, so apparently the recurring decimal 0.999... is 'equal' to the integer 1. Can you explain this?

**Exercise**

**1.3** Multiply 142857 by 53271483. If your calculator doesn't have a big enough display to provide the exact value of the product, break it up into smaller components such as $142857 \times 5327, 142857 \times 1483$ and do an appropriate simple addition by hand. Verify that the property described in (d) above holds.

Repeat this verification for the product $142857 \times 371890923546$.

**Exercise**  **1.4**  Obtain the recurring decimal expansion of 1/27. Multiply this result by $3, 6, 9, 12, 15, 18, 21, 24$. Explain what has happened.

**Exercise**  **1.5**  Express as rational numbers the following recurring decimals:

(a)  $0.543543543\ldots$

(b)  $1.07545454\ldots$

(c)  $1.07546546546\ldots$

**Exercise**  **1.6**  Multiply 76923 by the integers from 2 to 12 inclusive. Each of the products is a six-digit number. Separate these products into two sets, one of which consists of cyclic arrangements of the digits 076923. Identify the second set of products as another cyclic arrangement. Divide 1 by 13 without using a calculator, and relate it to the above.

### 1.1.3    Hailstone numbers

Think of a number, say 6, which is *even*. Halve it to obtain 3, which is odd. Triple 3 and add one to obtain 10, which is even. Now halve this to get 5; triple this and add one to get 16; continue doing this, so you'll get the sequence

6, 3, 10, 5, 16, 8, 4, 2, 1

Such numbers have been called **hailstone numbers** because their behaviour of rising and falling before ending at 1 is something like hailstones in a thundercloud. These repeatedly fall under gravity and rise on upcurrents of air, before eventually falling to the ground (of course, hailstones get bigger and bigger in size).

The rules for generating a sequence of hailstone numbers can be summarized as follows: if a number is even, halve it; if a number is odd, triple it and add 1. You can see that if the number 1 is reached then continuing to apply the rules gives an endless loop 142142... so once 1 is reached you can stop.

Try starting with a bigger number, say 15, to get:

15, 46, 23, 70, 35, 106, 53, 160, 80, 40, 20, 10, 5, 16, 8, 4, 2, 1

The rising and falling is very apparent, and although a high of 160 is attained, the sequence soon ends up at 1 again. Try generating some sequences yourself, with successively bigger starting values. Incidentally, you can see there's no point starting with an even number, as this gets halved at the first step. You'll keep getting back to 4, 2, 1 quite quickly, until you start with 27. This reaches a high of 9232, and doesn't reach 1 until the 112th number. Try it, and you'll find that there are a couple of intermediate highs of 1780 and 7288.

You can have fun trying different starting values, and seeing what highs are achieved, and how many steps are needed to end up back at 1. But nobody knows for sure (that is, no mathematical proof has been found) whether *every* such sequence ends up at 1. However, computer trials have been done for all starting values up to one trillion (that is, one million million, or one followed by twelve zeros, which is written $10^{12}$) and in *every* case the sequence eventually ends up at 1.

A surprising fact about each of these one trillion sequences can now be deduced by a simple but ingenious argument. This fact is that in every one of the sequences, all the numbers are different from each other. To see this, suppose that at some stage in a sequence a number $x$ (bigger than 1) is reached, and that further on $x$ is obtained again. Then the numbers following this second appearance of $x$ will be exactly the same as those following the original $x$, so continuing to compute the sequence will bring you to $x$ a third time, and so on. The set of numbers from one appearance of $x$ to the next appearance of $x$ will be endlessly repeated. This means that the number one will not be reached, which we know is wrong because (as the computer trials have shown) all the trillion sequences do indeed end up at 1. We can only conclude that it is *impossible* for any number (bigger than 1) to occur more than once in any particular sequence. Remember, though, that what happens for sequences beyond those trillion which have been evaluated on the computer remains uncertain.

**Exercise**   **1.7**   Prove that the highest value reached in any sequence of hailstone numbers must be even.

**Exercise**   **1.8**   Suppose that the rule for handling odd numbers in a hailstone sequence is changed to: if a number is odd, triple it and add 7 (even numbers are halved as before). Investigate what happens to sequences starting with 5 or 7 or 9.

## 1.2   Russian multiplication

The story goes that Russian peasants were not very good at arithmetic, and could only add and multiply or divide by 2. They therefore invented the method described below for multiplying together any two integers using only these operations. This explanation for the origin of the procedure is certainly not politically correct, but whatever its history the method is undoubtedly fascinating. It's best explained by means of an example, say the product $45 \times 27$.

The rules are:

(1)  Write down two columns, headed by 45 and 27.

(2)  The numbers in the left column are obtained by successively dividing by 2, *ignoring remainders*, until 1 is reached.

   The second entry is therefore 22, since $45 \div 2 = 22$ with remainder 1. The subsequent entries are: 11, 5, 2, 1.

(3) The numbers in the right column are obtained by successively multiplying by 2, until a column of the same length is obtained.

At this stage therefore the two columns are:

|  | 45 | | 27 |
|---|---|---|---|
| → | 22 | | 54 ← |
| | 11 | | 108 |
| | 5 | | 216 |
| → | 2 | | 432 ← |
| | 1 | | 864 |

obtained by halving↑     obtained by doubling↑

To obtain the value of the product, there are two final steps:

(4) Delete all *even* numbers in the left column, together with the corresponding numbers in the right column. The deleted numbers are indicated by arrows in the array above.

(5) Add up the numbers which remain in the right column. Their sum equals the required value of the product, as follows:

$$
\begin{array}{r}
27 \\
108 \\
216 \\
864 \\
\hline
1215 = 45 \times 27
\end{array}
$$

Notice that in step (1) it doesn't matter which way round you write the two original numbers. In this example, if you take 27 as heading the left column the calculation becomes:

| 27 | 45 |
|---|---|
| 13 | 90 |
| 6̸ | 1̸8̸0̸ |
| 3 | 360 |
| 1 | 720 |

$$1215 = 27 \times 45$$

where the deleted numbers are slashed through.

As a further illustration, the computations for $38 \times 117$ are set out below:

| 3̸8̸ | 1̸1̸7̸ |
|---|---|
| 19 | 234 |
| 9 | 468 |
| 4̸ | 9̸3̸6̸ |
| 2̸ | 1̸8̸7̸2̸ |
| 1 | 3744 |

$$4446 = 38 \times 117$$

So long as you can multiply and divide by 2, and add up, you can work out products even if you don't have a calculator handy! In view of the way you can ignore remainders in the divisions, and the apparently arbitrary deletion of even numbers in the left column, it's somewhat mysterious that the method is valid. However, the technique can be proved to work in all possible cases (see Section 2.2.2.1).

**Exercise**  |  **1.9** Evaluate the products $73 \times 44$ and $367 \times 189$ using Russian multiplication.

## 1.3  Divisibility

It can be useful to know whether one number can be exactly divided by another, without actually having to carry out the division. For example, you're certainly aware that if a number ends with a zero (for example, 8130) then it is divisible by 10 (a technical term is that 10 is a **factor** of 8130. You also probably know that if a number ends with an even digit (that is, $2, 4, 6, 8$) or zero then it is divisible by 2 (for example, 8136 is divisible by 2). The other single-digit divisors can be dealt with as follows.

### 1.3.1  Division by 3

A number is divisible by 3 if the sum of its digits is divisible by 3. For example, the sum of the digits of the number 58317 is $5 + 8 + 3 + 1 + 7 = 24$ which is divisible by 3, so 58317 is divisible by 3. However, for 17528 the sum of the digits is 23 which is not divisible by 3, so the original number is not divisible by 3.

As another example, take

893545268178903573

This 18-digit number is too large to be entered into most calculators, but you can use the calculator to add up the digits, giving a sum of 93. To confirm that this is divisible by 3, you can either spot that $93 = 31 \times 3$, or you can repeat the test by adding the digits $9 + 3 = 12$, which is clearly divisible by 3. In any case, this shows that the original 18-digit number is divisible by 3.

### 1.3.2  Division by 4

A number is divisible by 4 if the number consisting of the last two digits is divisible by 4 (this includes the case 00). For example, 51729 is not divisible by 4 since 29 is not divisible by 4; whereas 51728 *is* divisible by 4 since $28 = 7 \times 4$.

### 1.3.3 Division by 5

This is an easy one: a number is divisible by 5 if its last digit is 0 or 5. For example 51725 is divisible by 5, but 51726 is not.

### 1.3.4 Division by 6

Since $6 = 2 \times 3$, it's necessary to apply the divisibility tests for both 2 and 3. Thus for a number to be divisible by 6 it must be even and the sum of its digits must be divisible by 3. For example, 38520 is divisible by 6 because it is even and the sum of its digits is $3 + 8 + 5 + 2 + 0 = 18$ which is divisible by 3.

### 1.3.5 Division by 7

This case stands out as being a bit more awkward to deal with than any other single digit division. For numbers with three or fewer digits no simplification is possible – you have to divide by 7 and see if there is a remainder. For numbers with more than three digits, split the digits into groups of three, starting from the right. Alternately add and subtract these three-digit numbers to produce a 'test sum' (you can ignore any negative sign). If this is divisible by 7, then so is the original number. Two examples will make it clear how the test works:

(1) For 1412236, the number groups are $1, 412, 236$ so the test sum is

$$1 - 412 + 236 = -175$$

Since this is divisible by 7 (that is, $175 = 7 \times 25$), so is the original number.

(2) For 130747591, the test sum is

$$130 - 747 + 591 = -26$$

and since 26 is not divisible by 7, neither is the original number.

If the test sum has more than three digits, the procedure can be repeated until you end up with a new test sum which has three or fewer digits.

It's rather curious that this scheme can also be applied to test for divisibility by 11 or by 13. The test sum is calculated in exactly the same way as above, and if this is divisible by 11 or 13, then so is the original number. In the example (1) above, the test sum is not divisible by 11 or 13, so neither is 1412236; but in example (2) the test sum is $26 = 13 \times 2$, so the original nine-digit number is divisible by 13.

One useful consequence of this extension is that if the test sum is zero then the original number is divisible by the *three* factors 7, 11 and 13 (for a proof, see Problems 1.10 and 2.9). For example, it's easy to see that the number 251251 has this property. In fact, so far as divisibility by 11 is concerned, it is preferable to use a much simpler test which is described in Problem 1.8.

### 1.3.6 Division by 8

If the number consisting of the last three digits of the original number is divisible by 8, then the original number is divisible by 8. For example, 17880 is divisible by 8 since $880 = 8 \times 110$; but 29313 is not divisible by 8 because 313 is not divisible by 8.

### 1.3.7 Division by 9

This is similar to divisibility by 3 in Section 1.3.1: a number is divisible by 9 if the sum of its digits is divisible by 9. This follows from a more general result in Section 2.2.6. For example, the sum of the digits of the number 582147 is $5 + 8 + 2 + 1 + 4 + 7 = 27$ which is divisible by 9, and hence so is the original number. As in the case of division by 3, you can add up the digits of the test sum, if you wish. For example, using the 18-digit number in Section 1.3.1 the test sum was 93, and $9 + 3 = 12$ which is not divisible by 9, so the original number is not divisible by 9.

### 1.3.8 A mind-reading trick

You can use the result of the previous section to develop a 'mind reading' trick which will amaze your friends. Give a friend a calculator and ask him or her to select any two horizontal or vertical or diagonal lines on the keyboard, without your seeing what they are doing. For example, in Figure 1.2 your friend has selected 753 and 258 – it doesn't matter in which direction the numbers go. Then ask your friend to multiply together these two numbers – in the example chosen the product is 194274. Since you haven't seen what your friend has done, there is no way that you can guess what this product is – your friend can surely believe that he or she has generated a genuinely random number. Now ask your friend to concentrate hard on just one of the digits, say 7 in this example. Finally, ask your friend to tell you what the other digits are, in any order, apart from the one being concentrated on. You are then able to announce immediately what the missing digit is, thereby 'reading your friend's mind'!

$753 \times 258 = 194274$

*Figure 1.2* Random number generation

How is it done? The trick relies on the fact (which you may not have noticed) that the digits 1 to 9 on the keyboard of a calculator are always laid out in the same way, as shown in Figure 1.2, irrespective of make or model. There are three horizontal lines, three vertical lines and two diagonals on the keyboard, and every three-digit number they produce is divisible by 3. You can easily verify this property by using the test in Section 1.3.1, checking that in each case the sum of the digits is divisible by 3. For example, the two numbers selected in Figure 1.2 have $7 + 5 + 3 = 15$, $2 + 5 + 8 = 15$.

Since every three-digit number which can be selected from the calculator keyboard has a factor 3, the *product* of these two numbers will contain a factor $3 \times 3 = 9$. Because this product has a factor 9, the rule in Section 1.3.7 applies, namely that the sum of the digits of the product must be divisible by 9. This is the key property you use.

When your friend reads out all the digits except the one being kept back, you add them up. This is easily and quickly done in your head. Thus in the example we are using, your friend tells you 19424 (remember your friend is concentrating on the digit 7). The sum is $1 + 9 + 4 + 2 + 4 = 20$. You know that this sum must be divisible by 9, so you can immediately say that the number your friend is thinking of must be 7, since this gives $20 + 7 = 27$. Clearly 7 is the only digit which when added to 20 gives a multiple of 9. Your friend should be truly amazed! Moreover, the sum of the digits doesn't depend upon the order in which they are read out, so your friend could scramble the digits, and tell you, for example, 12449 and you would still give the correct answer 7 – even more amazing!

There is just one important point to emphasize. When you ask your friend to hold back one of the digits to concentrate on, you must insist that he or she does not select zero. You tell them that this is because you 'can't read their mind if they are thinking about nothing'(!) What is the real reason behind this? Suppose your friend chooses

$$123 \times 456 = 56088$$

If your friend holds back the digit zero, and tells you the others are 5688, you sum these digits to obtain 27, which is divisible by 9. You then have no way of telling whether the missing digit is 0 or 9, because in both cases you will get a total sum (27 or 36) which is divisible by 9. So the trick may fail, and in order to prevent this happening you tell your friend not to choose zero. There can never then be any ambiguity, and the 'mind-reading' trick will always work.

Of course, a convincing part of the trick is your insistence that the number your friend has generated on the calculator is 'random', so you cannot possibly guess it. In fact, you know that your friend's five- or six-digit number will always be divisible by 9, so is *not* random. Just hope your friend doesn't realize this!

**Exercise**  **1.10**  Use the tests of this section to find factors of the following numbers:

(a) 89364  (b) 2198868  (c) 275860088  (d) 1884843051.

**Exercise**  **1.11**  Use the test in Section 1.3.5 to show that any six-digit number which has all its digits the same is divisible by 7, 11 and 13.

Give some generalizations of this result for numbers:

(a) having six digits  (b) having five digits

(c) having more than six digits.

**Exercise**  **1.12**  Use the test in Section 1.3.5 to determine the smallest number divisible by 7 having

(a) 5 digits  (b) 6 digits  (c) 7 digits

(the first digit of each number must be non-zero).

**Exercise**  **1.13**  What is the test for divisibility by 12? Use it to verify that 16019376 is divisible by 12.

**Exercise**  **1.14**  A very simple trick is to ask a friend to write down a number consisting of three identical digits (not zero). You then tell him or her to add the three digits, divide the original number by this sum, and 'concentrate' on the answer. After a few seconds you announce he or she is thinking about 37. For example, if they choose 666 then $6 + 6 + 6 = 18$ and $666 \div 18 = 37$. Explain why this always works.

You may be thinking at this point that the divisibility tests are irrelevant now that pocket calculators are commonplace. However, remember that any calculator (or computer) can only work to a *finite* number of digits. An illustrative example involving divisibility by 3 of an 18-digit number, beyond the capacity of most calculators, was given in Section 1.3.1. It is easy to generate a profusion of such examples – for example, the 10-digit number in Exercise 1.10(d). As another example, you should be able to see quickly by using the rule at the end of Section 1.3.5 that the 24-digit number

123123123123123123123123

is divisible by 7, 11 and 13; try doing that on your calculator! It's interesting, therefore, that even in this age of electronic computations, the old methods still have their place.

## 1.4  Other types of number

So far attention has focused on integers, with a brief mention of recurring decimals and rational numbers in Section 1.1.2. Historically, 'whole numbers' (that is, integers) came first and 'fractions' (that is, rational numbers) were a natural development originating with the idea, which dates back to ancient Egyptian and Babylonian times (four or five thousand years ago), of dividing a whole into parts. The School of Pythagoras in ancient Greece (in the sixth century BC) built up a philosophy of the world based on integers and rational numbers. For example, they were deeply impressed by the relationships between numbers and musical harmony. It therefore came as a considerable blow to the Pythagoreans when they discovered that in order to describe something as simple as the length of the diagonal of a unit square, rational numbers would not suffice. It was necessary to invent a new kind of number, called non-rational, or 'irrational' – the name carries a connotation that these numbers are 'unreasonable', 'silly'! Even worse was to come, as from about the late sixteenth century onwards mathematicians increasingly began to realize that yet another type of number was needed to deal with square roots of negative numbers. These numbers were eventually called 'imaginary' to distinguish them from all the previous types which were now called 'real'. In fact, as the introduction to this chapter mentioned, all numbers are figments of the imagination, and no type of number is more or less 'real' than any of the others. Even though you can see four chairs round a table, this doesn't make the number 4 any more substantial than, say, the negative number $-4$. The terms 'rational', 'irrational', 'real', 'imaginary', 'complex', and so on, as applied to numbers must always be regarded purely as technical terms. Not even a hint of the everyday meaning of these words is relevant in a number sense.

### 1.4.1    Irrational numbers

As stated earlier, a rational number is the ratio of two integers, such as $3/4$, $-9/11$, and so on. A rational number has either a finite decimal expression, such as 0.75 for $3/4$; or as we saw in Section 1.1.2, there is a repeating decimal expansion which for the example of $-9/11$ is $-0.818181\ldots$ (as you should verify by dividing 11 into 9 by hand). A number which cannot be expressed as the ratio of two integers is called **irrational**, and has a non-repeating decimal expansion. The simplest example of an irrational number arises from trying to determine the length of the diagonal of a square whose side is one unit of distance. If $d$ denotes the length of the diagonal as shown in Figure 1.3, then from the right-angled triangle $d^2 = 1^2 + 1^2$, so $d = \sqrt{2}$. Why can't $\sqrt{2}$ be expressed as the ratio of two integers, say $a/b$? The following argument to show that this is impossible, so that $\sqrt{2}$ is indeed irrational, contains some subtleties, so you should go through it carefully. The first step is to ensure that the fraction $a/b$ is in its simplest possible form, that is to say any factor common to both numerator and denominator is removed. For example, if the fraction was $722/510$, then there is a factor 2 common to both 722 and 510, so this can be removed, leaving $361/255$. Next, assuming that it is possible to

*Figure 1.3* Diagonal of unit square

write $\sqrt{2} = a/b$ with this proviso in mind, and squaring both sides of the equality gives

$$2 = \frac{a^2}{b^2}$$

so that $a^2 = 2b^2$. This means that $a^2$ is even, because $b^2$ is an integer so $2b^2$ is an even integer. It now follows that the integer must itself be even, because the square of any odd number is always odd (see Exercise 1.15). Hence you can 'factor out' the 2, and write $a = 2c$, where $c$ is another integer. Therefore $a^2 = (2c)^2 = 4c^2$ and substituting this into $a^2 = 2b^2$ gives $4c^2 = 2b^2$, so that $b^2 = 2c^2$. This shows that $b^2$ is also even, so $b$ must also be even (by exactly the same argument as was used for $a$). So the conclusion has been reached that both $a$ and $b$ are even, that is, they have a common factor 2. But this is impossible, since it was agreed at the outset that any such factor common to $a$ and $b$ would be removed. An impasse has therefore been arrived at: assuming that $\sqrt{2} = a/b$ for $a$ and $b$ integers without any common factor leads to the *contradictory* conclusion that $a$ and $b$ *do* have a common factor 2. Because of this contradiction, you are forced to admit that it must be impossible to express $\sqrt{2}$ as $a/b$; in other words $\sqrt{2}$ is not a rational number.

Having established that $\sqrt{2}$ is irrational, you can then construct as many other irrational numbers as you like, for example $\sqrt{2} + 3$, $9 - \sqrt{2}$, $5\sqrt{2}/8$, and so on. The argument developed above is therefore remarkable in that it establishes a wholly new class of numbers, yet relies not on deep mathematical concepts but only on simple ideas of common factors and even numbers. Perhaps because of this, you may feel on a first reading that the proof is a bit of a swindle! Go through it again, and convince yourself that the proof is indeed completely watertight.

Most people find it helpful to see numbers represented along a straight line, as in Figure 1.4. One point represents zero, and it is usual to take positive

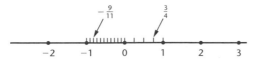

*Figure 1.4* Rational numbers

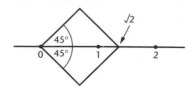

*Figure 1.5* Measuring out $\sqrt{2}$

numbers on the right and negative numbers on the left. The distance between 0 and 1 is the basic unit of length. Integers $2, 3, 4, \ldots$ and $-1, -2, -3, \ldots$ are represented by moving the appropriate number of unit distances to the right and left of 0. To find the point representing, for example, the rational number $3/4$, first divide up the unit distance into four equal parts, and then mark off three of these parts to the right of 0. Similarly for $-9/11$, move to the left of 0 by nine of the lengths equalling $1/11$. Irrational numbers can also be marked on the line. For example, $\sqrt{2}$ is the diagonal of the unit square shown in Figure 1.5.

**Exercise**  **1.15**  Every positive odd integer can be written as $2n + 1$, where $n$ takes the values $0, 1, 2, 3, \ldots$ Use this to show that the square of any odd integer is also odd.

**Exercise**  **1.16**  A proof that $\sqrt{3}$ is irrational can be developed through the following steps.

(a) Suppose $\sqrt{3} = a/b$, where $a$ and $b$ are positive integers having no factor in common. Show that $a^2 = 3b^2$.

(b) Show that the square of any positive integer must have as its last digit either $0, 1, 4, 5, 6$ or $9$ (consider $1^2, 2^2, 3^2, 4^2$, and so on).

(c) Show that three times the square of any positive integer must have as its last digit either $0, 2, 3, 5, 7$ or $8$.

(d) In part (a) the integers $a^2$ and $3b^2$ are equal, so must have the same last digit. Hence deduce that $a^2$ and $3b^2$ must both end in either $0$ or $5$.

(e) Explain why both $a$ and $b$ must end either in $0$ or $5$.

(f) Use Section 1.3.3 to deduce that $a$ and $b$ are each divisible by 5, and hence complete the proof that $\sqrt{3}$ is irrational.

**Exercise**  **1.17**  An alternative proof that $\sqrt{3}$ is irrational parallels the argument used for $\sqrt{2}$, as follows:

(a) Any positive integer $x$ can be written as $3n$, $3n + 1$ or $3n + 2$ where $n = 0, 1, 2, 3, \ldots$ Use this to show that if $x^2$ is a multiple of 3, then so is $x$.

(b) Use the above result to show that the assumption in part (a) of Exercise 1.16 produces a contradiction, from which it follows that $\sqrt{3} \neq a/b$.

### 1.4.2  Imaginary numbers

For anyone who has learned a little algebra, another way of looking at various types of numbers is helpful. For example, the rational number 3/4 can be regarded as the solution of an equation with integer coefficients, such as

$$4x + 1 = 4$$

In a similar way, the negative integer $-3$ is the solution of an equation

$$3x + 9 = 0$$

Again, this equation has coefficients which are positive integers but now its solution requires another kind of number, namely negative integers. Negative rational numbers can be handled similarly, so that overall *any* equation in the form

$$ax + b = 0$$

where $a$ and $b$ are positive or negative integers, can be solved using rational numbers or integers. Such equations involve only $x$ and no higher powers, so if $y = ax + b$ is plotted in the usual cartesian coordinates a straight line is obtained – for this reason such equations are called **linear**.

If you move up to 'quadratic' equations which contain $x^2$ as well as (possibly) terms in $x$, then for example the irrational number $\sqrt{2}$ is the solution of

$$x^2 - 2 = 0$$

Notice that this equation still has coefficients which are integers (regarding $x^2$ as $1x^2$) just like the linear equations, but its solution cannot be expressed in terms of integers or rational numbers – a new kind of number has to be invented. In an exactly similar way, to solve the equation

$$x^2 + 1 = 0$$

yet another kind of number is needed, even though the coefficients are again merely integers. The **imaginary number** $i$ is defined as the solution of this equation, so that $i^2 + 1 = 0$. In other words, $i$ satisfies $i^2 = -1$, so that $i = \sqrt{(-1)}$. Once again you are reminded that $i$ is no or more less imaginary than any other kind of number.

It's now possible to solve any quadratic equation

$$ax^2 + bx + c = 0$$

where $a, b, c$ are rational numbers. You may have come across the general formula

$$x = \frac{-b \pm \sqrt{(b^2 - 4ac)}}{2a}$$

If $b^2 - 4ac$ is negative, the solution will involve $i$. For example, the solution of

$$x^2 - 2x + 2 = 0$$

is

$$x = \frac{2 \pm \sqrt{(4 - 8)}}{2}$$
$$= 1 \pm \tfrac{1}{2}\sqrt{-4}$$
$$= 1 \pm i$$

the last step is obtained by using

$$\sqrt{-4} = \sqrt{(4)(-1)} = \sqrt{4}\sqrt{(-1)} = 2i$$

The roots $1 + i$ and $1 - i$ of the quadratic equation are examples of what are called **complex numbers**, which consist of the sum of a 'real' number (not involving $i$) and an imaginary number. Further examples of complex numbers are

$$\tfrac{1}{2} - \tfrac{3}{4}i, \quad \sqrt{2} + 5i, \quad -11 - i\sqrt{7}$$

The 'real parts' of these complex numbers are respectively $1/2$, $\sqrt{2}$, $-11$; the 'imaginary parts' are $-3/4$, $5$, $-\sqrt{7}$. Notice that the imaginary part is the term multiplying $i$ (which incidentally can be written before or after the multiplier). Pairs of complex numbers like $1 + i$ and $1 - i$, which differ only in the signs of their imaginary parts, are called **complex conjugate pairs**. You can see from the $\pm$ sign in front of the square root in the expression for the solution $x$ of the general quadratic equation that when a quadratic equation (with rational coefficients) has complex roots these always occur as a conjugate pair. It's impossible for such a quadratic equation to have one real and one complex root; such an equation either has two real roots (which may be equal) or a pair of complex conjugate roots.

The rules of arithmetic for complex numbers are just the same as those for real numbers which you are already familiar with; the only point to remember is that when $i^2$ occurs it is replaced by $-1$. For example,

$$(2 + 3i) + (7 - 11i) = 2 + 7 + (3 - 11)i = 9 - 8i$$
$$(2 + 3i) \times (7 - 11i) = 2 \times 7 + 2(-11i) + (3i)7 + (3i)(-11i)$$
$$= 14 - 22i + 21i - 33i^2$$
$$= 14 + (21 - 22)i + 33$$
$$= 47 - i$$

To handle division, for example $(2 + 3i)/(7 - 11i)$, you need first to realize that the product of a number and its conjugate is a real number; in this example

$$(7 - 11i) \times (7 + 11i) = 49 + 77i - 77i - 121i^2$$
$$= 49 + 121 = 170$$

The following example shows how to use this property. Multiply both numerator and denominator of the expression to be evaluated by the conjugate of the denominator. Then proceed as shown in the following steps:

$$\frac{2+3i}{7-11i} = \frac{(2+3i)(7+11i)}{(7-11i)(7+11i)}$$

$$= \frac{14+22i+21i+33i^2}{170}$$

$$= \frac{14+43i-33}{170}$$

$$= \frac{-19+43i}{170}$$

$$= \frac{-19}{170} + \frac{43}{170}i$$

If you like geometrical representations, then complex numbers can be displayed by using two lines, at right angles to each other (as with cartesian coordinates). Distances along the 'horizontal' axis correspond to the real part, and those along the 'vertical' axis refer to the imaginary part. Some examples are shown in Figure 1.6. Such a diagram is usually named after **Argand** (and sometimes after the famous mathematician, Gauss) but it was actually first discovered by a Norwegian called Wessel at the end of the eighteenth century. Unfortunately for Wessel his work was lost for almost a century in the

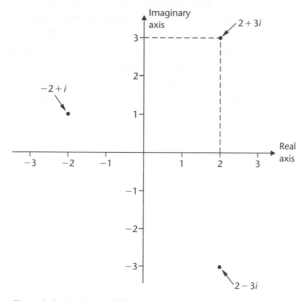

*Figure 1.6*   An Argand diagram

obscurity of the transactions of the Danish Academy of Sciences. So even in the apparently well-ordered world of mathematics, lasting fame can be a chance affair!

**Exercise** **1.18** Express the following in the form $x + iy$, where $x$ and $y$ are real.
(a) $(7 - 3i) + (-2 + 11i) - (6 - 9i)$
(b) $(-8 + 11i)(-4 + 17i)$
(c) $(5 + 3i)^2$
(d) $(7 - 3i)(-2 + 11i)(6 - 9i)$
(e) $(2 - 5i)/(7 - 2i)$
(f) $(5i - 3)/(3 + 5i)^2$.

**Exercise** **1.19** Solve the quadratic equations:
(a) $x^2 + 12x + 40 = 0$          (b) $2x^2 + 7x + 9 = 0$.

**Exercise** **1.20** Verify that the three roots of $x^3 = -1$ are $-1$, $(1 + i\sqrt{3})/2$, $(1 - i\sqrt{3})/2$. Plot these roots on an Argand diagram.

**Exercise** **1.21** Verify that the four roots of $x^4 = -1$ are $(1 + i)/\sqrt{2}$, $(1 - i)/\sqrt{2}$, $(-1 + i)/\sqrt{2}$, $(-1 - i)/\sqrt{2}$. Plot these roots on an Argand diagram.

### 1.4.3  What else?

You've seen the apparently inexorable necessity to keep on inventing new kinds of numbers in order to be able to solve new types of mathematical problem. It's natural to wonder whether this process ever stops – is there an endless list of new species of numbers waiting to be brought into play? Fortunately, for the mathematics you're likely to encounter, the answer is 'no', with the important exception of 'transcendental' numbers, which will be explained shortly. You can be confident that complex numbers are as far as you need to go (and in this book these won't actually be required again until Chapter 5). For example, in the preceding two exercises (1.20 and 1.21) you'll see that the cube roots and fourth roots of $-1$ can be evaluated as complex numbers, as indeed can expressions like $(2 + 3i)^{\frac{1}{3}}$. Going further, the roots of polynomial equations of any degree, like

$$x^{17} - \tfrac{7}{4}x^{15} + 3x^9 - \tfrac{2}{3}x^3 + \tfrac{7}{9}x - 2 = 0$$

are also complex numbers, and in fact this is true even if some of the coefficients in the equation are irrational or complex. Likewise, mathematical functions like $\log(-7)$ or $\sin(2 + 3i)$ can be evaluated using complex numbers. Incidentally, it's worth mentioning here that expressing numbers to different bases (for example, counting eggs in dozens instead of tens) does not constitute a different *type* of number (see Chapter 2).

Now let's turn as promised to 'transcendental' numbers which, like the numbers in the previous section, can be understood through the solution of equations. Consider a polynomial equation with **rational** numbers as coefficients, like the example above beginning with $x^{17}$. All the numbers you have encountered so far – positive and negative integers, rational and irrational numbers, imaginary and complex numbers – can be expressed as the roots of such an algebraic equation, and are called **algebraic** numbers. However, it was proved in 1844 that there are some irrational numbers which *cannot* be expressed in this way, and these were called **transcendental** from the idea that they 'transcend' (that is, 'go beyond') algebraic numbers. You've almost certainly encountered the two most important transcendental numbers: $\pi$ (pi) which is the ratio of the circumference (in Greek 'peripheria', beginning with the letter $\pi$) of a circle to its diameter; and e, the base of natural logarithms.

Both these numbers occur widely in mathematics. The number $\pi$, apart from geometrical applications such as the area of a circle and the volume of a sphere, arises in many areas of calculus and statistics. So far as e is concerned, a typical application is the idea of 'exponential growth' in economics and population science; and 'exponential decay' describes the behaviour of radioactive materials.

The two numbers have unending and non-repeating decimal representations, of which the first 20 places are

$\pi = 3.14159265358979323846\ldots$
$e = 2.71828182845904523536\ldots$

It's worth repeating that these numbers cannot arise as the root of a polynomial equation with rational coefficients, however high you may take the degree of the first term in the equation, or however you may try to select suitable coefficients.

Around 4000 years ago approximate values used for $\pi$ were $3\frac{1}{8}$ by the Babylonians, and $(16/9)^2 \simeq 3.1605$ by the Egyptians. This latter value is contained in the Rhind Papyrus, the oldest known mathematical document, discovered in Egypt by a Scot called Rhind in 1858 and now in the British Museum. The first Book of Kings in the Old Testament of the Bible gives the value of 3 for $\pi$, so the Jewish editors who compiled this book around 550 BC were not very up to date in their mathematical knowledge! The value of $\pi$ was known correct to six decimal places in China by the fifth century AD, 35 decimal places were computed in 1596 and this had grown to 500 decimals by 1855. It is surprising, then, that it was not until 1882 that $\pi$ was proved to be transcendental. The symbol e was introduced by the very prolific Swiss mathematician Euler (pronounced 'Oiler') in the eighteenth century, and the transcendence of e was established in 1873. With the development of electronic computers, both $\pi$ and e had been calculated to over 100,000 decimal places by 1961 – who needs more! (see

Exercise 1.22). In fact more recently several billion digits of $\pi$ have been computed.

The numbers $\pi$, e and $i$ are defined in such different and unrelated ways that it comes as a major surprise to find that they are connected by a very simple formula, due to Euler, namely

$$e^{i\pi} + 1 = 0$$

This is a truly mysterious formula – no-one really understands why 0, 1, the two transcendental numbers $\pi$ and e, and the square root of minus one, should be linked together like this. It's as though the numbers are conspiring with each other in some way we can't fathom, especially as it was shown in 1934 that $e^{\pi}$ is actually transcendental.

Don't get the idea that there are only a few special transcendental numbers. Any number of the form $a^b$, where $a$ is an algebraic number ($\neq 0$ or 1) and $b$ is an irrational algebraic number, is always transcendental. For example, $3^{\sqrt{2}}$, $(\sqrt{5})^{\sqrt{3}}$, $(2 + 3i)^{\sqrt{7}}$ are all transcendental numbers. Another transcendental number is the unending decimal

0.110001000000000000000001000 . . .

where all the digits are zero except for 1's in positions 1, 2, 6(=3!), 24(=4!), 120(=5!) and so on. However, some questions remain: for example it is not known whether $e + \pi$, $e \times \pi$ and $\pi^e$ are transcendental or algebraic.

The quest for new types of numbers still goes on. The latest of these are called **surreal numbers** (from the French 'sur', meaning 'above', real numbers) and research into them is still in its infancy. It is hoped that surreal numbers will eventually be applied to problems of theoretical physics involving very small or very large quantities. In view of the esoteric nature of modern physics, it is perhaps fitting that the dictionary meaning of 'surreal' is 'dream-like', and that this name first appeared in a novel by Knuth in 1974.

**Exercise** **1.22** The diameter of the Earth at the equator is 12757 km. Assuming the Earth is spherical, compute the equatorial circumference using $\pi$ correct to 9 decimal places. What is the difference if 10 decimal places of $\pi$ are used?

**Exercise** **1.23** How accurate are the approximations 355/113 for $\pi$ and 878/323 for e?

**Exercise** **1.24** An expression for $\pi$ due to Newton is

$$\pi = 6\left(x + \frac{1}{2}\frac{x^3}{3} + \frac{1 \times 3}{2 \times 4}\frac{x^5}{5} + \frac{1 \times 3 \times 5}{2 \times 4 \times 6}\frac{x^7}{7} + \ldots\right)$$

with $x = 1/2$. Compute an approximate value for $\pi$ by taking the first five terms of this (unending) series.

**Exercise**

**1.25** A standard expression for e is

$$e = 1 + \frac{1}{1!} + \frac{1}{2!} + \frac{1}{3!} + \frac{1}{4!} + \ldots$$

where the factorial notation ! means

$$2! = 2 \times 1, \quad 3! = 3 \times 2 \times 1, \quad 4! = 4 \times 3 \times 2 \times 1$$

and so on. How many terms do you need to take in order to compute e correct to five decimal places?

**Exercise**

**1.26** All except the simplest pocket calculators have a square root button. The calculator uses a repetitive procedure like the following. Suppose you wish to compute $\sqrt{3}$. Clearly the answer lies between 1 and 2 (since $1^2 = 1$, $2^2 = 4$), so start with, say, 1.5. A better approximation is

$$\frac{1}{2}\left(1.5 + \frac{3}{1.5}\right) = 1.75$$

The same calculation is repeated with this new value, giving as the next approximation

$$\frac{1}{2}\left(1.75 + \frac{3}{1.75}\right) = 1.7321$$

Doing this one more time gives

$$\frac{1}{2}\left(1.7321 + \frac{3}{1.7321}\right) = 1.7320508$$

which is correct to seven decimal places. The calculator chip does these computations in the twinkling of an eye.

Try this procedure yourself to obtain an approximate value of $\sqrt{5}$ (notice that the second term within the brackets will now have numerator 5 instead of 3).

This type of repetitive, or 'iterative' calculation will be studied in Chapter 5.

## Miscellaneous problems

**1.1** Intriguing arrays of numbers can be constructed, such as the following:

$$1 \times 9 + 2 = 11$$
$$12 \times 9 + 3 = 111$$
$$123 \times 9 + 4 = 1111$$
$$1234 \times 9 + 5 = 11111$$
$$12345 \times 9 + 6 = 111111$$
$$123456 \times 9 + 7 = 1111111$$
$$1234567 \times 9 + 8 = 11111111$$
$$12345678 \times 9 + 9 = 111111111$$
$$123456789 \times 9 + 10 = 1111111111$$

Use your calculator to build up a similar array that begins with $1 \times 8 + 1 = 9$, $12 \times 8 + 2 = 98$ and ends with

$$123456789 \times 8 + 9 = 987654321$$

In a similar way, try to construct an array that starts off with

$$9 \times 9 + 7 = 88$$
$$98 \times 9 + 6 = 888$$

**1.2**  Another amusing array is the following:

$$
\begin{aligned}
4^2 &= 16 \\
(34)^2 &= 1156 \\
(334)^2 &= 111556 \\
(3334)^2 &= 11115556 \\
(33334)^2 &= 1111155556 \\
(333334)^2 &= 111111555556
\end{aligned}
$$

The pattern can be continued indefinitely. There is only one other pair of digits like 1 and 6 which when set out in a pattern like that on the right produce squares on the left. Use your calculator to try and find them.

**1.3**  (a)  Apply the procedure in Section 1.1.1 to the five-digit numbers 62964, 61974, 53955. What can you conclude from your results?

(b)  Do the same thing for the six-digit numbers 631764, 549945. Can you infer anything this time?

Investigate what happens when you apply the procedure to 840852.

**1.4**  Divide 1 by 17 by hand (you will need some patience!) to obtain a recurring decimal with sixteen repeating digits. Multiply this set of digits by $2, 3, 4, \ldots, 16$ to reveal a cycle.

**1.5**  If you examine the multiplication $48 \times 159 = 7632$ you will see that all the nine digits are different. There are six other examples of this type involving the product of a two-digit number and a three-digit number. Use your calculator to try and find some of these.

**1.6**  Consider a 'hailstorm' like that in Section 1.1.3 but generated by a different rule for odd numbers: if a number is odd, multiply it by 5 and add 1. As before, if a number is even, then halve it. Compute sequences of these hailstone numbers starting with 3, 5 or 7.

**1.7**  The Russian multiplication method in Section 1.2 can be modified for division, requiring only multiplication by 2 and subtraction. For example, to evaluate $139 \div 12$, first multiply the divisor 12 repeatedly by 2, stopping before 139 is exceeded, to obtain 24, 48, 96. Subtract these numbers successively from 139, if possible, until a remainder less than 12 is obtained, as follows:

$$
\begin{array}{ll}
139 & \\
\underline{-96} & (= 8 \times 12) \\
43 & \\
\underline{-24} & (= 2 \times 12) \\
19 & \\
\underline{-12} & (= 1 \times 12) \\
7 & (= \text{remainder})
\end{array}
$$

Altogether $8 + 2 + 1 = 11$ multiples of 12 are subtracted, so $139 \div 12 = 11$ with a remainder of 7, that is, $139 \div 12 = 11\frac{7}{12}$.

Apply this method to evaluate $185 \div 8$ and $656 \div 58$.

**1.8** The objective of this problem is to determine a general test for divisibility by 11, simpler than that given in Section 1.3.5. Begin by writing

$$
\begin{array}{rl}
10 = & 11 - 1 \\
100 = & 99 + 1 = 11 \times 9 + 1 \\
1000 = & 1001 - 1 = 11 \times 91 - 1 \\
10000 = & 9999 + 1 = 11 \times 909 + 1 \\
100000 = & 100001 - 1 = 11 \times 9091 - 1
\end{array}
$$

and so on. Thus, for example, the number 23715 can be expressed as

$$
\begin{aligned}
23715 &= 2 \times 10000 + 3 \times 1000 + 7 \times 100 + 1 \times 10 + 5 \\
&= 2(11 \times 909 + 1) + 3(11 \times 91 - 1) + 7(11 \times 9 + 1) \\
&\quad + 1(11 - 1) + 5 \\
&= (2 \times 11 \times 909 + 3 \times 11 \times 91 + 7 \times 11 \times 9 + 1 \times 11) \\
&\quad + (2 - 3 + 7 - 1 + 5)
\end{aligned}
$$

The terms in the first bracket all contain a factor 11. Therefore the divisibility of 23715 by 11 depends solely on the terms within the second bracket. In this example $2 - 3 + 7 - 1 + 5 = 10$, which is *not* divisible by 11, so 23715 is not divisible by 11. Similarly, for 54219, the 'test sum' is $5 - 4 + 2 - 1 + 9 = 11$, so 54219 *is* divisible by 11.

Extend this idea to provide a general test for divisibility by 11. Hence test the numbers 1086362343 and 9397338985. If the number $27a6362$ is divisible by 11, what is the digit $a$?

**1.9** Using the rule for divisibility by 11 which you were asked to find in the previous problem, determine the *largest* possible nine-digit number which is divisible by 11, where none of the digits is repeated.

**1.10** The key to the divisibility test in Section 1.3.5 is the fact that $7 \times 11 \times 13 = 1001 = 10^3 + 1$. Show that $10^6 - 1$, $10^9 + 1$, $10^{12} - 1$, $10^{15} + 1$, and so on, are all divisible by 1001. Hence prove the validity of the test for divisibility by 7, 11 or 13.

**1.11**  Queen Elizabeth the Second was born on 21 April 1926. She seems perhaps to be well qualified as Queen, because the digits of her birthdate (211926) have an interesting property. Rewrite these digits in any order, say 911226, and subtract the smaller number from the larger:

$$911226 - 211926 = 699300$$

This difference is divisible by 9, since the sum of the digits is $6 + 9 + 9 + 3 = 27$, which is divisible by 9. Try your own birthdate to see if it has the same property.

**1.12**  In fact, neither the Queen's birthdate, nor your own, is really special in being divisible by 9, as found in the preceding problem.

Try to prove that if you take a number with several digits, scramble these in any order, and subtract the smaller number from the larger, then this difference is *always* divisible by 9 (provided that the two numbers are not identical, when their difference is zero).

You could develop this into another 'mind-reading' trick. Ask a friend to think up any six-digit number (whose digits are not all the same). Ask him or her to scramble these digits, subtract the smaller number from the larger, and tell you all the digits of the difference, except one digit (not zero) which he or she 'concentrates' on. By adding these digits just as in Section 1.3.8, you can immediately tell your friend which digit was held back.

**1.13**  Here is another simple game which you can use to mystify a friend. First ask a friend to enter 98765432 into a calculator, and divide by 8. The result (12345679) is rather surprising, since it consists of all the digits but now in ascending order, except for 8. You tell your friend this digit does not appear 'because it's already been used as divisor' (!).

But more is to come: ask your friend for his or her favourite digit – suppose it is 5. You immediately tell him or her to multiply the number on the calculator display (that is, 12345679) by 45, and your friend will be impressed to see that the result consists of the digit 5, repeated so that it fills the display. The multiplier you supply is always the product of 9 and the stated digit.

Can you explain why this works?

**1.14**  A number is called 'happy' if on computing the sum of the squares of its digits, and then computing the sum of the squares of the digits of this new sum, and so on, the process ends in 1. For example, 82 is a happy number because

$$
\begin{aligned}
82 &\longrightarrow 8^2 + 2^2 &&= 68 \\
68 &\longrightarrow 6^2 + 8^2 &&= 100 \\
100 &\longrightarrow 1^2 + 0^2 + 0^2 &&= 1
\end{aligned}
$$

Numbers which do not end in 1 after this procedure are called 'sad'.

(a)  Find the twenty happy numbers in the range from 1 to 100.

(b)  Investigate what happens to sad numbers.

**1.15**   Prove that the number $a_n a_{n-1} \dots a_3 a_2 a_1$ is divisible by $2^k$, for any positive integer $k$, if the number $a_k a_{k-1} \dots a_2 a_1$, consisting of the last $k$ digits, is divisible by $2^k$. Apply this to test 72148368.

Notice that this result generalizes those in Section 1.3.2 for division by 4, and in Section 1.3.6 for division by 8.

**1.16**   Show that the result of the preceding problem also applies to division by 5. That is, a number is divisible by $5^k$, for $k = 1,2,3,\dots$, if the number consisting of the last $k$ digits is divisible by $5^k$. Apply this to test 5412875.

**1.17**   When you buy a pad of ruled paper you probably select size A4. If you measure a sheet you'll find its dimensions are 210 mm by 297 mm. This seems rather odd, why isn't the size something like 200 mm by 300 mm, which would seem more convenient? The reasons are found if you fold an A4 sheet in half to produce dimensions 148 mm by 210 mm – this is precisely the size A5. Similarly, if you fold A5 in half you get A6, and so on up to the smallest size A10. All the rectangular sheets are 'similar' in shape, which means that the ratio of the longer to the shorter side is always the same. Prove that this ratio is $\sqrt{2}$.

For example, for sizes A4 and A5 the ratios are

$$297/210 = 1.414, \quad 210/148 = 1.419$$

Slight discrepancies from $\sqrt{2}(=1.414)$ occur because paper sizes are rounded to the nearest millimetre. The largest size is A0, which has an area of 1 square metre, and this determines the sizes of the whole of the range. Construct a table of paper sizes from A0 to A10.

A larger size of paper used for items such as posters or wall charts uses the same principles but begins with size B0 having dimensions 1000 mm by 1414 mm. Construct the table of dimensions for sizes B0 to B10.

**1.18**   Verify that the three roots of the equation $x^3 = 1$ are 1, $(-1 + i\sqrt{3})/2$, $(-1 - i\sqrt{3})/2$. If either of the complex roots is denoted by $a$, show that the other root is $a^2$, and verify that $1 + a + a^2 = 0$. Can you prove this directly, without knowing $a$?

Plot the three roots on an Argand diagram.

**1.19**   You can easily verify that the four roots of the equation $x^4 = 1$ are 1, $-1$, $i$, $-i$. Plot these roots on an Argand diagram.

Compare this diagram with the one in the previous problem. Can you spot any pattern for the location of the roots?

Try to guess the location of the roots of the equation $x^6 = 1$. Convert these roots into the form $a + ib$ and test whether they satisfy the equation.

**1.20**   Evaluate $(2 + i)^2$ and hence deduce that $(3 + 4i)^{1/2} = \pm(2 + i)$. Plot the two values of the square root on an Argand diagram.

Similarly, show that one value of $(2 + 11i)^{1/3}$ is $2 + i$. Can you use the idea of the previous problem to plot the other two cube roots of $2 + 11i$ on the diagram?

**1.21**    Suppose you have £1 which you decide to invest, and you're lucky enough to find a bank which offers 100% annual interest. At the end of one year you would have £2. If instead the bank paid interest every six months, then the rate would be $100/2 = 50\%$. After the first six months you would have £1.50. This sum would then gain interest, so at the end of a further six months (that is, one year in total) you would have

$$1.50 + 1.50 \times \tfrac{50}{100} = 1.50(1 + 0.5)$$
$$= £(1.5)^2 = £2.25$$

which is more than the first case.

Similarly, if the interest is compounded quarterly (that is, four times per year) then the rate is $100/4 = 25\%$. At the end of the first quarter you would have

$$1 + \tfrac{25}{100} = £1.25$$

After a second quarter this becomes

$$1.25 + 1.25 \times \tfrac{25}{100} = 1.25(1 + 0.25)$$
$$= £(1.25)^2 = £1.56$$

Show that at the end of the year you would have

$$£(1.25)^4 = £2.44$$

This is more than if the interest is added annually or semi-annually. Does this mean that if the interest is added more and more frequently, then the sum you will have after one year will get bigger and bigger?

To see what happens, use your calculator to build up the following table. For example, the entry when interest is added monthly is

$$\left(1 + \tfrac{1}{12}\right)^{12} = 2.61$$

| Number of times per year interest is compounded | Total sum at end of one year in £ |
|:---:|:---:|
| 1 | 2 |
| 2 | 2.25 |
| 4 | 2.44 |
| 12 | 2.61 |
| 24 | 2.66 |
| 52 | 2.69 |
| 500 | 2.7156 |
| 1000 | 2.7169 |
| 10000 | 2.7181 |

The table suggests that you're not going to get more than about £2.72 even if the interest were to be added every few minutes!

In fact, what you have been computing is the expression

$$\left(1 + \frac{1}{n}\right)^n$$

for increasing values of $n$, which is the number in the left hand column. It can be shown that as $n$ gets larger and larger, the number in the right hand column gets closer and closer to e.

**Student project**

**1.22**  Try to develop a proof of the property of the number 6174 which was discussed in Section 1.1.1, via the following argument.

> *Step 1*:  Show that there are 9990 different four-digit numbers whose digits are not all the same.

> *Step 2*:  Recall that the operation applied to any four-digit number is to arrange the digits in decreasing order, then in increasing order, and subtract the smaller number from the larger one.

Show that when applied to the 9990 numbers in Step 1, then only 54 different four-digit numbers are obtained. Note that it is very tedious to do this computationally – try to find an algebraic argument.

> *Step 3*:  Of the 54 numbers in Step 2, remove any which contain the same quadruple of digits, to end up with the following thirty numbers:

> | | | | | | | |
> |---|---|---|---|---|---|---|
> | 5544 | 5553 | 6444 | 6543 | 6552 | 6642 | 7443 | 7533 |
> | 7551 | 7632 | 7641 | 7731 | 8442 | 8532 | 8550 | 8622 |
> | 8640 | 8721 | 8730 | 8820 | 9441 | 9531 | 9621 | 9711 |
> | 9810 | 9954 | 9963 | 9972 | 9981 | 9990 | | |

> *Step 4*:  Apply the operation described in Step 2 to each of the thirty numbers in Step 3, to verify that in each case 6174 is reached after at most six more subtractions.

**Student project**

**1.23**  Using appropriate sources, write a detailed essay on generalizations of numbers, including quaternions, Cayley numbers and surreal numbers.

## Further reading

Section 1.1.1  Honsberger R. (1970). *Ingenuity in Mathematics*. New York: Random House, p.79
Roberts J. (1992). *Lure of the Integers*. Mathematical Association of America, p.240

Section 1.1.2  Lines M.E. (1986). *A Number for your Thoughts*. Bristol: Adam Hilger, p.53

Reichmann W.J. (1958). *The Fascination of Numbers*. London: Methuen, p.94

Wells D. (1986). *The Penguin Dictionary of Curious and Interesting Numbers*. London: Penguin, p.179

Section 1.1.3    Lines M.E. (1990). *Think of a Number*. Bristol: Adam Hilger, p.20

Section 1.2    Reichmann W.J. (1958). op.cit., p.74

Sections 1.3.1–1.3.7    Holt, E. (1990). *The Big Book of Numbers, The Man Made World*. London: Pan, p.87

Reichmann W.J. (1958). op.cit., p.64

Section 1.3.8    Gardner R.M. (1956). *Mathematics, Magic and Mystery*. New York: Dover, p.164

Sections 1.4.1, 1.4.2    Cajori F. (1980). *A History of Mathematics*, New York: Chelsea

Campbell D.M. (1976). *The Whole Craft of Number*. Boston MA: Prindle, Weber and Schmidt

Flegg, G. (1984). *Numbers, Their History and Meaning*. London: Pelican

Gardner, M. (1989). Negative Numbers. In *Penrose Tiles to Trapdoor Ciphers*. New York: Freeman, p.151

Section 1.4.3    Beckmann P. (1971). *A History of Pi*. New York: St. Martin's Press

Crandall R.E. (1997). The challenge of large numbers. *Scientific American*, **276**(2), 58–62

Gardner M. (1966). The transcendental number pi. In *New Mathematical Diversions*. New York: Simon and Schuster, p.91

Gardner M. (1970). The transcendental number e. In *Further Mathematical Diversions*. London: George Allen and Unwin, p.34

Gardner, M. (1989). Conway's surreal numbers. In *Penrose Tiles to Trapdoor Ciphers*. New York: Freeman, p.49

Knuth D.E. (1974). *Surreal Numbers*. Reading MA: Addison-Wesley

Maor E. (1993). *e: The Story of a Number*. Princeton: Princeton University Press

Matthews R. (1995). The man who played God with infinity. *New Scientist*, **147**(1993), 36–40

General    For general background material the following are worth reading in addition to the books quoted above.

Blocksma M. (1989). *Reading the Numbers*. New York: Penguin

Conway, J.H. and Guy, R.K. (1996). *The Book of Numbers*. New York, Springer-Verlag

Holt E. (1990). *The Big Book of Numbers, The Natural World*. London: Pan

Humez A., Humez N. and Maguire J. (1993). *Zero to Lazy Eight, The Romance of Numbers*. New York: Simon and Schuster

Menninger K. (1969). *Number Words and Number Symbols, A Cultural History of Numbers*. Cambridge MA: MIT Press

Room A. (1989). *The Guinness Book of Numbers*. London: Guinness Publishing

# Interlude: Proof by Induction

## I.1    Introduction to the method

**Example I.1**    **Dividing up a circle**

Suppose a circle (perhaps a circular cake) is to be cut up into pieces by marking points on its circumference, and joining together pairs of points in all possible ways. The points need not be equally spaced, and cases with two and three points are shown in Figure I.1, where you can see that there are two and four pieces respectively. The case when there are four points results in eight pieces, as shown in Figure I.2. You should draw the case when there are five points and confirm that this produces 16 pieces. You can therefore draw up the following table:

| Number of points | 2 | 3 | 4 | 5 |
|---|---|---|---|---|
| Number of pieces | 2 | 4 | 8 | 16 |

You have probably noticed that the number of pieces doubles each time, and you would therefore guess that for six points there would be 32 pieces. However, the actual situation is shown in Figure I.3, where there are only 31

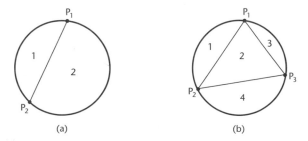

        (a)                                          (b)

*Figure I.1*    Dissecting a circle: (a) two points; (b) three points

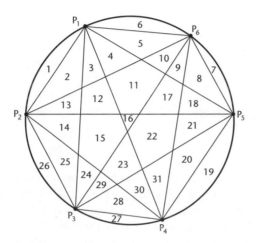

*Figure I.2* Four points

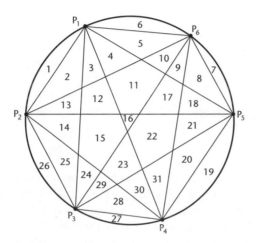

*Figure I.3* Six points

pieces: what has happened to the expected 32nd piece? The answer is that the guess was wrong – the number of pieces does *not* double every time an extra point is added. This example illustrates the danger of assuming that because a result is true for small values of a variable $n$ (here the number of points) then it is true for all values of $n$ (for the correct formula, see Exercise I.1).

### Example I.2    The sum of consecutive odd numbers

Consider the following sums of consecutive odd numbers:

$$1 + 3 = 4, \quad 1 + 3 + 5 = 9, \quad 1 + 3 + 5 + 7 = 16$$

You probably recognize the sums as $2^2$, $3^2$, $4^2$. It therefore looks as though the sums of consecutive odd numbers are equal to squares of consecutive even numbers. Expressing this in formal terms, denote the sum of $n$ consecutive odd numbers by $S_n$. Then $S_1 = 1$, $S_2 = 1 + 3$, $S_3 = 1 + 3 + 5$, and in general the last number in the $n$th sum is $2n - 1$, so that

$$S_n = 1 + 3 + 5 + \ldots + (2n - 1) \tag{I.1}$$

The results given above can be tabulated as

| $n$ | 1 | 2 | 3 | 4 |
|-----|---|---|---|---|
| $S_n$ | $1^2$ | $2^2$ | $3^2$ | $4^2$ |

(I.2)

It's tempting to guess that for $n = 5$ then $S_5 = 5^2$, and this is indeed correct since

$$S_5 = 1 + 3 + 5 + 7 + 9 = 25$$

However, in view of what happened in the previous example you should be wary of assuming that $S_n = n^2$ for *all* positive values of the integer $n$. What is needed is a guaranteed way of establishing whether the guess is indeed correct for all values of $n$. This is provided by the **method of induction**. To see how it works in this example, first write down $S_{n+1}$ by adding on the next term on the right-hand side in (I.1), so that

$$S_{n+1} = 1 + 3 + 5 + \ldots + (2n - 1) + (2n + 1)$$
$$= S_n + (2n + 1) \tag{I.3}$$

If the guess $S_n = n^2$ is correct, then (I.3) becomes

$$S_{n+1} = n^2 + (2n + 1)$$
$$= (n + 1)^2 \tag{I.4}$$

The crucial point you must now appreciate is that (I.4) is the *same formula* as the guess, with $n$ replaced by $n + 1$. This means that if the guess $S_n = n^2$ is correct for some value $n$, then it is also correct for the next value $n + 1$. But $S_1 = 1^2$, so the result is certainly true for $n = 1$. Hence by the above argument it is true for $n = 1 + 1 = 2$; and therefore it is true for $n = 2 + 1 = 3$; and by repeating the argument over and over again the result is true for $n = 4$, $n = 5$, $n = 6, \ldots$. This process can be continued for ever, so the guessed result $S_n = n^2$ is indeed correct for all positive integers $n$.

---

Many students are puzzled by their first encounter with the method of induction, as illustrated by the previous example. It's therefore important to be absolutely clear what's going on. You want to establish that a formula which works for small values of the positive integer $n$ is correct for *all* values of $n$. The method of induction consists of three steps.

*Step 1:* Either on the basis of looking at a pattern like that in (I.2), or some other way, guess a formula which is valid for at least one value $a$ of $n$.

*Step 2:* The crucial fact you have to prove is that if the formula is true for $n$, then it is also true for $n + 1$. Provided this can be done, then:

*Step 3:* It follows that since the result is true for $n = a$, it must also be true for $n = a + 1$; hence it is true for $n = (a + 1) + 1 = a + 2$; then for $n = (a + 2) + 1 = a + 3$; and so on for *all* values of $n \geqslant a$.

**Example I.3**    **The sum of consecutive positive integers**

Let $S_n$ now denote the sum of the first $n$ positive integers, so that

$$S_1 = 1, \quad S_2 = 1 + 2 = 3, \quad S_3 = 1 + 2 + 3 = 6, \quad S_4 = 1 + 2 + 3 + 4 = 10$$

A pattern emerges if you write

$$S_1 = \frac{1 \times 2}{2}, \quad S_2 = \frac{2 \times 3}{2}, \quad S_3 = \frac{3 \times 4}{2}, \quad S_4 = \frac{4 \times 5}{2}$$

so it looks as though

$$S_n = \tfrac{1}{2}n(n+1) \tag{I.5}$$

This is the guessed formula which you use in Step 1 of the induction procedure. The guess (I.5) is certainly true for $n = 1$ (and indeed for $n = 2, 3, 4$).

Moving on to Step 2, write

$$S_{n+1} = 1 + 2 + 3 + \ldots + n + (n+1)$$
$$= S_n + (n+1)$$

If the formula (I.5) is true for $n$, then

$$S_{n+1} = \tfrac{1}{2}n(n+1) + (n+1)$$
$$= (n+1)(\tfrac{1}{2}n + 1)$$
$$= \tfrac{1}{2}(n+1)(n+2)$$

which is precisely (I.5) with $n$ replaced by $n + 1$. In other words, (I.5) is also true for $n + 1$. Since (I.5) is true for $n = 1$, it follows from Step 3 that it is therefore true for *all* values of $n \geqslant 1$.

**Example I.4**    **Sum of powers of 2**

Consider the expressions

$$1 + 2 = 2^2 - 1, \; 1 + 2 + 4 = 2^3 - 1, \; 1 + 2 + 4 + 8 = 2^4 - 1$$

*Step 1*:    On the basis of these results, if you define

$$S_n = 1 + 2 + 2^2 + 2^3 + \ldots + 2^n$$

then you can guess that $S_n = 2^{n+1} - 1$. This formula is certainly correct for $n = 1$ (and $n = 2, 3$).

*Step 2*:    From the definition

$$S_{n+1} = 1 + 2 + \ldots + 2^n + 2^{n+1}$$
$$= S_n + 2^{n+1}$$

If the guessed formula is true for $n$, then

$$S_{n+1} = (2^{n+1} - 1) + 2^{n+1}$$
$$= 2 \times 2^{n+1} - 1$$
$$= 2^{n+2} - 1$$

This is the same as the guessed formula with $n$ replaced by $n + 1$, so the result is true for $n + 1$.

*Step 3:* Hence $S_n = 2^{n+1} - 1$ for all $n \geqslant 1$.

**Exercise**  **1.1** It can be shown that when there are $n$ points on the circle in Example 1.1 then there are $\frac{1}{24}(n^4 - 6n^3 + 23n^2 - 18n + 24)$ pieces. Evaluate this expression for $n = 2, 3, 4, 5, 6, 7$.

**Exercise**  **1.2** Let $S_n$ be the sum of the first $n$ even numbers, that is

$$S_n = 2 + 4 + 6 + \ldots + 2n$$

By evaluating $S_1$, $S_2$, $S_3$, $S_4$ and factorising the results, guess an expression for $S_n$.

**Exercise**  **1.3** Prove your result in the preceding exercise is true for all values of $n \geqslant 1$ by the method of induction.

**Exercise**  **1.4** What is wrong with the following 'proof' by induction?

*Step 1:* It is required to prove that all positive integers are small.

*Step 2:* If $n$ is small, so is $n + 1$.

*Step 3:* The result is true for $n = 1$. Hence it is true for all values of $n \geqslant 1$.

## 1.2  The sigma notation for summation

Before presenting some more applications of the method of induction it's useful to introduce a standard notation for sums. This uses the Greek capital letter $\Sigma$ ('sigma'). For example, the sum of the squares of the first $n$ positive integers is written as

$$\sum_{i=1}^{n} i^2 = 1^2 + 2^2 + 3^2 + \ldots + n^2 \tag{1.6}$$

The term on the left in (I.6) means that the quantity covered by the sigma is summed for consecutive integer values of the **index** $i$ going from 1 to $n$. For example, with $n = 4$ in (I.6)

$$\sum_{i=1}^{4} i^2 = 1^2 + 2^2 + 3^2 + 4^2$$

Similarly, the sum of the first $n$ positive integers in Example I.3 is simply

$$\sum_{i=1}^{n} i = 1 + 2 + 3 + \ldots + n$$

and the sum of $n$ consecutive odd integers in (I.1) can be written as

$$\sum_{i=1}^{n} (2i - 1) = 1 + 3 + 5 + \ldots + (2n - 1)$$

In general if $x_i$ is some expression depending on $i$ then

$$\sum_{i=1}^{n} x_i = x_1 + x_2 + x_3 + \ldots + x_{n-1} + x_n$$

There is no need for the index to start at $i = 1$, for example

$$\sum_{i=5}^{8} i^2 = 5^2 + 6^2 + 7^2 + 8^2$$

and similarly

$$\sum_{i=5}^{8} x_i = x_5 + x_6 + x_7 + x_8$$

The initial value of $i$ (5 in the last two examples) is called the **lower limit** of the sum, and the final value of $i$ (8 in the examples) is the **upper limit**. In a sum

$$\sum_{i=a}^{b} x_i = x_a + x_{a+1} + x_{a+2} + \ldots + x_b$$

you must have $a$ and $b$ integers with $b \geqslant a$. If $b = a$ then the sum consists of a single term $x_a$. Some properties of the sigma notation are as follows:

(1) The variable $i$ can be replaced by any other symbol without affecting the result. For example,

$$\sum_{j=1}^{n} j^2, \quad \sum_{k=1}^{n} k^2, \quad \sum_{r=1}^{n} r^2$$

are all the same as (I.6), and similarly

$$\sum_{i=a}^{b} x_i = \sum_{k=a}^{b} x_k = \sum_{r=a}^{b} x_r$$

For this reason, the index is said to be a **dummy variable**, since it plays no role in the final expression.

(2) The expression to be evaluated for each value of $i$ may contain more than one term. For example

$$\sum_{i=1}^{n}(i^2 + 4i) = (1^2 + 4 \times 1) + (2^2 + 4 \times 2)$$
$$+ (3^2 + 4 \times 3) + \ldots + (n^2 + 4 \times n)$$

(3) If a term in the summation does not involve the index $i$, then it is simply added to itself an appropriate number of times. For example,

$$\sum_{i=1}^{5} x = x + x + x + x + x = 5x$$

$$\sum_{i=1}^{n}(i^2 - 2) = (1^2 - 2) + (2^2 - 2) + (3^2 - 2) + \ldots + (n^2 - 2)$$
$$= 1^2 + 2^2 + 3^3 + \ldots + n^2 - 2n$$

The brackets within $\Sigma$ in the above example are important, since

$$\sum_{i=1}^{n} i^2 - 2 = (1^2 + 2^2 + \ldots + n^2) - 2$$

(4) Rearrangement of terms covered by $\Sigma$ is allowable, for example the sum in (2) can be written

$$\sum_{i=1}^{n}(i^2 + 4i) = \sum_{i=1}^{n} i^2 + \sum_{i=1}^{n} 4i$$
$$= \sum_{i=1}^{n} i^2 + 4 \sum_{i=1}^{4} i$$

Notice that the common factor 4 can be taken outside $\Sigma$.

(5) Sometimes changing the dummy variable is useful. For example, to simplify

$$\sum_{i=4}^{8} 2(i - 7) \tag{I.7}$$

write $j = i - 7$. When $i = 4$ then $j = -3$, and when $i = 8$ then $j = 1$. Hence the sum (I.7) can be written as

$$\sum_{j=-3}^{1} 2j = 2 \sum_{j=-3}^{1} j = 2(-3 - 2 - 1 + 0 + 1) \tag{I.8}$$
$$= -10$$

The expression in (I.8) is merely an alternative way of writing (I.7). Both sums have the same value $-10$. Incidentally, (I.8) illustrates that the limits of the sum need not be positive.

**Exercise**    **I.5**   Write out the following sums and evaluate them:

(a) $\displaystyle\sum_{i=1}^{6} 2(i+2)$   (b) $\displaystyle\sum_{j=1}^{4} j^3$   (c) $\displaystyle\sum_{k=-1}^{4} (k^2 - 2k)$   (d) $\displaystyle\sum_{i=1}^{3} x_{3i-1}.$

**Exercise**    **I.6**   Express the following in sigma notation:

(a) the sum of the first 200 odd integers

(b) the sum of the even integers from 20 to 200 inclusive

(c) $7 + 14 + 21 + 28 + \ldots + 126$

(d) $x_7 + x_{11} + x_{15} + x_{19} + \ldots + x_{39}$

(e) $7 + 28 + 63 + 112 + \ldots + 343.$

**Exercise**    **I.7**   Simplify the following expressions:

(a) $\displaystyle\sum_{i=1}^{14} 7x$   (b) $\displaystyle\sum_{j=3}^{20} (4j - 3)$   (c) $\displaystyle\sum_{k=10}^{10} (2k - 8)$   (d) $\displaystyle\sum_{i=25}^{50} 6(i - 30).$

## I.3   Further examples

**Example I.5**    **Incorrect guess**

It's instructive to see what happens if at Step 1 of the induction procedure you make an incorrect guess for the formula. Go back to Example I.2, where you wanted to find a formula for the sum $S_n$ of the first $n$ odd numbers. Suppose that for some reason you guessed that

$$S_n = 2n^2 - 3n + 2 \tag{I.9}$$

This certainly is correct for $n = 1$ and $n = 2$, since substituting $n = 1$ and $n = 2$ into (I.9) gives

$$S_1 = 2 \times 1^2 - 3 \times 1 + 2 = 1, \quad S_2 = 2 \times 2^2 - 3 \times 2 + 2 = 4$$

which agrees with the values in the table (I.2). To perform Step 2 of the induction procedure, recall from (I.3) that

$$S_{n+1} = S_n + (2n + 1)$$

If you assume that (I.9) is true for $n$, then

$$S_{n+1} = 2n^2 - 3n + 2 + (2n + 1)$$
$$= 2n^2 - n + 3 \tag{I.10}$$

However, replacing $n$ by $n + 1$ in (I.9) gives

$$S_{n+1} = 2(n+1)^2 - 3(n+1) + 2$$
$$= 2(n^2 + 2n + 1) - 3(n+1) + 2$$
$$= 2n^2 + n + 1$$

which is *not* the same as (I.10) except when $n = 1$. This shows that if (I.9) is true for $n (>1)$ it is *not* true for $n + 1$. Hence Step 3 of the induction method cannot be applied: although (I.9) is valid for $n = 2$ it does *not* hold for $n = 3$ or any subsequent values of $n$. The fact that (I.9) works for $n = 1$ and $n = 2$ is just a fluke, like the guess for the number of pieces of a dissected circle in Example I.1.

The method of induction is not just for proving the correctness of formulae, as the following two examples illustrate.

**Example I.6**   **Divisibility**

It is required to prove that for all positive integers $n$:

(a)  $n^2 + n + 2$ is even

(b)  $n(n^2 + 5)$ is divisible by 6.

In part (a) write $x_n = n^2 + n + 2$. The 'guess' in Step 1 of the induction method has been made for you, and is correct for $n = 1$ since $x_1 = 1 + 1 + 2 = 4$ which is even. For Step 2, assume that $x_n$ is even, so it can be written as $x_n = 2p$, where $p$ is a positive integer. Then replacing $n$ by $n + 1$ gives

$$x_{n+1} = (n+1)^2 + (n+1) + 2$$
$$= (n^2 + 2n + 1) + n + 3$$
$$= n^2 + n + 2 + 2n + 2$$
$$= x_n + 2(n+1)$$
$$= 2p + 2(n+1)$$

This shows that $x_{n+1}$ is also even, since it is the sum of two even numbers. Hence by Step 3 the result is true for all values of $n \geqslant 1$.

In part (b) write $y_n = n(n^2 + 5)$. The required result is certainly true for $n = 1$ since $y_1 = 1(1 + 5)$ which is divisible by 6. Assume that $y_n$ is divisible by 6, so that $y_n = 6r$, where $r$ is a positive integer. Then replacing $n$ by $n + 1$ gives

$$y_{n+1} = (n+1)\left[(n+1)^2 + 5\right]$$
$$= (n+1)(n^2 + 2n + 6)$$
$$= n^3 + 3n^2 + 8n + 6$$
$$= (n^3 + 5n) + 3n^2 + 3n + 6$$
$$= y_n + 3(n^2 + n + 2)$$
$$= y_n + 3x_n \qquad\qquad\text{(I.11)}$$

where $x_n$ is defined in the solution to part (a). Since it was shown that $x_n = 2p$, it follows from (I.11) that $y_{n+1} = y_n + 6p$ is divisible by 6, since it is the sum of two numbers each divisible by 6. Hence the result is true for $n + 1$, so it follows from Step 3 that it holds for all $n \geqslant 1$.

---

**Example I.7**   **An inequality**

It is required to prove that $2^{n-2} > 2n - 3$ for all positive integers $n \geqslant 5$. Once again the 'guess' in Step 1 has been made for you. When $n = 5$ then $2^{n-2} = 2^3 = 8$ and $2n - 3 = 7$, so the result is true for $n = 5$. Assuming it is true for $n$, this means that $2^{n-2} - (2n - 3)$ is positive, that is, for $n \geqslant 5$

$$2^{n-2} = 2n - 3 + p \tag{I.12}$$

where $p > 0$. To show the result is true for $n + 1$, replace $n$ by $n + 1$ in the given inequality. Consider the difference between the left and right sides, namely

$$\begin{aligned}
2^{(n+1)-2} - [2(n + 1) - 3] &= 2^{n-1} - 2n + 1 \\
&= 2 \times 2^{n-2} - 2n + 1 \\
&= 2(2n - 3 + p) - 2n + 1 \qquad \text{using (I.12)} \\
&= 2n + 2p - 5 \tag{I.13}
\end{aligned}$$

Since $n \geqslant 5$ and $p > 0$, the expression in (I.13) is positive, so the required inequality does hold for $n + 1$. Hence by Step 3 the result is true for all $n \geqslant 5$. Note that the result is **not** true for $n = 2, 3, 4$.

---

The method of induction will be used on a number of occasions throughout the chapters which follow.

**Exercise**   **I.8**   Prove the following by the method of induction:

(a) $\displaystyle\sum_{i=1}^{n} i^2 = \frac{n(n + 1)(2n + 1)}{6}$

(b) $\displaystyle\sum_{i=1}^{n} i^3 = \frac{n^2(n + 1)^2}{4}.$

**Exercise**   **I.9**   By considering the values of

$$S_n = \sum_{i=1}^{n} \frac{1}{i(i + 1)}$$

for $n = 1, 2, 3$, guess a general formula for $S_n$ and prove it by induction.

**Exercise**    **I.10**    Prove by induction that $n^3 + 2n$ is divisible by 3 for all integers $n \geqslant 1$.

**Exercise**    **I.11**    Let $x_n = 2^{2n} - 1$, and by considering $x_1$, $x_2$ and $x_3$ guess a result for the divisibility of $x_n$ by an integer. Prove this result by induction.

**Exercise**    **I.12**    Prove by induction that for all positive integers $n$:

(a) $1^2 + 3^2 + 5^2 + \ldots + (2n-1)^2 = \frac{1}{3}n(4n^2 - 1)$

(b) $1^3 + 3^3 + 5^3 + \ldots + (2n-1)^3 = n^2(2n^2 - 1)$.

**Exercise**    **I.13**    Prove by induction that:

(a) $n^2 - 2n - 1 > 0$,    $n \geqslant 3$      (b) $2^n > n^2$,    $n \geqslant 5$.

**Exercise**    **I.14**    Consider the identities

$$\left(1 - \frac{1}{2^2}\right) = \frac{3}{4}$$

$$\left(1 - \frac{1}{2^2}\right)\left(1 - \frac{1}{3^2}\right) = \frac{4}{6}$$

$$\left(1 - \frac{1}{2^2}\right)\left(1 - \frac{1}{3^2}\right)\left(1 - \frac{1}{4^2}\right) = \frac{5}{8}$$

Guess a general formula based on these, and prove it by induction.

## Miscellaneous problems

**I.1**    Given that

$$\sum_{i=1}^{n} i = \frac{n(n+1)}{2}$$

it follows from Exercise I.8(b) that

$$\sum_{i=1}^{n} i^3 = \left(\sum_{i=1}^{n} i\right)^2$$

Prove this result directly by induction.

**I.2**    What is the flaw in the following 'proof' by induction?

*Step 1:*    It is required to prove that $b^n = 1$ for any number $b \neq 0$ and all integers $n \geqslant 0$. The result is certainly true for $n = 0$, since $b^0 = 1$ by definition.

*Step 2:*    Assume true for $n$, that is, assume $b^n = 1$. Then

$$b^{n+1} = \frac{b^n b^n}{b^{n-1}} = \frac{1.1}{1} = 1$$

so the result is also true for $n + 1$.

*Step 3:*    Hence the result is true for all $n \geqslant 0$.

**I.3** Whilst playing with your calculator you discover the following:

$$2 + 3 + 4 = 1 + 8, \quad 5 + 6 + 7 + 8 + 9 = 8 + 27,$$
$$10 + 11 + 12 + 13 + 14 + 15 + 16 = 27 + 64$$

Guess a general formula based on these results, and prove it by induction.

## Student project

**I.4** The **harmonic numbers** are defined by

$$H_1 = 1, \quad H_2 = 1 + \frac{1}{2}, \quad H_3 = 1 + \frac{1}{2} + \frac{1}{3}, \quad \ldots, \quad H_n = \sum_{i=1}^{n} \frac{1}{i}$$

Prove by induction that for all integers $n \geqslant 1$:

(a) $H_m \leqslant 1 + n,$ where $m = 2^n$

(b) $\displaystyle\sum_{k=1}^{n} H_k = (n+1)H_n - n$

(c) $\displaystyle\sum_{k=1}^{n} kH_k = \frac{1}{2}n(n+1)H_{n+1} - \frac{1}{4}n(n+1)$

(d) $H_{2n+1} - \frac{1}{2}H_n = \displaystyle\sum_{i=0}^{n} \frac{1}{2i+1}.$

## Further reading

Dierker P.F. and Voxman W.L. (1986). *Discrete Mathematics*. San Diego: Harcourt Brace Jovanovich, p.12

Finkbeiner II D.T. and Lindstrom W.D. (1987). *A Primer of Discrete Mathematics*. New York: W.H. Freeman, p.72

Grimaldi R.P. (1994). *Discrete and Combinatorial Mathematics, An Applied Introduction* 3rd edn. Reading MA: Addison-Wesley, Section 4.1

Molluzzo J.C. and Buckley F. (1986). *A First Course in Discrete Mathematics*. Belmont CA: Wadsworth, p.111

Wheeler R.F. (1981). *Rethinking Mathematical Concepts*. Chichester: Ellis Horwood, Chapter 18

# 2 Number Systems

Only older people now remember the pre-decimal British currency which went out of use in 1971. The system consisted of the pound which was broken up into 20 shillings, each of which was worth 12 pence (or 'pennies'). To do money calculations you had therefore to count in twelves and twenties. The old imperial system of weights and measures, still in partial use in Britain, is even worse. There are 12 inches to the foot, three feet to the yard, and 1760 yards to the mile. For weights there are 16 ounces to the pound, 14 pounds to the stone, and 2240 pounds to the ton. No wonder the metric system was invented! Usually, then, you count in tens using the **decimal** system, this having arisen because we have ten **digits** on our two hands (notice that 'digit' denotes finger or thumb as well as a numeral from 0 to 9). The number ten is called the **base** of the number system. Counting using bases other than ten is also important, for example there are 24 hours in a day and 360 degrees in a circle. The **binary** system of counting in twos is all-powerful in the world of computers. An introduction to some ideas of different systems of counting is given in Section 2.1. The remainder of this chapter then investigates properties of important number systems and the appropriate arithmetic involved.

## 2.1    Clocks, compasses and computers

The most common everyday situation where you count using a non-decimal system is in measuring time. On a clock face the hours are counted in twelves or twenty fours; and minutes and seconds in sixties. On a calendar, days are counted in sevens. A variation on the base 60 arises with measurement of angles, where 360 degrees constitute a complete circle. Bases larger than ten are certainly useful, but despite the ability of computers to work extremely quickly they count only in twos, using the 'binary system'. If you can do even simple arithmetic in your head then you can feel superior to computers, which are so stupid that they rely on binary arithmetic involving only the digits 0 and 1!

### 2.1.1 Telling the time

Most children learn to read the time from a clock face at quite an early age, but this is not an easy task because of the illogicality and difficulty of the system in general use. A standard clock or watch usually has three pointers called 'hands'. The fastest moving hand counts out the seconds and makes one complete revolution per minute, which consists of 60 seconds. The minute hand completes a revolution in 60 minutes, which constitute an hour. The hour hand, which is the shortest of the three, makes a complete revolution in 12 hours. An immediate difficulty arises because there are 24 hours in a complete day. There are two ways round this problem. One is to use a.m. (an abbreviation for the Latin *ante meridiem*, meaning 'before noon') to indicate the time period from midnight to noon; and p.m. for the period from noon to midnight, this being an abbreviation for the Latin *post meridiem* 'after noon'. The second scheme, which is widely used throughout most of Europe, is the '24-hour clock', whereby counting continues after noon with 13.00 standing for 1.00 p.m., 14.00 for 2.00 p.m. and so on, ending up with 24.00 for midnight. Digital clocks and watches usually offer the options either to show a.m. and p.m. symbols, or to use a 24-hour display The 24-hour clock is certainly more logical – an unfortunate convention which is used with the a.m./p.m. system is to refer to a time like 30 minutes past noon as 12.30 p.m., when strictly it should be 00.30 p.m. (the same holds for the first hour after midnight – for example, 12.30 a.m. should really be 00.30 a.m.).

What's of particular relevance to this chapter is the curious way in which counting time is done in sixties and twelves (or twenty-fours). The use of 60 as a number base goes back to the Babylonians, who used it for calculations with fractions. The base 60 may have been preferred because it has many factors, being divisible by 2, 3, 4, 5, 6, 10, 12, 15, 20 and 30. The division of day and night into 12 hours each, giving a total duration of 24 hours, is due to the ancient Egyptians. It's remarkable that the same method of measuring time is used everywhere in the world. Despite the (almost) universal use of the metric system of weights and measures, no one has seriously suggested dividing a day up into 10 or 20 hours, an hour into 100 minutes, or a minute into 100 seconds!

Anyone who can tell the time with a conventional clock is therefore familiar with using 60 as a base for counting. For example, if the time is 2 o'clock, then 75 minutes later the minute hand shows 15 minutes past the hour, and the time will be 3.15 since 1 hour and 15 minutes has elapsed. In other words, you read the minutes on a clock face by adding (or subtracting) suitable multiples of 60.

| Example 2.1 | **Time calculations** |
|---|---|

Suppose that a conventional clock shows 3.00. The problem is to work out the positions of the hands when given periods of time have elapsed.

(a) Consider first the hour hand. This is in the same position (that is, it points at 3) every 12 hours. For example, since $19 = 12 + 7$ then after 19 hours the hour hand will advance through 7 hours, so the clock will show 10.00 (see Figure 2.1(b)). In general, since multiples of 12-hour periods do not affect the position of the hour hand, you divide the number of elapsed hours by 12 and use only the *remainder* to find by how many hours the hour hand has advanced. For example, since 100 divided by 12 gives a remainder of 4 (that is, $100 = 8 \times 12 + 4$) then after 100 hours the hour hand will have advanced by 4 and the clock will show 7.00.

The same applies when going backwards in time. For example, 4 hours ago the clock showed 11.00 (see Figure 2.1(c)). To see how this is computed, note that $3 - 4 = -1$, but there are no negative numbers on the clock face! However, the negative sign can be interpreted as going in the anticlockwise direction, so that $-1$ represents one hour before noon, as shown in Figure 2.1(c). Alternatively, since the position of the hour hand is unaltered by 12-hour periods you can convert $-1$ to a positive time by adding 12 to it, that is, $-1 + 12 = 11$. This shows once again that the hour hand points at 11. Similarly, 19 hours ago the hour hand was in the same position as 7 hours ago, and so was pointing at $3 - 7 = -4$. This converts to $-4 + 12 = 8$, so 19 hours earlier the clock showed 8.00. In each case you use the remainder after division by 12; as a further example, the time showing 100 hours ago was the same as that 4 hours ago (that is, 11.00).

If your digital clock or watch is set to the 24-hour cycle, then the calculations are similar except that the display repeats itself every 24 hours. For example, if the display shows 15:10 (that is, 3.10 p.m.) then what will it show 110 hours later? The clock will record a total time of 125:10. The number of hours remaining after division by 24 is 5 (since $125 = 5 \times 24 + 5$) so the actual display will be 5:10 (that is, 5.10 a.m.).

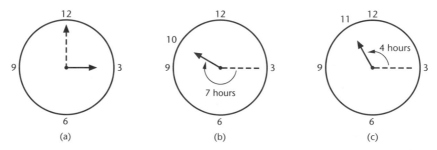

*Figure 2.1*  Clock hour hand: (a) 3 o'clock; (b) 7 hours later; (c) 4 hours earlier

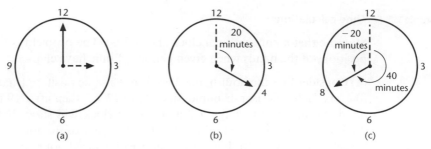

*Figure 2.2*  Clock minute hand: (a) 3 o'clock; (b) 320 minutes later; (c) 200 minutes earlier

(b) Now consider the minute hand. Its position is unaffected by 60-minute periods, backwards or forwards. You therefore now use remainders after division by 60. For example, where will the minute hand point after 320 minutes have elapsed? Since $320 = 5 \times 60 + 20$, the minute hand will have moved through 20 minutes and so will be pointing at 4 (see Figure 2.2(b)).

Similarly, since $200 = 3 \times 60 + 20$, then 200 minutes earlier the minute hand pointed at $-20$ minutes, as shown in Figure 2.2(c). To convert this to a positive time (that is, in the clockwise direction) you now add 60, thus $-20 + 60 = 40$. Either way, 200 minutes ago the minute hand pointed at 8.

If the clock has a third hand, counting seconds, then since it makes one complete revolution in 60 seconds, computing its position follows exactly the same rules as for the minute hand.

(c) If the clock also displays the day of the week, then this follows a 7-day cycle. For example, if the clock shows Monday then after 7 complete days have elapsed it will again show Monday. The remainders after division by the base 7 are now used in the same way as base 12 for hours and base 60 for minutes (and seconds). For example, what day will the clock show after 26 days have passed? Since $26 = 3 \times 7 + 5$, the overall effect is to advance the day by 5. Hence the clock will record Saturday, as shown schematically below.

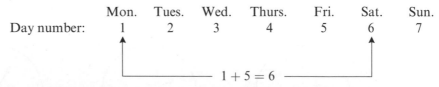

This numbering can also be used to determine days in the past. For example, what day was displayed 17 days ago? Here the remainder after division by 7 is 3 (since $17 = 2 \times 7 + 3$), so the day recorded was $1 - 3 = -2$. Adding 7 gives 5 which corresponds to Friday, and this was displayed on the clock 17 days ago.

Even if you prefer to use digital watches and clocks, you'll still have learned to tell the time with a conventional timepiece having hands. So without realizing it, you've also learned how to do calculations involving base 12 or base 60. Using a calendar has also taught you to compute using base 7. Arithmetic of this type is called 'modular', and will be studied in detail in Section 2.4. The properties of different bases for counting will be explored in Section 2.2.

**Exercise**   **2.1**   What time does a conventional clock (that is, with hands) show

(a) 23 hours after it shows 4 o'clock?

(b) 118 hours after it shows 11 o'clock?

(c) 82 hours before it shows noon?

**Exercise**   **2.2**   A conventional clock shows 2.40. What does the minute hand show

(a) 1000 minutes later?

(b) 2030 minutes earlier?

**Exercise**   **2.3**   If your digital watch shows 15:30, what time will it show in 205 hours' time?

**Exercise**   **2.4**   If today is Tuesday, what day of the week will it be in 107 days from now? What day was it 81 days ago?

**Exercise**   **2.5**   If 8 January is Monday, what day of the week will 25 January be?

**Exercise**   **2.6**   If your birthday this year falls on a Saturday, what day of the week will your birthday be in two years' time (assuming there are no intervening leap years)?

### 2.1.2   Finding your way

If you're lost on the moors or in the woods, then a compass is very useful in pointing you back onto the correct path. It's marked in degrees, there being 360 in a complete circle. Each degree is divided into 60 minutes, and each minute into 60 seconds of arc. Once again, as for counting time, the Babylonian base of 60 is used. One reason for having 360° in a circle may be that the circle was divided up into six equal parts, each subtending 60° at the centre. Another possibility arises from the primitive astronomical belief that the year consisted of 360 days, made up of 12 lunar months each 30 days long. Although the French counting system is a combination of tens and twenties, it is interesting that the structure of words for multiples of ten changes after 60. Thus, *dix* = 10, *vingt* = 20, *trente* = 30, *quarante* = 40, *cinquante* = 50, *soixante* = 60, which are all independent number words. But a compound structure is then used: 70 = *soixante-dix* (sixty-ten), 80 = *quatre-vingts* (four-twenties), 90 = *quatre-vingt-dix* (four-twenty-ten).

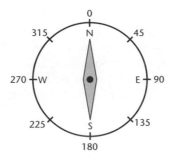

*Figure 2.3* Points of the compass

You read a compass clockwise; north is 0°, east is 90°, south is 180° and west is 270°. Negative angles can be regarded as going in the anticlockwise direction, so for example −90° is the same as 270°.

---

**Example 2.2**   **Using a compass**

If the needle points due north and is rotated through 1530°, where is it pointing? To answer this, you proceed as for 'clock arithmetic', but now use 360 as the base. Since $1530 = 4 \times 360 + 90$, you use only the *remainder* 90° so the needle is pointing due east. Similarly, for a rotation of −1170°, then $-1170 = -3 \times 360 - 90$, so the net effect is a rotation through −90°, that is, the needle is pointing due west.

Another way of measuring angles is in terms of radians instead of degrees. One **radian** is defined to be the angle subtended at the centre of a circle by an arc of length $r$, where $r$ is the radius of the circle. Since the total circumference is $2\pi r$, the total angle round the circle is $2\pi$ radians. It is customary to use the superfix $^c$ to denote radians. Thus $2\pi^c = 360°$ so the conversion from radians to degrees is $\pi^c = 180°$, or $1^c = 180/\pi° \approx 57.3°$. The trigonometric functions sine, cosine and tangent are all periodic with period $2\pi$, for example, $\sin(x + 2\pi k) = \sin x$, for any integer $k$ and angle $x$ (in radians). If $x$ is in degrees, then $\sin(x + 360k) = \sin x$.

---

**Exercise**   **2.7**   If a compass needle points due east and then rotates through either 3000° or −3000°, where does it end up pointing in each case?

**Exercise**   **2.8**   Given that $\sin 30° = 0.5$, $\sin 60° = 0.866$, $\sin 90° = 1$, $\sin 120° = 0.866$, $\sin 150° = 0.5$, $\sin 180° = 0$, determine the values of

(a) $\sin 1110°$   (b) $\sin 2970°$   (c) $\sin 4500°$   (d) $\sin(-7770°)$.

## 2.1.3    Computer reckoning

A humble electric light switch has two positions, 'on' or 'off'. Despite the popular conception of a computer as an 'electronic brain', in essence a computer is a very large collection of on/off switches, connected together in exceedingly intricate ways. It is natural to use the numbers 0 and 1 to represent the two states of an on/off switch. Since computers count in twos we can fancifully imagine that a computer has only one finger on each hand! The digits 0 and 1 are called **bits** (an abbreviation for **binary digits**). A string of bits like 10011011 is called a **byte** and is the basic unit of information in the computer. A byte usually consists of eight bits, and since each bit has two states the total number of different 8-bit bytes is $2^8 = 256$.

Most computers use a system called the **American Standard Code for Information Interchange** (ASCII) for representing information in numerical form. The ASCII code uses seven bits so there are $2^7 = 128$ possibilities, and the eighth bit in the byte is used in various ways, for example detecting errors (see Section 4.1.2). The 128 characters represented by the ASCII code include the digits 0 to 9, upper- and lower-case letters of the alphabet, and some mathematical and other symbols. A sample listing is given in Table 2.1. For example, the word 'bad' in ASCII code is 110001011000011100100.

A computer memory stores information using one byte per character. A **kilobyte** (or kb) consists of $2^{10} = 1024$ bytes, and a **megabyte** (Mb) is 1024 kb, or $2^{20} = 1,048,576$ bytes. A typical typed page of around 500 words contains about 3000 characters (including spaces) and so requires 3 kb of memory. The Read Only Memory (ROM) is a computer's permanent internal memory which cannot be altered. The Random Access Memory (RAM) is the accessible memory which holds data and programs temporarily. Personal computers can have as much as 16 Mb (or more) RAM, and so can store vast amounts of information and run very powerful software. Information in RAM can be saved on 'floppy disks' or 'hard disks' which can have capacities up to hundreds of Mb.

The manipulation of binary numbers using binary arithmetic will be dealt with in Section 2.2.2.

*Table 2.1* Part of seven-bit ASCII code

| Character | Seven-bit symbol |
|-----------|------------------|
| A | 1000001 |
| B | 1000010 |
| C | 1000011 |
| a | 1100001 |
| b | 1100010 |
| c | 1100011 |
| d | 1100100 |
| 1 | 0110001 |
| 2 | 0110010 |
| 3 | 0110011 |

**Exercise**  **2.9**  There are four two-bit bytes, which can be numbered as follows:

00  01  10  11
0   1   2   3

This representation can be used to halve the number of digits in the ASCII code. Put an extra zero onto the left-hand end of each seven-bit symbol in Table 2.1, so for example the capital letter A becomes 01000001. This can be replaced by the four-digit symbol 1001 because of the correspondence:

01  00  00  01
1   0   0   1

Write out the four-digit representations for the rest of the entries in Table 2.1.

## 2.2  Number bases

The mathematical properties of decimal, binary and other number systems are developed in this section. The key property for any representation is that the *position* of a digit gives its value in terms of the number base being used. Contrast this with the system of Roman numerals where X stands for ten, V for five and I for 1, so that XXXVI represents 36. Although the three X's occupy different positions in the block of numerals, each only counts for the same value 10. For this reason it's impossible to perform even addition in a standardized way, never mind multiplication and division. The 'place-value' system is essential for the development of systematic arithmetical procedures, and is also economical in the use of symbols – for example, compare the Roman CCCLXXII with the decimal equivalent 372.

### 2.2.1  Decimal system

The decimal system uses the base ten, and as already mentioned derives from counting on the fingers (and thumbs) of both hands. There are counting systems based on 20, involving toes as well as fingers – indeed the word 'digit' can also mean a toe. Relics of a base-20 system can be seen in the word 'score' ('three-score years and ten' = 70, the expected human lifespan according to the King James Bible); in the French '*quatre-vingts*' = four twenties; and in the Welsh '*ugeint*' = 20, '*de-ugeint*' = 40, '*tri ugeint*' = 60, '*pedwar-ugeint*' = 80 (other Celtic languages have a similar pattern).

The number three hundred and seventy two means just what it says: three hundreds, seven tens and two units. In symbols,

$$372 = 3 \times 100 + 7 \times 10 + 2 \times 1$$
$$= 3 \times 10^2 + 7 \times 10 + 2 \times 1$$

where, reading from right to left, the first digit is the number of units, the second is the number of tens, the third is the number of hundreds. Thus the *position* of the digit tells you what power of ten it corresponds to. For example, the decimal number $x_4 x_3 x_2 x_1$ corresponds to

$$x_4 x_3 x_2 x_1 = x_4 \times 10^3 + x_3 \times 10^2 + x_2 \times 10 + x_1$$

where each of the $x$'s is one of the ten digits $0, 1, 2, \ldots, 9$.

Everybody quickly learns how to multiply a number by ten – just add a zero onto the end. For example, if you multiply the number 372 above by 10, this corresponds to

$$(3 \times 10^2 + 7 \times 10 + 2 \times 1) \times 10 = 3 \times 10^3 + 7 \times 10^2 + 2 \times 10$$
$$= 3720$$

Each digit is moved one place to the left, so 3 hundreds become 3 thousands, 7 tens become 7 hundreds and 2 units become 2 tens. In exactly the same way, multiplying a decimal number by $10^k$, where $k$ is a positive integer, moves each digit $k$ places to the left, and puts $k$ zeros onto the right-hand end of the number – for example,

$$372 \times 10^4 = 3720000$$

The converse clearly applies: if a number ends with $k$ zeros then it is divisible by $10^k$. Otherwise, division by 10 produces a non-integer quotient. For example $372 \div 10 = 37.2$, which is obtained by inserting a **decimal point** to the left of the last digit. Similarly, to divide by $10^k$ insert a decimal point $k$ places to the left, for example $1372 \div 10^3 = 1.372$. Digits to the *right* of the decimal point correspond to *negative* powers of 10, for example

$$x_4 x_3 x_2 x_1 \cdot y_1 y_2 y_3 = x_4 10^3 + x_3 10^2 + x_2 10 + x_1 + y_1 10^{-1} + y_2 10^{-2} + y_3 10^{-3}$$

The generalisation to number systems with base $b$, where $b$ is a positive integer, proceeds in an identical way with ten replaced by $b$. For example, a number $x_4 x_3 x_2 x_1$ expressed in base $b$ is

$$x_4 x_3 x_2 x_1 = x_4 b^3 + x_3 b^2 + x_2 b + x_1$$

where each of the $x$'s is one of the $b$ digits $0, 1, 2, 3, \ldots, b-1$. Notice that if $b$ is greater than ten, then special symbols are needed for the numbers ten, eleven, $\ldots, b-1$. For example, if $b$ is eleven then the Roman numeral X can be used for $b-1$ (see Exercise 2.104 for a practical application of this to the International Standard Book Number).

To avoid confusion, the notation $(x_4 x_3 x_2 x_1)_b$ will be used to denote a number $x_4 x_3 x_2 x_1$ when the base is $b$, provided $b$ is *not* ten. Thus from now on in this chapter *only* decimal numbers will be used without a suffix. In particular notice that $(10)_b = 1 \times b + 0 = b$ for any base $b$, so the suffix notation is essential to avoid a confusing statement like $10 = b$, when $b$ is not ten. The symbol $(10)_b$ is read as 'one-zero to base $b$', *not* 'ten to base $b$'.

Just as for decimal numbers, multiplying a number to base $b$ by the base itself has the effect of moving each digit one place to the left. For example, $(x_4 x_3 x_2 x_1)_b \times b$ corresponds to

$$(x_4 b^3 + x_3 b^2 + x_2 b + x_1)b = x_4 b^4 + x_3 b^3 + x_2 b^2 + x_1 b$$
$$= (x_4 x_3 x_2 x_1 0)_b$$

so multiplication by $b$ corresponds to putting a zero onto the right-hand end of the number. Again, just as for decimal numbers, multiplying by $b^k$ has the effect of putting $k$ zeros onto the right-hand of the number. A particular example of this is $(10)_b = b$, $(100)_b = b^2$, $(1000)_b = b^3$, and so on. The 'point' notation can also be used, where digits to the right of the point correspond to negative powers of $b$ – see Exercise 2.12, and Section 2.2.5.1.

**Exercise**  **2.10**  Express the following as decimal numbers:

(a) $(231)_4$  (b) $(3021)_5$  (c) $(94X)_{11}$.

**Exercise**  **2.11**  The **point** notation for any base $b$ is defined in the same way as for decimal numbers. For example

$$(2.34)_6 = 2 + 3 \times 6^{-1} + 4 \times 6^{-2}$$
$$(x_1.y_1 y_2 y_3)_b = x_1 + y_1 b^{-1} + y_2 b^{-2} + y_3 b^{-3}$$

Express the following in terms of base 10, giving answers involving fractions:

(a) $(1.101)_2$  (b) $(23.14)_5$  (c) $(XX.XX)_{11}$.

**Exercise**  **2.12**  To square a decimal number ending in 5, that is, $x_n x_{n-1} x_{n-2} \ldots x_2 5$, a simple rule is to evaluate the product $x_n x_{n-1} \ldots x_2 \times [(x_n x_{n-1} \ldots x_2) + 1]$ and then append the digits 25 onto the right-hand end of this product. For example, to evaluate $(2735)^2$, first compute $273 \times 274 = 74802$ and then append 25 to get $(2735)^2 = 7480225$. Prove the validity of this rule.

### 2.2.2  Binary system

#### 2.2.2.1  Binary arithmetic

The only digits in the binary system are the bits 0 and 1. Numbers are expressed as sums of powers of 2, for example

$$(x_3 x_2 x_1)_2 = x_3 2^2 + x_2 2 + x_1$$

**Example 2.3**  **Binary to decimal conversion**

The number $(1101011)_2$ can be converted to decimal by writing

$$(1101011)_2 = 1 \times 2^6 + 1 \times 2^5 + 0 \times 2^4 + 1 \times 2^3 + 0 \times 2^2 + 1 \times 2 + 1 \times 1$$
$$= 64 + 32 + 0 + 8 + 0 + 2 + 1$$
$$= 107$$

It's easy to build up the following table.

*Table 2.2* Decimal–binary equivalents

| Decimal | 0 | 1 | 2 | 3 | 4 | 5 |
|---------|---|---|---|---|---|---|
| Binary | $(0)_2$ | $(1)_2$ | $(10)_2$ | $(11)_2$ | $(100)_2$ | $(101)_2$ |

| Decimal | 6 | 7 | 8 | 9 | 10 |
|---------|---|---|---|---|----|
| Binary | $(110)_2$ | $(111)_2$ | $(1000)_2$ | $(1001)_2$ | $(1010)_2$ |

To convert a decimal number to binary form, proceed as follows.

## Algorithm to convert decimal to binary

**Step 1:**  Divide the number by 2, retain the quotient and record the remainder.

**Step 2:**  If the quotient in Step 1 is 0, stop.

**Step 3:**  If the quotient in Step 1 is not 0 repeat Step 1, using the quotient as the number which is divided by 2.

The required binary number is given by the remainders in the *reverse* order to which they are obtained.

**Example 2.4**  **Decimal to binary conversion**

To convert 45 to binary form, the algorithm produces the following array. At the first application of Step 1 the quotient is 22 and the remainder is 1. Step 3 then requires 22 to be divided by 2 giving quotient 11, remainder 0, and so on.

$$
\begin{array}{cc}
 & \text{remainder} \\
2\,\overline{)45} & 1 \\
\phantom{2)}22 & \\
2\,\overline{)22} & 0 \\
\phantom{2)}11 & \\
2\,\overline{)11} & 1 \\
\phantom{2)}5 & \\
2\,\overline{)5} & 1 \\
\phantom{2)}2 & \\
2\,\overline{)2} & 0 \\
\phantom{2)}1 & \\
2\,\overline{)1} & 1 \qquad \uparrow\text{read upwards} \\
\phantom{2)}0 &
\end{array}
$$

Reading upwards the required binary number is $(101101)_2$. In practice you can write the calculations in the following more compact form:

$$
\begin{array}{lc}
 & \text{remainder} \\
2\,\overline{)45} & 1 \\
2\,\overline{)22} & 0 \\
2\,\overline{)11} & 1 \\
2\,\overline{)5} & 1 \\
2\,\overline{)2} & 0 \\
2\,\overline{)1} & 1 \quad \uparrow \\
0 &
\end{array}
$$

A proof of the algorithm will be given in Section 2.2.5.3.

To perform the basic operations of addition, subtraction, multiplication and division on binary numbers, you follow the same procedures that you should be familiar with from decimal arithmetic. In some ways it's much simpler, since instead of having to remember addition and multiplication tables, like $3 + 4 = 7$ or $3 \times 4 = 12$, you only need $1 + 1 = 2$ and $1 \times 1 = 1$. In binary notation the complete set of rules for addition is

$$(0)_2 + (0)_2 = (0)_2, \quad (0)_2 + (1)_2 = (1)_2, \quad (1)_2 + (1)_2 = (10)_2 \tag{2.1}$$

and for multiplication

$$(0)_2 \times (0)_2 = (0)_2, \quad (0)_2 \times (1)_2 = (0)_2, \quad (1)_2 \times (1)_2 = (1)_2 \tag{2.2}$$

Armed with these, you can now tackle binary arithmetic problems. This is best explained through examples.

**Example 2.5**    **Binary addition**

First recall how you add decimal numbers. For example, $43 + 19 = 62$ is written as:

$$
\begin{array}{r}
① \\
4\ 3 \\
1\ 9 \\
\hline
6\ 2
\end{array}
$$

You first add 3 and 9 in the units column to get 12, and the 1 is 'carried over' to the next column, shown within a circle. The second (tens) column then gives $1 + 4 + 1 = 6$.

In a similar way you can add together two binary numbers, say $(1101)_2 + (1011)_2$. The first step is

$$
\begin{array}{@{}cccc}
 & \textcircled{1} & & \\
1 & 1 & 0 & 1 \\
1 & 0 & 1 & 1 \\
\hline
 & & & 0
\end{array}
$$

using $(1)_2 + (1)_2 = (10)_2$ from (2.1), and carrying over the 1, again put inside a circle. Move to the next column, which is $(1)_2 + (0)_2 + (1)_2 = (10)_2$ so that

$$
\begin{array}{@{}cccc}
\textcircled{1} & \textcircled{1} & & \\
1 & 1 & 0 & 1 \\
1 & 0 & 1 & 1 \\
\hline
 & & 0 & 0
\end{array}
$$

Continuing similarly:

$$
\begin{array}{@{}cccc}
\textcircled{1} & \textcircled{1} & \textcircled{1} & \\
1 & 1 & 0 & 1 \\
1 & 0 & 1 & 1 \\
\hline
 & 0 & 0 & 0
\end{array}
$$

and finally the last column gives

$$(1)_2 + (1)_2 + (1)_2 = 3 = (11)_2$$

so the complete sum is

$$
\begin{array}{@{}ccccc}
 & \textcircled{1} & \textcircled{1} & \textcircled{1} & \\
 & 1 & 1 & 0 & 1 \\
 & 1 & 0 & 1 & 1 \\
\hline
1 & 1 & 0 & 0 & 0
\end{array}
$$

In decimals, $(1101)_2 = 13$, $(1011)_2 = 11$, and $13 + 11 = 24 = (11000)_2$.

When adding more than two binary numbers, add the bits in each column in decimal and then convert to binary. There may well be two (or more digits) to carry over to the next two (or more) columns. As an illustration, consider $(11011)_2 + (10010)_2 + (11100)_2 + (1111)_2$. The first step is

$$
\begin{array}{@{}ccccc}
 & & \textcircled{1} & & \\
1 & 1 & 0 & 1 & 1 \\
1 & 0 & 0 & 1 & 0 \\
1 & 1 & 1 & 0 & 0 \\
 & 1 & 1 & 1 & 1 \\
\hline
 & & & & 0
\end{array}
$$

In the second column, $1 + 1 + 1 + 0 + 1 = 4 = (100)_2$ so the left and middle bits 1 and 0 are carried over to the next *two* columns as follows:

$$\begin{array}{ccccc} \textcircled{1} & \textcircled{0} & \textcircled{1} & & \\ 1 & 1 & 0 & 1 & 1 \\ 1 & 0 & 0 & 1 & 0 \\ 1 & 1 & 1 & 0 & 0 \\ & 1 & 1 & 1 & 1 \\ \hline & & & 0 & 0 \end{array}$$

You should now be able to follow through the remaining steps set out below:

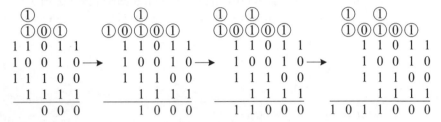

In decimals, $(11011)_2 = 27$, $(10010)_2 = 18$, $(11100)_2 = 28$, $(1111)_2 = 15$, and $27 + 18 + 28 + 15 = 88 = (1011000)_2$.

| Example 2.6 | **Binary subtraction** |

Begin by recalling how you do decimal subtraction. For example $43 - 19 = 24$ is written as:

$$\begin{array}{cc} & \textcircled{1} \\ 4 & 3 \\ 1 & 9 \\ \textcircled{-1} & \\ \hline 2 & 4 \end{array}$$

In the units column, because 9 is bigger than 3, you must 'borrow' 1 from the tens column, shown within the circle at the top of the units column. This makes the 3 into 13, so for the units column $13 - 9 = 4$. In order to balance the 'borrowed' 1, a $-1$ is needed, shown circled at the bottom of the tens column, which becomes $4 - 1 - 1 = 2$.

Now see how this works for a binary subtraction like $(1101)_2 - (1011)_2$, which in decimals is $13 - 11 = 2 = (0010)_2$. It's important to remember that when 'borrowing' a 1 from a column this represents 2 (compared with 10 in the decimal case). The steps in this example are as follows:

$$\begin{array}{cccc} 1\ 1\ 0\ 1 & 1\ 1\ 0\ 1 & 1\ 1\ 0\ 1 & 1\ 1\ 0\ 1 \\ 1\ 0\ 1\ 1 & 1\ 0\ 1\ 1 & 1\ 0\ 1\ 1 & 1\ 0\ 1\ 1 \\ \hline \quad 0 & \ \ 1\ 0 & 0\ 1\ 0 & 0\ 0\ 1\ 0 \end{array}$$

Notice that at the second step the 'borrowed' 1 gives the result $2 - 1 = 1$ for the second column.

Another example $(110010)_2 - (11011)_2$ should help you to understand the method:

$$
\begin{array}{cccccc}
 & & & & \textcircled{1} & \\
1 & 1 & 0 & 0 & 1 & 0 \\
1 & 1 & 0 & 1 & 1 & \\
 & & & & \textcircled{-1} & \\
\hline
 & & & & & 1
\end{array}
\rightarrow
\begin{array}{cccccc}
 & & & & \textcircled{1}\textcircled{1} & \\
1 & 1 & 0 & 0 & 1 & 0 \\
1 & 1 & 0 & 1 & 1 & \\
 & & & \textcircled{-1} & \textcircled{-1} & \\
\hline
 & & & & 1 & 1
\end{array}
\rightarrow
\begin{array}{cccccc}
 & & \textcircled{1} & & \textcircled{1} & \textcircled{1} \\
1 & 1 & 0 & 0 & 1 & 0 \\
1 & 1 & 0 & 1 & 1 & \\
 & & \textcircled{-1} & \textcircled{-1} & \textcircled{-1} & \\
\hline
 & & & 1 & 1 & 1
\end{array}
$$

$$
\rightarrow
\begin{array}{cccccc}
 & \textcircled{1} & \textcircled{1} & & \textcircled{1} & \textcircled{1} \\
1 & 1 & 0 & 0 & 1 & 0 \\
1 & 1 & 0 & 1 & 1 & \\
\textcircled{-1} & \textcircled{-1} & \textcircled{-1} & \textcircled{-1} & & \\
\hline
 & 0 & 1 & 1 & 1
\end{array}
\rightarrow
\begin{array}{cccccc}
\textcircled{1} & \textcircled{1} & \textcircled{1} & & \textcircled{1}\textcircled{1} \\
1 & 1 & 0 & 0 & 1 & 0 \\
1 & 1 & 0 & 1 & 1 & \\
\textcircled{-1} & \textcircled{-1} & \textcircled{-1} & \textcircled{-1} & \textcircled{-1} & \\
\hline
1 & 0 & 1 & 1 & 1
\end{array}
\rightarrow
\begin{array}{cccccc}
\textcircled{1} & \textcircled{1} & \textcircled{1} & & \textcircled{1} & \textcircled{1} \\
1 & 1 & 0 & 0 & 1 & 0 \\
1 & 1 & 0 & 1 & 1 & \\
\textcircled{-1} & \textcircled{-1} & \textcircled{-1} & \textcircled{-1} & \textcircled{-1} & \\
\hline
0 & 1 & 0 & 1 & 1 & 1
\end{array}
$$

In decimals, $(110010)_2 = 50$, $(11011)_2 = 27$, and $50 - 27 = 23 = (10111)_2$.

| Example 2.7 | **Binary multiplication** |
| --- | --- |

Again you set out the calculation by analogy with decimal products. For example, $231 \times 124$ is done by breaking up 124 as $100 + 20 + 4$, multiplying 231 by each of these three numbers, and adding as follows:

$$
\begin{array}{l}
\phantom{0}231 \\
\underline{\phantom{0}124} \\
23100 \longleftarrow 231 \times 100 \\
\phantom{0}4620 \longleftarrow 231 \times \phantom{0}20 \\
\phantom{00}\underline{924} \longleftarrow 231 \times \phantom{00}4 \\
28644 \phantom{0}= 231 \times 124
\end{array}
$$

Recall from Section 2.2.1 that multiplying by $(10)_2$, $(100)_2$, $(1000)_2$, ... has the same effect for binary numbers as does multiplication by 10, 100, 1000, ... for decimal numbers – you put one, two, three, ... zeros onto the end. Thus for example the product of $(1101)_2$ by $(1011)_2$ is done by writing $(1011)_2 = (1000)_2 + (10)_2 + (1)_2$, then multiplying $(1101)_2$ by each of these numbers using the rules in (2.2), and finally adding together the three parts as follows:

$$
\begin{array}{cccccccc}
 & & & & 1 & 1 & 0 & 1 \\
 & & & & 1 & 0 & 1 & 1 \\
\hline
1 & 1 & 0 & 1 & 0 & 0 & 0 & \longleftarrow (1101)_2 \times (1000)_2 \\
 & & 1 & 1 & 0 & 1 & 0 & \longleftarrow (1101)_2 \times \phantom{0}(10)_2 \\
 & & & 1 & 1 & 0 & 1 & \longleftarrow (1101)_2 \times \phantom{00}(1)_2 \\
 & \textcircled{1} & \textcircled{1} & \textcircled{1} & & & & \\
\hline
1 & 0 & 0 & 0 & 1 & 1 & 1 & 1 & = \phantom{0}(1101)_2 \times (1011)_2
\end{array}
$$

The addition of the three product terms is carried out as explained in Example 2.5, the 'carried over' 1's again being circled. In decimals $(1101)_2 = 13$, $(1011)_2 = 11$ and $13 \times 11 = 143 = (10001111)_2$.

Another example $(11001)_2 \times (11011)_2$ is now given to help your understanding of the procedure.

$$
\begin{array}{r}
1\ 1\ 0\ 0\ 1 \\
1\ 1\ 0\ 1\ 1 \\
\hline
\end{array}
$$

$$
\begin{array}{l}
1\ \ 1\ \ 0\ \ 0\ \ 1\ \ 0\ \ 0\ \ 0\ \ 0 \leftarrow (11001)_2 \times (10000)_2 \\
\quad 1\ \ 1\ \ 0\ \ 0\ \ 1\ \ 0\ \ 0\ \ 0 \leftarrow (11001)_2 \times \ \ (1000)_2 \\
\qquad\quad 1\ \ 1\ \ 0\ \ 0\ \ 1\ \ 0 \leftarrow (11001)_2 \times \quad\ \ (10)_2 \\
\qquad\qquad\quad 1\ \ 1\ \ 0\ \ 0\ \ 1 \leftarrow (11001)_2 \times \qquad (1)_2 \\
\qquad ①\ ①\ ①\ ⓪\ ① \\
\hline
1\ 0\ 1\ 0\ 1\ 0\ 0\ 0\ 1\ 1 = (11001)_2 \times (11011)_2
\end{array}
$$

In decimals, $(11001)_2 = 25$, $(11011)_2 = 27$ and $25 \times 27 = 675 = (1010100011)_2$.

---

**Example 2.8**  **Binary division**

Once again there is a direct analogy with 'long division' for decimal numbers, so it's best to begin with an example of this, say $12953 \div 13$.

**Step 1:**

$$
\begin{array}{r}
9\phantom{0000} \\
13\,)\overline{12953} \\
117\phantom{00} \\
\hline
12\phantom{00}
\end{array}
$$

13 does not divide into 12, so it must be divided into 129. It goes 9 times, with remainder 12.

**Step 2:**

$$
\begin{array}{r}
99\phantom{000} \\
13\,)\overline{12953} \\
117\phantom{00} \\
\hline
125\phantom{0} \\
117\phantom{0} \\
\hline
8\phantom{0}
\end{array}
$$

Bring down the next digit 5, and $125 \div 13$ gives divisor 9 with remainder 8.

**Step 3:**

$$
\begin{array}{r}
996 \\
13\,)\overline{12953} \\
117\phantom{00} \\
\hline
125\phantom{0} \\
117\phantom{0} \\
\hline
83 \\
78 \\
\hline
5
\end{array}
$$

Bring down the next digit 3; $83 \div 13$ gives divisor 6, remainder 5.

Hence $12953 \div 13 = 996$, with a remainder of 5, that is, $12953 = 996 \times 13 + 5$.

The main difference for binary division is that at each step the divisor can only be 1 or 0. First consider the example $(100111)_2 \div (11)_2$.

**Step 1:**

$$\begin{array}{r} 1 \\ 11\,)\overline{100111} \\ \underline{11} \\ 1 \end{array}$$

$(11)_2$ does not divide into $(10)_2$, so it must be divided into $(100)_2$, and $(100)_2 - (11)_2 = (1)_2$.

**Step 2:**

$$\begin{array}{r} 11 \\ 11\,)\overline{100111} \\ \underline{11} \\ 11 \\ \underline{11} \\ 0 \end{array}$$

Bring down the next bit 1, divide $(11)_2$ by $(11)_2$, remainder 0.

**Step 3:**

$$\begin{array}{r} 110 \\ 11\,)\overline{100111} \\ \underline{11} \\ 11 \\ \underline{11} \\ 01 \end{array}$$

Bring down the next bit 1, $(01)_2$ does not divide by $(11)_2$.

**Step 4:**

$$\begin{array}{r} 1101 \\ 11\,)\overline{100111} \\ \underline{11} \\ 11 \\ \underline{11} \\ 011 \\ \underline{11} \\ 00 \end{array}$$

Bring down the next bit 1, divide $(011)_2$ by $(11)_2$ with remainder 0.

Since the final remainder is 0, $(100111)_2 \div (11)_2 = (1101)_2$. In decimals this is $39 \div 3 = 13$.

---

You should now work through the following steps (1) to (6) for $(110111001)_2 \div (101)_2$. In decimals this is $441 \div 5 = 88$ with remainder 1.

$$\begin{array}{r} 1 \\ 101\,)\overline{110111001} \\ \underline{101} \\ 1 \end{array}$$

(1) $\longrightarrow$
$$\begin{array}{r} 10 \\ 101\,)\overline{110111001} \\ \underline{101} \\ 11 \end{array}$$

(2) $\longrightarrow$
$$\begin{array}{r} 101 \\ 101\,)\overline{110111001} \\ \underline{101} \\ 111 \\ \underline{101} \\ 10 \end{array}$$

(3) $\longrightarrow$
$$\begin{array}{r} 1011 \\ 101\,)\overline{110111001} \\ \underline{101} \\ 111 \\ \underline{101} \\ 101 \\ \underline{101} \\ 0 \end{array}$$

$$
\begin{array}{r}
10110 \\
(4) \quad 101\,\overline{)110111001} \\
101 \\
\hline
111 \\
101 \\
\hline
101 \\
101 \\
\hline
101 \\
101 \\
\hline
00
\end{array}
\qquad
\begin{array}{r}
101100 \\
(5) \quad 101\,\overline{)110111001} \\
101 \\
\hline
111 \\
101 \\
\hline
101 \\
101 \\
\hline
000
\end{array}
$$

$$
\begin{array}{r}
1011000 \\
(6) \quad 101\,\overline{)110111001} \\
101 \\
\hline
111 \\
101 \\
\hline
101 \\
101 \\
\hline
0001
\end{array}
$$

The quotient is $(1011000)_2 = 88$, and there is a remainder $(1)_2$.

It's interesting that binary numbers can be used to give an explanation of the method of 'Russian multiplication', described in Section 1.2. An example for the product $45 \times 27 = 1215$ was set out as follows:

| 45 | 27 |
|----|-----|
| ~~22~~ | ~~54~~ |
| 11 | 108 |
| 5 | 216 |
| ~~2~~ | ~~432~~ |
| 1 | 864 |
| | 1215 |

The number 45 is successively divided by 2 (with remainders ignored) and the number 27 is successively multiplied by 2. The desired product is the sum of terms in the right column which do *not* correspond to even numbers in the left column. The use of the factor 2 should make you suspect that binary numbers are somehow involved, and in fact if you write $45 = (101101)_2$ then the way the method works can be understood. The product is

$$
\begin{aligned}
45 \times 27 &= (101101)_2 \times 27 \\
&= (1 \times 2^5 + 0 \times 2^4 + 1 \times 2^3 + 1 \times 2^2 + 0 \times 2 + 1) \times 27 \\
&= 2^5 \times 27 + 2^3 \times 27 + 2^2 \times 27 + 1 \times 27 \\
&= 864 + 216 + 108 + 27
\end{aligned}
$$

which agrees with the tabular array above using decimal numbers. The zero bits in the binary form of 45 do not contribute to the product, and this explains why certain terms are struck out in the two columns. To be precise, the successive divisions of 45 by 2 correspond exactly to the scheme for converting 45 to binary, and the array in Example 2.4 is repeated below:

$$
\begin{array}{r l}
2\,)\overline{45} & 1 \\
2\,)\overline{22} & 0 \\
2\,)\overline{11} & 1 \\
2\,)\overline{\ \,5} & 1 \\
2\,)\overline{\ \,2} & 0 \\
2\,)\overline{\ \,1} & 1 \\
& 0
\end{array}
\qquad \uparrow \quad 45 = (101101)_2
$$

The *zero* bits in the binary form of 45 occur where there are *even* numbers being divided by 2. This is why terms in both columns of the Russian multiplication array are deleted when the left-hand number is even. The non-zero bits in the binary form of 45 give products of 27 with the appropriate powers of 2, and these are the terms which are added up to give the value of the product.

You may have a calculator which works in a binary mode, in which case you could use it to do some of the computations of this section. However, it is easy to run out of space on the calculator display: for example, the binary form of $10^4$ contains 14 bits (see Exercise 2.16(d)). You should attempt the following exercises without using the binary mode on a calculator.

**Exercise**    **2.13**   (a) Using the four-digit code in Exercise 2.9, determine the binary representation of 1123.

(b) If *two* extra zeros are prefixed to the ASCII symbols in Table 2.1, so for example 'a' becomes 001100001, this can be reduced to a three-digit symbol 141 using

$$
\begin{array}{ccc}
(001)_2 & (100)_2 & (001)_2 \\
1 & 4 & 1
\end{array}
$$

Write out the three-digit representations for the rest of the entries in Table 2.1.

**Exercise**    **2.14**   Convert the following binary numbers to decimal form:

(a) $(101111)_2$   (b) $(11110010)_2$   (c) $(101110100111)_2$.

**Exercise**    **2.15**   Extend Table 2.2 to give the binary equivalents from 11 to 20.

**Exercise**    **2.16**   Convert the following decimal numbers to binary form:

(a) 187   (b) 121   (c) 424   (d) 10000.

**Exercise**    **2.17**   Evaluate the following binary additions:

(a) $(10111)_2 + (10011)_2$      (b) $(100101)_2 + (110110)_2$

(c) $(10101)_2 + (1111)_2 + (10011)_2$

(d) $(100010)_2 + (10001)_2 + (1100101)_2 + (11010010)_2$.

In each case give the decimal equivalent.

**Exercise** | **2.18** Evaluate the following:

(a) $(10110)_2 - (1110)_2$ (b) $(1100101)_2 - (1001111)_2$

(c) $(11010100)_2 - (10111111)_2 + (101101)_2$.

In each case give the decimal equivalent.

**Exercise** | **2.19** Evaluate the following binary products:

(a) $(10101)_2 \times (101000)_2$ (b) $(11110011)_2 \times (101111)_2$

(c) $(1011101)_2 \times (100011)_2 \times (10001)_2$.

In each case give the decimal equivalent.

**Exercise** | **2.20** Evaluate the following using binary arithmetic:

(a) $512 \times 16$ (b) $21(45 - 17)$ (c) $(83 + 51)(79 - 62)$.

**Exercise** | **2.21** Evaluate the following:

(a) $(1100011)_2 \div (11)_2$ (b) $(110111)_2 \div (100)_2$

(c) $(100000011101)_2 \div (11111)_2$ (d) $(100101000101)_2 \div (10010)_2$.

In each case give the decimal equivalent.

**Exercise** | **2.22** Using the notation in Exercise 2.12, convert the following to decimal form:

(a) $(11.011)_2$ (b) $(0.000111)_2$ (c) $(1.10111)_2$.

**Exercise** | **2.23** Show that

$$2^k = (-1)^k + 3t_k$$

is valid for $k = 1$, 2, 3, where $t_k$ is a positive integer. Use the method of induction (see the Interlude) to prove that this result is true for any positive integer $k$.

Hence prove that the binary number $(a_n a_{n-1} \ldots a_2 a_1)_2$ is divisible by 3 if and only if

$$a_1 - a_2 + a_3 - a_4 + \ldots + (-1)^{n-1} a_n$$

is divisible by 3.

**Exercise** | **2.24** Apply the result of the preceding exercise to test for divisibility by 3 the binary numbers

(a) $(111001101)_2$ (b) $(1001101101)_2$.

Check your results by finding the decimal equivalents and using the test in Section 1.3.1.

## 2.2.2.2 Computer arithmetic

In order to carry out a subtraction of binary numbers $x - y$ in a computer, the minus sign must be represented in terms of bits, as is any other piece of information stored in a computer. One frequently used procedure is called the '2's-complement method'. Suppose an 8-bit representation is being used, so that for example $27 = (00011011)_2$. The **1's-complement** of a binary number is obtained by replacing 0 by 1 and 1 by 0, so for example the 1's-complement of 27 is $(11100100)_2$. To represent a negative number, add 1 to the 1's-complement to produce the **2's-complement**. For example, the 2's-complement of $-27$ is

$$\underset{\substack{\text{1's-complement} \\ \text{of 27}}}{(11100100)_2} + (00000001)_2 = \underset{\substack{\text{2's-complement} \\ \text{of } -27}}{(11100101)_2}$$

and this is how $-27$ is stored in the computer. You can confirm that the decimal equivalent of $(11100101)_2$ is 229, which when added to 27 gives $256 = 2^8$. In general if $y = (y_8 y_7 \ldots y_2 y_1)_2$ then the 1's-complement of $y$ is $\bar{y} = (\bar{y}_8 \bar{y}_7 \ldots \bar{y}_2 \bar{y}_1)_2$ where $y_i + \bar{y}_i = 1$ for $i = 1, 2, \ldots, 8$, so that

$$y + \bar{y} = (11111111)_2$$
$$= 2^8 - 1$$

The 2's-complement of $-y$, which can be denoted by $\bar{\bar{y}}$, is defined by $\bar{\bar{y}} = \bar{y} + 1$, so that

$$y + \bar{\bar{y}} = y + \bar{y} + 1 = 2^8$$

which accounts for the name '2's-complement'.

Using eight bits a total of $2^8 = 256$ numbers can be represented. The negative numbers range from $-128$, which has 2's-complement

$$\bar{\bar{y}} = 2^8 - 128 = 128 = (10000000)_2$$

up to $-1$, which has 2's-complement

$$\bar{\bar{y}} = 2^8 - 1 = (11111111)_2$$

The left-most bit of the 2's-complement of all these negative numbers is 1. The remaining numbers range from 0 to 127 and are represented by their normal binary forms, from $(00000000)_2$ up to $(01111111)_2$, and in each case the left-most bit is 0.

Suppose that $x \leq 127$ and $x > y$. Then using $y = 2^8 - \bar{\bar{y}}$ gives

$$x - y = x + \bar{\bar{y}} - 2^8$$
$$= x + \bar{\bar{y}} - (100000000)_2$$

which shows that $x - y$ is equal to the **sum** of $x$ and 2's-complement of $-y$; the final term indicates that a (ninth) left-most bit is discarded. This is because the computer is working with eight bits.

**Example 2.9** **Binary subtraction**

Repeat the subtraction of $50 - 27 = 23$ carried out in Example 2.6, now using 8-bit representations so that $50 = (00110010)_2$. The subtraction is done as

$$50 + (-27) = \underset{\substack{\text{binary form} \\ \text{of 50}}}{(00110010)_2} + \underset{\substack{\text{2's-complement} \\ \text{of 27}}}{(11100101)_2}$$

The binary addition is carried out in the standard way, as described in Section 2.2.2.1, that is,

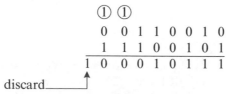

discard

Discarding the left-most (ninth) bit produces the answer to the subtraction

$(00010111)_2 = 23$

and the arrowed left-most bit is 0, showing that the answer is positive.

The procedure is similar if the answer is negative. For example, consider $27 - 50 = -23$ which will be evaluated as $27 + (-50)$. The 1's-complement of 50 is $(11001101)_2$, and the 2's-complement of $-50$ is obtained by adding 1 to this, giving $(11001110)_2$. The sum of 27 and $(-50)$ is therefore

```
① ① ① ①
0 0 0  1  1  0 1 1     binary form of 27
1 1 0  0  1  1 1 0     2's-complement of −50
1 1 1  0  1  0 0 1
↑
```

This time there is no ninth bit, and the arrowed left-most bit is 1, indicating that the answer is negative. To obtain the numerical value of the answer the 2's-complement construction must be reversed as follows:

```
11101001 ─────────────→ 00010110
            1's-complement
                             ↓ add
                        00000001
                             ↓
                        00010111
```

and $(00010111)_2 = 23$. Hence the required answer is $-23$.

The procedure illustrated in Example 2.9 can be summarized as follows.

## Algorithm: $x - y$ using 8-bit representations

    **(1)** Replace 0 with 1 and 1 by 0 in the binary representation of $y$ (gives $\bar{y}$).

    **(2)** Add 1 to the result of (1) (gives $\bar{\bar{y}}$).

    **(3)** Add the result of (2) to the binary representation of $x$ (gives $x + \bar{\bar{y}}$).

    **(4)** If the sum has nine bits, discard the left-most bit.

    **(5)** If the left-most bit after applying (4) is 0, the answer is positive.

    **(6)** If the left-most bit after applying (4) is 1 the answer is negative; apply (1) and (2) to obtain the binary number whose negative is the required answer.

The method works in precisely the same way if more than eight bits are used to represent numbers. However, no matter how many bits are used there is a limit to the size of integers which can be handled. For example, with eight bits

$$108 + 45 = (01101100)_2 + (00101101)_2$$
$$= (10011001)_2$$
$$\uparrow$$

The arrowed left-most bit is 1, indicating a negative answer to the sum of two positive numbers! This is because the sum is 153, which exceeds the maximum size of positive numbers (that is, 127) allowable with eight bits. The spurious left-most bit 1 indicates there has been an **overflow error**. Similarly, if two negative numbers are added via their 2's-complements, an overflow error is detected if the left-most bit of their sum is 0.

**Exercise**     **2.25** Determine the 5-bit representations of all the integers from $-16$ to 15.

**Exercise**     **2.26** Use 2's-complement arithmetic with five bits, together with your results in the preceding exercise, to compute the following:

(a) $3 - 5$    (b) $8 - 3$    (c) $9 + 10$    (d) $14 - 7$    (e) $15 - 15$.

**Exercise**     **2.27** Use 2's-complement arithmetic with eight bits to compute the following:

(a) $75 - 19$    (b) $19 - 75$    (c) $-19 - 75$    (d) $75 + 65$.

**Exercise**     **2.28** One computer uses 2's-complement arithmetic with 16 bits, another with 32 bits. What is the range of negative and positive integers which they can represent? In each case determine the representations of 128 and $-64$.

Exercise **2.29** Verify that the second part of step (6) of the algorithm for computing $x - y$ is valid.

Exercise **2.30** Verify that using 2's-complement arithmetic with eight bits gives $x + (-x) = 0$ for $-127 \leq x \leq 127$.

Exercise **2.31** With 8-bit arithmetic, show that if $x < 0$ has 2's-complement $\bar{x}$, then the 2's-complement of $\bar{x}$ is $-x$, with one exception. What is this exception?

### 2.2.3 Octal system

The base 8 is used in computer applications, when numbers are expressed as powers of 8 using the digits 0, 1, 2, 3, 4, 5, 6, 7. These are sometimes referred to as **octits** (short for octal digits). For example,

$$(237)_8 = 2 \times 8^2 + 3 \times 8 + 7$$
$$= 2 \times 64 + 3 \times 8 + 7$$
$$= 159$$

and the first 16 decimal numbers are given in Table 2.3. Notice that in counting you 'jump' from $(7)_8$ to $(10)_8$ (read as 'one-zero to base eight'), from $(17)_8$ to $(20)_8$, and so on.

Table 2.3 Decimal–octal equivalents

| Decimal | 0 | 1 | 2 | 3 | 4 | 5 | 6 | 7 | 8 |
|---|---|---|---|---|---|---|---|---|---|
| Octal | $(0)_8$ | $(1)_8$ | $(2)_8$ | $(3)_8$ | $(4)_8$ | $(5)_8$ | $(6)_8$ | $(7)_8$ | $(10)_8$ |

| Decimal | 9 | 10 | 11 | 12 | 13 | 14 | 15 | 16 |
|---|---|---|---|---|---|---|---|---|
| Octal | $(11)_8$ | $(12)_8$ | $(13)_8$ | $(14)_8$ | $(15)_8$ | $(16)_8$ | $(17)_8$ | $(20)_8$ |

Decimal to octal conversion uses the same algorithm as for decimal to binary in Section 2.2.2.1 except that in Step 1 division is by 8 instead of 2.

Example 2.10 **Decimal to octal conversion**

To convert 583 to octal involves the following steps using the algorithm in Section 2.2.2.1.

```
              remainder
8 )583         7
8 ) 72         0
8 )  9         1
8 )  1         1       ↑ read upwards
     0
```

Reading upwards gives $583 = (1107)_8$.

Because $8 = 2^3$, conversions between octal and binary are very simple, in either direction. These rely on representing the eight octits as the following 3-bit binaries.

*Table 2.4* Octits and 3-bit binary equivalents

| Octit | $(0)_8$ | $(1)_8$ | $(2)_8$ | $(3)_8$ | $(4)_8$ | $(5)_8$ | $(6)_8$ | $(7)_8$ |
|---|---|---|---|---|---|---|---|---|
| Binary | $(000)_2$ | $(001)_2$ | $(010)_2$ | $(011)_2$ | $(100)_2$ | $(101)_2$ | $(110)_2$ | $(111)_2$ |

To convert octal to binary, simply expand each octit to its binary equivalent in Table 2.4. Conversely, to convert binary to octal, compress each set of three bits to its octal equivalent in Table 2.4, starting at the right-hand end.

**Example 2.11**     **Octal to binary conversions**

To convert $(3506)_8$ to binary, use Table 2.4:

$$
\begin{array}{cccc}
3 & 5 & 0 & 6 \\
011 & 101 & 000 & 110
\end{array}
$$

so the binary equivalent is $(11101000110)_2$, since the left-most 0 can be dropped. It's easy to explain why this method works. Expanding $(3506)_8$ gives

$$
\begin{aligned}
(3506)_8 &= 3 \times 8^3 + 5 \times 8^2 + 6 \\
&= 3 \times (2^3)^3 + 5 \times (2^3) + 6 \\
&= 3 \times 2^9 + 5 \times 2^6 + 6 \\
&= (011)_2 \times 2^9 + (101)_2 \times 2^6 + (110)_2 \\
&= (2 + 1)2^9 + (2^2 + 1)2^6 + (2^2 + 2) \\
&= 2^{10} + 2^9 + 2^8 + 2^6 + 2^2 + 2 \\
&= (11101000110)_2
\end{aligned}
$$

As an example of the reverse procedure, convert $(1101011)_2$ into octal. First separate the bits into blocks of three, starting from the right, and put the octal equivalents from Table 2.4 underneath:

$$
\begin{array}{ccc}
001 & 101 & 011 \\
1 & 5 & 3
\end{array}
$$

The octal equivalent is therefore $(153)_8$. Notice that in this example two leading zeros have to be inserted so as to produce a block of three bits.

In general the number of octits is about one-third the number of bits required to represent a given decimal number.

| Example 2.12 | **Octal addition** |

'Carrying over' is done in the same way as for binary or decimal addition, but the sums for each column must be converted via Table 2.3. For example, $(636)_8 + (355)_8$ has the first step

$$
\begin{array}{ccc}
\textcircled{1} & & \\
6 & 3 & 6 \\
3 & 5 & 5 \\
\hline
 & & 3
\end{array}
\qquad 6 + 5 = 11 = (13)_8, \text{ the 1 is carried over}
$$

The next two steps are as follows, where as usual 'carried over' numbers are circled.

$$
\begin{array}{ccc}
\textcircled{1} \textcircled{1} & & \\
6 & 3 & 6 \\
3 & 5 & 5 \\
\hline
 & 1 & 3
\end{array}
\qquad
\begin{array}{l}
\textit{Column sum} \\
\textcircled{1} + 3 + 5 = 9 = (11)_8
\end{array}
$$

$$
\begin{array}{ccc}
\textcircled{1} & \textcircled{1} & \\
6 & 3 & 6 \\
3 & 5 & 5 \\
\hline
1 & 2 & 1 & 3
\end{array}
\qquad \textcircled{1} + 6 + 3 = 10 = (12)_8
$$

The answer is $(1213)_8$. The decimal equivalent is $414 + 237 = 651$.

| Example 2.13 | **Octal subtraction** |

Here the only difference from the procedure in Example 2.6 is that a 'borrowed' 1 now represents 8 (instead of 2 in binary and 10 in decimal). The example $(714)_8 - (237)_8 = (455)_8$ proceeds as follows:

$$
\begin{array}{ccc}
 & \textcircled{1} & \\
7 & 1 & 4 \\
2 & 3 & 7 \\
 & \ominus{1} & \\
\hline
 & & 5
\end{array}
\qquad
\begin{array}{l}
\textit{Column operation} \\
(10)_8 + 4 - 7 = 8 + 4 - 7 = 5
\end{array}
$$

$$
\begin{array}{ccc}
\textcircled{1} & \textcircled{1} & \\
7 & 1 & 4 \\
2 & 3 & 7 \\
\ominus{1} & \ominus{1} & \\
\hline
 & 5 & 5
\end{array}
\qquad
\begin{array}{l}
(10)_8 + 1 - 3 \ominus{1} = 8 + 1 - 3 - 1 \\
\qquad\qquad\qquad = 5
\end{array}
$$

$$
\begin{array}{ccc}
\textcircled{1} & \textcircled{1} & \\
7 & 1 & 4 \\
2 & .3 & 7 \\
\ominus{1} & \ominus{1} & \\
\hline
4 & 5 & 5
\end{array}
\qquad 7 - 2 \ominus{1} = 4
$$

The decimal equivalent is $460 - 159 = 301$.

**Example 2.14** **Octal multiplication**

This is done in a similar way to binary multiplication in Example 2.7, except that you need to use decimal-to-octal conversions as given in Table 2.3 and its extension (see Exercise 2.32). Recall once again that multiplying an octal number by $(10)_8$, $(100)_8$, $(1000)_8$, ... has the effect of putting one, two, three, ... zeros onto the right-hand end, so for example, $(127)_8 \times (10)_8 = (1270)_8$. Thus to evaluate, for example, $(127)_8 \times (354)_8$ the second number is written as $(300)_8 + (50)_8 + (4)_8$, and the three parts are computed separately, as shown below.

$$
\begin{array}{cccc}
 & & & \textit{Intermediate steps} \\
1 & 2 & 7 & 3 \times 7 = 21 = (25)_8 \\
3 & 0 & 0 & 3 \times 2 + ② = 8 = (10)_8 \\
① \ ② & & & 3 \times 1 + ① = (4)_8 \\
\hline
4 \ 0 \ 5 \ 0 \ 0 & & &
\end{array}
$$

$$
\begin{array}{ccc}
1 & 2 & 7 & 5 \times 7 = 35 = (43)_8 \\
 & 5 & 0 & 5 \times 2 + ④ = 14 = (16)_8 \\
① \ ④ & & & 5 \times 1 + ① = (6)_8 \\
\hline
6 \ 6 \ 3 \ 0 & & &
\end{array}
$$

$$
\begin{array}{ccc}
1 & 2 & 7 & 4 \times 7 = 28 = (34)_8 \\
 & & 4 & 4 \times 2 + ③ = 11 = (13)_8 \\
① \ ③ & & & 4 \times 1 + ① = (5)_8 \\
\hline
5 \ 3 \ 4 & & &
\end{array}
$$

The overall product is the sum of these parts, that is,

$$
\begin{array}{ccccc}
 & & & & \textit{Column sums} \\
4 & 0 & 5 & 0 & 0 & 3 + 3 = (6)_8 \\
 & 6 & 6 & 3 & 0 & 5 + 6 + 5 = 16 = (20)_8 \\
 & & 5 & 3 & 4 & 6 + ② = 8 = (10)_8 \\
 & ① & ② & & & 4 + ① = (5)_8 \\
\hline
5 & 0 & 0 & 6 & 4 &
\end{array}
$$

so $(127)_8 \times (354)_8 = (50064)_8$. The decimal equivalent is $87 \times 236 = 20532$.

**Example 2.15** **Octal division**

This follows the same principles as decimal or binary division, but is rather tedious as it relies on using a multiplication table for octal numbers (see Exercise 2.38). Only a simple example $(4416)_8 \div (7)_8$ will be given, therefore, to illustrate the basics of the procedure.

**Step 1:**

$$
\begin{array}{r}
5 \\
7\,)\overline{4416} \\
\underline{43} \\
1
\end{array}
$$

$(44)_8 \div (7)_8$ gives divisor $(5)_8$, remainder 1

[since $(7)_8 \times (5)_8 = 35 = (43)_8$]

Step 2: $\dfrac{51}{7)\overline{4416}}$ Bring down 1, $(11)_8 \div (7)_8$ gives divisor $(1)_8$, remainder $(2)_8$
$\underline{43}$ [since $(11)_8 = 9$]
$11$
$\underline{7}$
$2$

Step 3: $\dfrac{513}{7)\overline{4416}}$ Bring down 6, $(11) \div (7)_8$ gives divisor $(3)_8$, remainder 1
$\underline{43}$ [since $(7)_8 \times (3)_8 = 21 = (25)_8$]
$11$
$\underline{7}$
$26$
$\underline{25}$
$1$

Hence $(4416)_8 \div (7)_8 = (513)_8$ with remainder 1, that is
$(4416)_8 = (513)_8 \times (7)_8 + (1)_8$.

---

**Exercise** **2.32** Extend Table 2.3 to give the octal equivalents of the decimal numbers 17 to 49.

**Exercise** **2.33** Convert the following octal numbers to decimal form:

(a) $(5612)_8$ (b) $(76104)_8$.

**Exercise** **2.34** Convert the two octal numbers in the preceding exercise to binary form. Add these two numbers together in both octal and binary form. Check using decimal arithmetic.

**Exercise** **2.35** Convert the following binary numbers to octal:

(a) $(110111111110100)_2$ (b) $(1101010100011)_2$.

Then convert each octal number to decimal form, and compare your answers with direct conversion of the binary numbers to decimal form.

**Exercise** **2.36** Evaluate the following octal additions:

(a) $(7654)_8 + (1230)_8$

(b) $(743)_8 + (276)_8 + (450)_8 + (674)_8$

(c) $(73425)_8 + (60376)_8 + (41263)_8$.

In each case give the decimal equivalent.

**Exercise** **2.37** Evaluate the following:

(a) $(743)_8 - (356)_8$ (b) $(2170)_8 - (613)_8 + (452)_8$ (c) $(70421)_8 - (26573)_8$

In each case give the decimal equivalent.

**Exercise** **2.38** Complete the following addition and multiplication tables for octal numbers:

| + | 1 | 2 | 3 | 4 | 5 | 6 | 7 |
|---|---|---|---|---|---|---|---|
| 1 | $(2)_8$ | $(3)_8$ | $(4)_8$ | $(5)_8$ | $(6)_8$ | $(7)_8$ | $(10)_8$ |
| 2 | | $(4)_8$ | $(5)_8$ | $(6)_8$ | $(7)_8$ | $(10)_8$ | $(11)_8$ |
| 3 | | | $(6)_8$ | $(7)_8$ | $(10)_8$ | $(11)_8$ | $(12)_8$ |
| 4 | | | | | | | |
| 5 | | | | | | | |
| 6 | | | | | | | |
| 7 | | | | | | | |

| × | 1 | 2 | 3 | 4 | 5 | 6 | 7 |
|---|---|---|---|---|---|---|---|
| 1 | $(1)_8$ | $(2)_8$ | $(3)_8$ | $(4)_8$ | $(5)_8$ | $(6)_8$ | $(7)_8$ |
| 2 | | $(4)_8$ | $(6)_8$ | $(10)_8$ | $(12)_8$ | $(14)_8$ | $(16)_8$ |
| 3 | | | $(11)_8$ | $(14)_8$ | $(17)_8$ | $(22)_8$ | $(25)_8$ |
| 4 | | | | | | | |
| 5 | | | | | | | |
| 6 | | | | | | | |
| 7 | | | | | | | |

**Exercise** **2.39** Evaluate the following:

(a) $(37)_8 \times (25)_8$   (b) $(246)_8 \times (315)_8$   (c) $(3745)_8 \div (5)_8$.

In each case give the decimal equivalent.

**Exercise** **2.40** Evaluate the following using octal arithmetic:

(a) $32768 \times 64$   (b) $(153 + 201)(77 - 60)$.

**Exercise** **2.41** Using the notation in Exercise 2.12, convert the following to decimal form:

(a) $(5.21)_8$   (b) $(0.436)_8$.

**Exercise** **2.42** What is the condition for an octal number to be exactly divisible (that is, without remainder) by $8^k$, where $k$ is a positive integer?

What is the corresponding result for binary numbers?

**Exercise** **2.43** Verify that

$$8^k = 1 + 7t_k$$

is valid for $k = 1, 2, 3$ where $t_k$ is a positive integer. Use the method of induction to prove that this result is true for all positive integers $k$.

Hence prove that the octal number $(a_n a_{n-1} \ldots a_2 a_1)_8$ is divisible by 7 if and only if the sum of its octits $a_1 + a_2 + \ldots + a_n$ is divisible by 7.

**Exercise** | **2.44** Apply the result in the preceding exercise to test the following octal numbers for divisibility by 7:

(a) $(12004)_8$  (b) $(166241)_8$.

**Exercise** | **2.45** Verify that

$$8^k = (-1)^k + 9s_k$$

is valid for $k = 1$, 2, 3, where $s_k$ is a positive integer. Use the method of induction to prove that this result is true for all positive integers $k$.

Hence prove that the octal number $(a_n a_{n-1} \dots a_2 a_1)_8$ is divisible by 9 if and only if

$$a_1 - a_2 + a_3 - a_4 + \dots + (-1)^{n-1} a_n$$

is divisible by 9.

**Exercise** | **2.46** Apply the result in the preceding exercise to test the following octal numbers for divisibility by 9:

(a) $(11515)_8$  (b) $(706674)_8$.

Find the decimal equivalents, and check your results using the test in Section 1.3.7.

## 2.2.4 Hexadecimal system

The base 16 is also used in computer applications. Since 16 is bigger than 10, the first ten hexadecimal digits are 0, 1, 2, 3, 4, 5, 6, 7, 8, 9 and the other six are usually denoted by the letters $A$, $B$, $C$, $D$, $E$, $F$ where $A$ is ten, $B$ is eleven and so on up to fifteen for $F$. The digits are called **hexits**. Conversion from hexadecimal to decimal is done in the usual way by expanding in powers of the base. For example,

$$\begin{aligned}(2C7E)_{16} &= 2 \times 16^3 + C \times 16^2 + 7 \times 16 + E \\ &= 2 \times 16^3 + 12 \times 16^2 + 7 \times 16 + 14 \\ &= 11390\end{aligned}$$

The first 32 decimal numbers are given in Table 2.5. In a similar fashion to octal counting, you now 'jump' from $(F)_{16}$ to $(10)_{16}$, read as 'one-zero to base 16'.

*Table 2.5* Decimal–hexadecimal equivalents

| Decimal | 0 | 1 | 2 | 3 | 4 | 5 | 6 | 7 | 8 |
|---|---|---|---|---|---|---|---|---|---|
| Hexadecimal | $(0)_{16}$ | $(1)_{16}$ | $(2)_{16}$ | $(3)_{16}$ | $(4)_{16}$ | $(5)_{16}$ | $(6)_{16}$ | $(7)_{16}$ | $(8)_{16}$ |
| Decimal | 9 | 10 | 11 | 12 | 13 | 14 | 15 | 16 | 17 |
| Hexadecimal | $(9)_{16}$ | $(A)_{16}$ | $(B)_{16}$ | $(C)_{16}$ | $(D)_{16}$ | $(E)_{16}$ | $(F)_{16}$ | $(10)_{16}$ | $(11)_{16}$ |
| Decimal | 18 | 19 | 20 | 21 | 22 | 23 | 24 | 25 | 26 |
| Hexadecimal | $(12)_{16}$ | $(13)_{16}$ | $(14)_{16}$ | $(15)_{16}$ | $(16)_{16}$ | $(17)_{16}$ | $(18)_{16}$ | $(19)_{16}$ | $(1A)_{16}$ |
| Decimal | 27 | 28 | 29 | 30 | 31 | 32 | | | |
| Hexadecimal | $(1B)_{16}$ | $(1C)_{16}$ | $(1D)_{16}$ | $(1E)_{16}$ | $(1F)_{16}$ | $(20)_{16}$ | | | |

The various operations using hexadecimal numbers are all similar to those using octal numbers, so some representative examples should suffice to give you the general ideas.

**Example 2.16**  **Decimal to hexadecimal conversion**

You again use the algorithm for decimal to binary in Section 2.2.2.1 but with division by 16 instead of 2. For example, conversion of 603 gives

$$
\begin{array}{ll}
& \text{remainder} \\
16\,\overline{)603} & 11 = B \\
16\,\overline{)\ 37} & 5 \\
16\,\overline{)\ \ 2} & 2 \qquad \uparrow \text{read upwards} \\
\qquad\ 0 &
\end{array}
$$

Reading upwards shows that $603 = (25B)_{16}$.

**Example 2.17**  **Hexadecimal to binary conversions**

Because $16 = 2^4$ you now need the binary representations shown in Table 2.6 using *four* bits.

*Table 2.6* Hexits and 4-bit binary equivalents

| Hexit | $(0)_{16}$ | $(1)_{16}$ | $(2)_{16}$ | $(3)_{16}$ | $(4)_{16}$ | $(5)_{16}$ | $(6)_{16}$ | $(7)_{16}$ |
|---|---|---|---|---|---|---|---|---|
| Binary | $(0000)_2$ | $(0001)_2$ | $(0010)_2$ | $(0011)_2$ | $(0100)_2$ | $(0101)_2$ | $(0110)_2$ | $(0111)_2$ |

| Hexit | $(8)_{16}$ | $(9)_{16}$ | $(A)_{16}$ | $(B)_{16}$ | $(C)_{16}$ | $(D)_{16}$ | $(E)_{16}$ | $(F)_{16}$ |
|---|---|---|---|---|---|---|---|---|
| Binary | $(1000)_2$ | $(1001)_2$ | $(1010)_2$ | $(1011)_2$ | $(1100)_2$ | $(1101)_2$ | $(1110)_2$ | $(1111)_2$ |

For example, to convert $(7DA1)_{16}$ to binary, simply expand each hexit using Table 2.6 to obtain

| 7 | D | A | 1 |
|---|---|---|---|
| 0111 | 1101 | 1010 | 0001 |

The left-most 0 is dropped to give $(111110110100001)_2$.

Similarly, to do a reverse conversion, collect the bits into groups of four, starting from the right, and insert extra leading zeros if necessary. For example, $(11010010111001)_2$ gives

| 0011 | 0100 | 1011 | 1001 |
|---|---|---|---|
| 3 | 4 | B | 9 |

so the hexadecimal equivalent is $(34B9)_{16}$.

To convert from hexadecimal to octal, and conversely, it's best to go via binary – see Exercise 2.53.

**Example 2.18** **Hexadecimal arithmetic**
(a) Addition $(A49)_{16} + (EFD)_{16}$

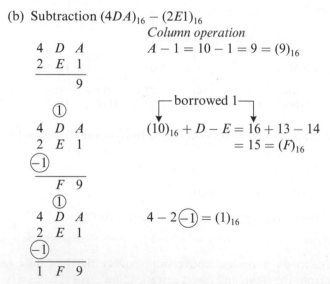

|   |   |   | *Column sum* |
|---|---|---|---|
| $①$ | | | $(9)_{16} + (D)_{16} = 9 + 13 = 22$ |
| $A$ | $4$ | $9$ | $= (16)_{16}$ |
| $E$ | $F$ | $D$ | |

The first step shows:
$A$ $4$ $9$ / $E$ $F$ $D$ with result $6$, and column sum $(9)_{16} + (D)_{16} = 9 + 13 = 22 = (16)_{16}$

$①$ $①$
$A$ $4$ $9$
$E$ $F$ $D$
————
$4$ $6$

$① + (4)_{16} + (F)_{16} = 1 + 4 + 15 = 20 = (14)_{16}$

$①$ $①$
$A$ $4$ $9$
$E$ $F$ $D$
————
$1$ $9$ $4$ $6$

$① + (A)_{16} + (E)_{16} = 1 + 10 + 14 = 25 = (19)_{16}$

The final sum is $(1946)_{16}$. The decimal equivalent is $2633 + 3837 = 6470$. Notice that it's convenient to do intermediate sums with decimal arithmetic, for example in the first step to evaluate $(9)_{16} + (D)_{16}$, with the help of Table 2.5. This practice is continued for the other operations.

(b) Subtraction $(4DA)_{16} - (2E1)_{16}$

*Column operation*
$A - 1 = 10 - 1 = 9 = (9)_{16}$

$4$ $D$ $A$
$2$ $E$ $1$
————
$9$

┌── borrowed 1 ──┐

$①$
$4$ $D$ $A$
$2$ $E$ $1$
$⊖①$
————
$F$ $9$

$(10)_{16} + D - E = 16 + 13 - 14 = 15 = (F)_{16}$

$①$
$4$ $D$ $A$
$2$ $E$ $1$
$⊖①$
————
$1$ $F$ $9$

$4 - 2 ⊖① = (1)_{16}$

The answer is $(1F9)_{16}$, and the decimal equivalent is $1242 - 737 = 505$.

(c) Multiplication $(BAE)_{16} \times (C5)_{16}$
Remember that multiplication by $(10)_{16}$ has the effect of putting a zero onto the right-hand end, so write

$$(BAE)_{16} \times (C5)_{16} = [(BAE)_{16} \times (C0)_{16}] + [(BAE)_{16} \times (5)_{16}]$$

and evaluate each part separately as follows. The first product is

$$\begin{array}{ccc} B & A & E \\ & C & 0 \end{array}$$

*Intermediate steps*

$$C \times E = 12 \times 14 = 168$$
$$= (A8)_{16}$$

$$\begin{array}{cc} Ⓐ & \\ \hline 8 & 0 \end{array}$$

$$\begin{array}{ccc} B & A & E \\ & C & 0 \end{array}$$

$$Ⓐ + C \times A = 10 + 12 \times 10$$
$$= 130$$
$$= (82)_{16}$$

$$\begin{array}{ccc} ⑧ & Ⓐ & \\ \hline 2 & 8 & 0 \end{array}$$

$$\begin{array}{ccc} B & A & E \\ & C & 0 \end{array}$$

$$⑧ + C \times B = 8 + 12 \times 11$$
$$= 140$$
$$= (8C)_{16}$$

$$\begin{array}{cccc} ⑧ & Ⓐ & & \\ \hline 8 \quad C & 2 & 8 & 0 \end{array}$$

So $(BAE)_{16} \times (C0)_{16} = (8C280)_{16}$.
  Similarly, the second product is

$$\begin{array}{ccc} B & A & E \\ & & 5 \end{array}$$

$$\begin{array}{cc} ③ & ④ \\ \hline 3 \quad A & 6 \quad 6 \end{array}$$

The required product is the sum of these two individual products, worked out as follows:

*Column sums*

$$\begin{array}{ccccc} 8 & C & 2 & 8 & 0 \\ & 3 & A & 6 & 6 \\ \hline 8 & F & C & E & 6 \end{array}$$

$$(8)_{16} + (6)_{16} = 14 = (E)_{16}$$
$$(2)_{16} + (A)_{16} = 2 + 10 = 12 = (C)_{16}$$
$$(C)_{16} + (3)_{16} = 12 + 3 = 15 = (F)_{16}$$

so the answer is $(8FCE6)_{16}$. The decimal equivalent is $2990 \times 197 = 589030$.

---

**Exercise**   **2.47**  Extend Table 2.5 to give the hexadecimal equivalents of the decimal numbers 33 to 60.

**Exercise**   **2.48**  Convert to decimal form the following hexadecimal numbers:
(a) $(2A8E)_{16}$   (b) $(B13F40)_{16}$.

**Exercise**   **2.49**  Convert the following decimal numbers to hexadecimal form:
(a) 7813   (b) 48925.

**Exercise** **2.50** Express in binary form each of the hexadecimal numbers you found in the previous exercise. Find the sum and difference of the two numbers using both binary and hexadecimal arithmetic.

**Exercise** **2.51** Convert the following binary numbers to hexadecimal form:

(a) $(11011100001100011)_2$ (b) $(11101101001001010000)_2$.

**Exercise** **2.52** Express $943 \times 256$ as hexadecimals, and hence evaluate the product.

**Exercise** **2.53** The best way of converting from octal to hexadecimal is to first convert to binary. Use this to convert to hexadecimal:

(a) $(714)_8$ (b) $(1625)_8$.

Similarly, hexadecimal is converted to octal via binary. Apply this to convert to octal:

(c) $(AD3)_{16}$ (d) $(B13F40)_{16}$.

**Exercise** **2.54** Evaluate the following using hexadecimal arithmetic:

(a) $292(1508 - 539)$ (b) $(42213)^2$ (c) $(8DE7)_{16} \div (6)_{16}$.

**Exercise** **2.55** Using the notation in Exercise 2.12, convert the following to decimal form:

(a) $(0.FFF)_{16}$ (b) $(A.3CE)_{16}$.

**Exercise** **2.56** Look at the result in Section 1.3.7 for divisibility of decimal numbers by 9, and the result in Exercise 2.43 for divisibility of octal numbers by 7. Guess a corresponding result for divisibility of hexadecimal numbers, and try to prove it.

Hence test for divisibility by 15 the hexadecimal numbers:

(a) $(22F0B)_{16}$ (b) $(140A98)_{16}$.

**Exercise** **2.57** In a similar fashion to the preceding exercise, look at the results of divisibility of

(i) decimal numbers by 11 (Problem 1.8)
(ii) binary numbers by 3 (Exercise 2.23)
(iii) octal numbers by 9 (Exercise 2.45).

Guess a corresponding result for hexadecimal numbers, and try to prove it.
Hence test for divisibility by 17 the hexadecimal numbers:

(a) $(161C45)_{16}$ (b) $(10E8FE9)_{16}$.

## 2.2.5 General base

It's now useful to repeat the introduction to an arbitrary number base $b$ given at the end of Section 2.2.1. Here $b$ is a positive integer, and

$$(x_n x_{n-1} \ldots x_3 x_2 x_1)_b = x_n b^{n-1} + x_{n-1} b^{n-2} + \ldots + x_3 b^2 + x_2 b + x_1 \qquad (2.3)$$

where each of $x_1, \ldots, x_n$ is one of the numbers $0, 1, 2, \ldots, b-1$. As for the case of hexadecimal numbers (when $b = 16$) if $b > 9$ then special symbols are needed for the numbers ten, eleven, $\ldots, b-1$. The results developed in the preceding sections for binary, octal and hexadecimal numbers can now be generalized.

### 2.2.5.1  Point notation

Insert a point in the expression for a number $X_b$ so that digits to the left of it correspond to positive powers of $b$, and digits to the right to negative powers of $b$. This idea was introduced in Exercise 2.12, and in general

$$X_b = (x_n x_{n-1} \ldots x_3 x_2 x_1 . y_1 y_2 y_3 \ldots y_m)_b$$
$$= x_n b^{n-1} + x_{n-1} b^{n-2} + \ldots + x_3 b^2 + x_2 b + x_1$$
$$+ y_1 b^{-1} + y_2 b^{-2} + \ldots + y_m b^{-m}$$

Multiplying both sides of this expression by $b$ produces

$$X_b \times b = x_n b^n + \ldots + x_1 b + y_1 + y_2 b^{-1} + \ldots + y_m b^{-m+1}$$
$$= (x_n \ldots x_1 y_1 . y_2 \ldots y_m)_b$$

so all the digits are shifted one place to the left. Similarly, multiplying by $b^k$ for any positive integer $k$ moves all the digits $k$ places to the left. In particular, if $X_b$ is an integer (that is, all the $y$'s are zero) then $X_b \times b^k$ is $X_b$ with $k$ zeros put onto the right-hand end. Conversely, if $(x_n x_{n-1} \ldots x_1)_b$ ends in $k$ zeros then it is divisible by $b^k$.

In Problem 2.32 you are asked to investigate a method for expressing a rational number less than 1 in the form $(0.y_1 y_2 y_3 \ldots)_b$.

### 2.2.5.2  Conversion to decimal form

In the previous sections, conversion of binary, octal or hexadecimal numbers to decimal form was done by direct evaluation of an expression involving powers of 2, 8 or 16 – see Example 2.3 for the binary case. In fact there is a simpler method of doing the conversion to decimal form. This is best illustrated with an example, say $(x_5 x_4 x_3 x_2 x_1)_b$. To convert this to decimal form requires the evaluation of

$$x_5 b^4 + x_4 b^3 + x_3 b^2 + x_2 b + x_1$$

This is actually a polynomial in $b$. To compute its value, construct the following sequence $z_1, z_2, z_3, z_4$ using decimal arithmetic:

$$z_1 = x_5 b + x_4$$
$$z_2 = z_1 b + x_3$$
$$\phantom{z_2} = (x_5 b + x_4)b + x_3$$
$$\phantom{z_2} = x_5 b^2 + x_4 b + x_3$$

Similarly $z_3$ and $z_4$ are

$$z_3 = z_2 b + x_2 = x_5 b^3 + x_4 b^2 + x_3 b + x_2$$
$$z_4 = z_3 b + x_1 = x_5 b^4 + x_4 b^3 + x_3 b^2 + x_2 b + x_1$$

The last expression shows that $z_4$ is the required decimal number.

This scheme is known as **Horner's method**. In general to convert $(x_n x_{n-1} \ldots x_2 x_1)_b$ to decimal form the algorithm is:

*Step 1*: Set $z_0 = x_n$
*Step 2*: Compute $z_i = z_{i-1} b + x_{n-i}$, for $i = 1, 2, \ldots, n-1$.
*Step 3*: $z_{n-1}$ is the decimal equivalent of $(x_n \ldots x_1)_b$.

The total number of arithmetical 'operations' (that is, additions and multiplications) required to compute $z_n$ is $2n - 2$, since each $z_i$ in Step 2 requires one multiplication and one addition. Compare this with direct evaluation of the polynomial of degree $n - 1$ in $b$ in (2.3): this requires $n - 2$ multiplications to compute $b^2$, $b^3$, $\ldots$, $b^{n-1}$, then $n - 1$ multiplications to form the products $x_2 b$, $x_3 b^2$, $\ldots$, $x_n b^{n-1}$, and finally $n - 1$ additions to form the sum in (2.3), so in total there are $3n - 4$ operations. For large values of $n$, Horner's method therefore is considerably more efficient. If you have a calculator with binary, octal and hexadecimal modes, then its conversions to decimals will actually employ Horner's method, performed very quickly.

**Example 2.19**  **Horner's algorithm**

To convert $(FB62A3)_{16}$ to decimal form, first identify $b = 16$, $n = 6$, $x_1 = 3$, $x_2 = (A)_{16} = 10$, $x_3 = 2$, $x_4 = 6$, $x_5 = (B)_{16} = 11$, $x_6 = (F)_{16} = 15$. Step 1 gives $z_0 = x_6 = 15$, and for Step 2 the calculations are

$$z_1 = z_0 16 + x_5 = 15 \times 16 + 11 = 251$$
$$z_2 = z_1 16 + x_4 = 251 \times 16 + 16 = 4022$$
$$z_3 = z_2 16 + x_3 = 4022 \times 16 + 2 = 64354$$
$$z_4 = z_3 16 + x_2 = 64354 \times 16 + 10 = 1029674$$
$$z_5 = z_4 16 + x_1 = 1029674 \times 16 + 3 = 16474787$$

and $z_5$ is the required decimal equivalent. You should compute the expansion (2.3) for this example, and notice the extra effort involved.

Horner's method is especially simple for binary numbers. You should verify that $(1101011)_2$ in Example 2.3 produces the sequence

$$z_0 = 1, z_1 = 3, z_2 = 6, z_3 = 13, z_4 = 26, z_5 = 53, z_6 = 107$$

2.2.5.3    Conversion from decimal form

The conversion algorithm from decimal to binary in Section 2.2.2.1 requires repeated division by the base 2 until a zero quotient is obtained. The remainders, taken in the opposite order to which they are generated, give the binary number. This algorithm was then used for conversion to octal in Section 2.2.3 and to hexadecimal in Section 2.2.4, the only difference being that the repeated division is by the respective bases 8 or 16.

The verification of this procedure for *any* base $b$ can now be described. For convenience, take a simple case: suppose the decimal number $x = x_4x_3x_2x_1$ is to be converted to base $b$, and that on the fourth division a zero quotient is obtained. This means that at the first division by $b$, if the quotient is $q_1$ and the remainder is $r_1$ then

$$x = bq_1 + r_1 \qquad (2.4)$$

The next step is to divide $q_1$ by $b$, giving a quotient $q_2$ and remainder $r_2$, that is,

$$q_1 = bq_2 + r_2 \qquad (2.5)$$

Similarly the next two divisions by $b$ correspond to

$$q_2 = bq_3 + r_3 \qquad (2.6)$$

$$q_3 = b.0 + r_4 \qquad (2.7)$$

since it's assumed that the algorithm terminates at this point. Notice that decimal arithmetic is used throughout. To express $x$ in terms of the base $b$, you need to use the expressions (2.5), (2.6) and (2.7) to eliminate the $q$'s. Firstly, substitute for $q_1$ in (2.4) using (2.5) to get

$$\begin{aligned} x &= b(bq_2 + r_2) + r_1 \\ &= b^2q_2 + br_2 + r_1 \end{aligned}$$

Next, substitute for $q_2$ from (2.6) to obtain

$$\begin{aligned} x &= b^2(bq_3 + r_3) + br_2 + r_1 \\ &= b^3q_3 + b^2r_3 + br_2 + r_1 \end{aligned}$$

Finally, since $q_3 = r_4$ in (2.7) this becomes

$$x = b^3r_4 + b^2r_3 + br_2 + r_1$$

Hence the expression of $x$ to base $b$ is $(r_4r_3r_2r_1)_b$. Notice that the order in which the remainders are computed gives the digits in the base-$b$ number reading from right to left. The proof in the general case of a decimal number with $n$ digits proceeds in exactly the same way.

At each division the quotient and remainder are *unique* integers – this fact will be established in Section 2.3.1.

| Example 2.20 | **Decimal to base 11 conversion** |

It was pointed out in Section 2.2.1 that the Roman numeral X can be used to denote ten when the base $b$ is eleven. To convert the decimal number 6627 to base 11, repeatedly divide by 11, as follows:

$$
\begin{array}{rl}
 & \text{remainder} \\
11 \overline{)6627} & 5 \\
11 \overline{)\ 602} & 8 \\
11 \overline{)\ \ 54} & 10 = X \quad \uparrow \\
11 \overline{)\ \ \ 4} & 4 \\
 \ \ \ \ \ 0 &
\end{array}
$$

Reading upwards, the base eleven equivalent is $(4X85)_{11}$.

Base eleven, in conjunction with appropriate 'modular' arithmetic (see Section 2.4) is used for the International Standard Book Number (ISBN) which identifies all published books worldwide (see Exercise 2.104 and Section 4.4.1, Chapter 4).

Base twelve (called **duodecimal**) has already been encountered as the number of hours in a day, the number of inches in a foot, or the number of old British pennies in a shilling. None of these cases is a true duodecimal system, however, since the next level of counting does not involve $(12)^2$ (there are seven days in a week, three feet in a yard and twenty shillings in a pound).

Armed with the experience gained from binary, octal and hexadecimal arithmetic, you should be able to do arithmetic for any number base.

Exercise

**2.58** Numbers to base three form the **ternary system**. Evaluate the following:

(a) decimal equivalent of $(12022110)_3$ (use Horner's method)

(b) ternary equivalent of 7940

(c) $(21021)_3 + (10212)_3 + (220)_3$

(d) $(212211)_3 - (1022)_3$

(e) $(122022)_3 \times (2101)_3$.

Exercise

**2.59** Use Horner's algorithm to convert the following to decimals:

(a) $(32012)_4$

(b) $(X189X)_{11}$

(c) $(101110100111)_2$ (see Exercise 2.14(c)).

Exercise

**2.60**

(a) Convert 2967 to base 7.

(b) Convert $(62134)_7$ to decimal.

| Exercise | **2.61** Apply Horner's method to evaluate the polynomial |

$$f(x) = 2x^5 - 3x^4 + 7x^3 - x^2 + 5x + 1$$

when $x = 8$ and when $x = 10$.

| Exercise | **2.62** Eggs are usually counted in twelves, with 1 **dozen** = 12, 1 **gross** = $(12)^2$. Use duodecimal arithmetic to solve the following problems: |

(a) (9 gross, 4 dozen and 7) + (2 gross, 7 dozen and 8)

(b) (15 gross, 7 dozen and 2) − (4 gross, 8 dozen and 6)

(c) A consignment of eggs totalling 74 gross, 10 dozen and 9 is to be divided equally amongst three supermarkets. How many eggs will go to each store?

| Exercise | **2.63** In the duodecimal system let $A$ denote ten and $B$ eleven. Express 17243 to base 12. |

| Exercise | **2.64** Express 1158319 in the Babylonian base 60. |

| Exercise | **2.65** Go back to the results in Exercise 2.56. Hence try to state and prove a result for a ternary number (that is, a number to base 3) to be even. Use it to test whether $(12022110)_3$ and $(211112002)_3$ are even. |

| Exercise | **2.66** Explain how to convert a number from base 3 to base 9, or from base 9 to base 3. Use your results to convert $(220112)_3$ and $(782)_9$. |

| Exercise | **2.67** Write out the proof of the method in Section 2.2.5.3 for conversion of the decimal number $x_n x_{n-1} \ldots x_1$ to $(r_m r_{m-1} \ldots r_1)_b$. |

## 2.2.6 Divisibility

In Chapter 1, Section 1.3 was devoted to rules for divisibility of decimal numbers – remember that 'divisibility' means exactly divisible without a remainder. In several places in this present Section 2.2 you have been asked to determine divisibility tests for numbers to base 2, 8 or 16. Thus Exercise 2.43 required you to prove the test for divisibility of octal numbers by 7 (= 8 − 1), and Exercise 2.56 asked you for the test for divisibility of hexadecimal numbers by 15 (= 16 − 1). You probably realized that in both cases the divisor was *one less* than the base; indeed the nature of the results was the same as divisibility of decimal numbers by 9 (= 10 − 1), requiring divisibility of the sum of the digits. Similarly, when the divisor is *one more* than the base, the divisibility result for decimal numbers is in Problem 1.11; for binary numbers, Exercise 2.23; octal numbers, Exercise 2.45; and hexadecimal numbers, Exercise 2.57. In fact all these results are special cases of two general theorems which can now be stated and proved.

## Theorem: Divisibility by $b - 1$

The number $a = (a_n a_{n-1} \ldots a_2 a_1)_b$ is divisible by $b - 1$ if and only if the sum of its digits $S = a_1 + a_2 + \ldots a_n$ is divisible by $b - 1$.

**Proof** This follows exactly the same lines as indicated for the base $b = 8$ in Exercise 2.43. First notice that

$$b = 1 + (b - 1)$$
$$b^2 = 1 + (b - 1)(b + 1)$$

The next step is to prove that

$$b^k = 1 + (b - 1)t_k \tag{2.8}$$

for all values of the positive integer $k$, where the $t_k$ are positive integers. Since (2.8) is true for $k = 1$ and $k = 2$, it is natural to try and prove it by the method of induction. Assume that (2.8) holds for some particular value of $k$. Then

$$
\begin{aligned}
b^{k+1} &= b(b^k) \\
&= b[1 + (b - 1)t_k] \qquad \text{by assumption} \\
&= b + b(b - 1)t_k \\
&= 1 + (b - 1) + b(b - 1)t_k \\
&= 1 + (b - 1)(1 + bt_k) \\
&= 1 + (b - 1)t_{k+1}
\end{aligned}
$$

This shows that if (2.8) holds for a certain value of $k$, then it is also true for $k + 1$. However, since (2.8) is true for $k = 1$ and $k = 2$, it follows that it *must* be true for *all* values of $k = 1, 2, 3, 4, \ldots$.

The expression (2.8) can now be substituted into the expansion

$$a = (a_n a_{n-1} \ldots a_2 a_1)_b = a_n b^{n-1} + a_{n-1} b^{n-2} + \ldots + a_2 b + a_1$$

to produce

$$
\begin{aligned}
a = a_n[1 + (b - 1)t_{n-1}] + a_{n-1}[1 + (b - 1)t_{n-2}] + \ldots + a_2[1 + (b - 1)t_1] \\
+ a_1 = (a_n + a_{n-1} + \ldots a_2 + a_1) + (b - 1)p
\end{aligned}
\tag{2.9}
$$

where $p$ is also a positive integer (actually equal to $a_n t_{n-1} + a_{n-1} t_{n-2} + \ldots + a_2 t_1$). It follows from (2.9) that since the term $(b - 1)p$ is divisible by $b - 1$, then $a$ is itself divisible by $b - 1$ if and only if the first term on the right in (2.9), namely $S$, is divisible by $b - 1$.

---

**Example 2.21** | **Divisibility by $b - 1$**

(a) In Exercise 2.44 you were asked to test the octal numbers $(12004)_8$ and $(166241)_8$ for divisibility by 7. In the first case the sum of the octits is $1 + 2 + 4 = 7$ so $(12004)_8$ is divisible by 7. In the second case

$1 + 6 + 6 + 2 + 4 + 1 = 20$, which is not divisible by 7, so $(166241)_8$ is not divisible by 7. Incidentally, the decimal equivalents are respectively 5124 and 60577, but if you try the test for divisibility of these by 7 in Section 1.3.5, you'll find it's more complicated.

(b)  In Exercise 2.56 you were asked to test two hexadecimal numbers for divisibility by 15. The first was $(22F0B)_{16}$, whose hexits sum is

$$(2 + 2 + F + 0 + B)_{16} = 2 + 2 + 15 + 0 + 11 = 30$$

which is divisible by 15, so $(22F0B)_{16}$ is divisible by 15.

The second number was $(140A98)_{16}$, for which

$$(1 + 4 + 0 + A + 9 + 8)_{16} = 1 + 4 + 0 + 10 + 9 + 8 = 32$$

showing that $(140A98)_{16}$ is not divisible by 15.

Notice that the test provided by the Theorem is carried out for convenience using decimal arithmetic.

---

**Theorem: Divisibility by $b + 1$**

The number $a = (a_n a_{n-1} \ldots a_2 a_1)_b$ is divisible by $b + 1$ if and only if $a_1 - a_2 + a_3 - \ldots + (-1)^{n-1} a_n$ is divisible by $b + 1$.

**Proof**   Like the previous theorem, this follows along the same lines as the proof for the case $b = 8$, as indicated in Exercise 2.45. First notice that

$$b = -1 + (b + 1), \quad b^2 = 1 + (b + 1)(b - 1)$$

and then prove by induction that

$$b^k = (-1)^k + (b + 1)s_k, \qquad k = 3, 4, 5, \ldots$$

where the $s_k$ are positive integers. Substituting into

$$a = (a_n \ldots a_2 a_1)_b = a_n b^{n-1} + \ldots + a_2 b + a_1$$

gives

$$a = a_n[(-1)^{n-1} + (b+1)s_{n-1}] + \ldots + a_2[-1 + (b+1)s_1] + a_1$$
$$= a_1 - a_2 + \ldots + (-1)^{n-1} a_n + (b+1)q$$

where $q$ is a positive integer. The stated result now follows using an argument like that for the previous theorem.

---

**Example 2.22**   **Divisibility by $b + 1$**

(a)  In Exercise 2.24 you were asked to test the binary numbers $b_1 = (111001101)_2$ and $b_2 = (10011001101)_2$ for divisibility by 3. In the first case $n = 9$, and the sum of bits in the theorem is

$$(1 - 0 + 1 - 1 + 0 - 0 + 1 - 1 + 1)_2 = 2$$

which is not divisible by 3, so neither is $b_1$. In the second case $n = 10$, and the sum is

$$(1 - 0 + 1 - 1 + 0 - 1 + 1 - 0 + 0 - 1)_2 = 0$$

showing $b_2$ is divisible by 3. (The decimal equivalents are $b_1 = 461$, $b_2 = 621$.)

(b)　In Exercise 2.46 your were asked to test the octal numbers $c_1 = (11515)_8$ and $c_2 = (706674)_8$ for divisibility by 9. The sums of octits are respectively

$$(5 - 1 + 5 - 1 + 1)_8 = 9, \quad (4 - 7 + 6 - 6 + 0 - 7)_8 = -10$$

showing that $c_1$ is divisible by 9, but $c_2$ is not.

(c)　In Exercise 2.57 you were asked to test for divisibility by 17 the hexadecimal numbers $d_1 = (161C45)_{16}$ and $d_2 = (10E8FE9)_{16}$. The sums of hexits are respectively

$$(5 - 4 + C - 1 + 6 - 1)_{16} = 5 - 4 + 12 - 1 + 6 - 1 = 17$$
$$(9 - E + F - 8 + E - 0 + 1)_{16} = 9 - 14 + 15 - 8 + 14 - 0 + 1 = 17$$

so both $d_1$ and $d_2$ are divisible by 17.

---

**Exercise** **2.68**　Write out the details of the induction proof for the expression $b^k$ in the theorem on divisibility by $b + 1$.

**Exercise** **2.69**　Test the following ternary numbers for divisibility by 2 and 4:

(a) $(2102111)_3$　(b) $(20220110)_3$　(c) $(2210011)_3$.

**Exercise** **2.70**

(a) What are the conditions for an octal number to be divisible by 2 or 4?

(b) What are the conditions for a hexadecimal number to be divisible by 2 or 4 or 8?

**Exercise** **2.71**　Apply the theorems of this section to the following:

(a) $(4302551)_7$　(b) $(188180848)_9$　(c) $(58388XX)_{11}$.

Check by finding the decimal equivalents.

**Exercise** **2.72**　Express the following decimal numbers in terms of the base 1000, and hence test whether they are divisible by 999:

(a) 21,461,517　(b) 869,385,744　(c) 312,963,802,151.

**Exercise** **2.73**　Express the following decimal numbers in terms of the base 1000, and hence test whether they are divisible by 1001:

(a) 8,522,515　(b) 9,860,023,173　(c) 719,237,505,814,828.

See Problem 2.9 for a development of this idea.

**Exercise** **2.74**　Prove that a number to base 5 is even if and only if the sum of its digits is even. This result can be generalized – see Problem 2.10.

**Exercise** | **2.75** Prove by induction that

$$5^k = \begin{cases} 1 + 3s_k & k \text{ even} \\ -1 + 3t_k & k \text{ odd} \end{cases}$$

where $s_k$ and $t_k$ are positive integers. Hence prove that $(a_n \ldots a_2 a_1)_5$ is divisible by 3 if and only if $a_1 - a_2 + a_3 - \ldots + (-1)^{n-1} a_n$ is divisible by 3. Hence test $(401134)_5$ for divisibility by 3 and by 6.

## 2.3 Greatest common divisor

### 2.3.1 Definitions

Suppose $a$ and $b$ are positive integers, and there is an integer $c$ such that $a = bc$. Then $b$ is said to **divide** $a$, or to be a **divisor** or **factor** of $a$, written $b|a$. If $b$ does not divide $a$ then the notation used is $b \nmid a$. In this case there is a **remainder** $r$, that is,

$$a = bq + r \tag{2.10}$$

where $q$ is the **quotient**.

---

**Division theorem**

There exist unique integers $q > 0$ and $b > r \geq 0$ satisfying (2.10).

This theorem is simply a formal mathematical statement of what you are already familiar with from 'long division'. You go on dividing $b$ into $a$ until you can't continue because the remaining integer $r$ is less than $b$. This result in the theorem therefore seems obvious, but as usual a **proof** is necessary: suppose that (2.10) is satisfied by two *different* pairs of quotient and remainder, that is,

$$a = bq_1 + r_1, \qquad b > r_1 \geq 0$$
$$a = bq_2 + r_2, \qquad b > r_2 \geq 0$$

Subtracting the second identity from the first gives

$$0 = b(q_1 - q_2) + r_1 - r_2$$

so that

$$r_1 - r_2 = b(q_2 - q_1) \tag{2.11}$$

However, if $q_2 \neq q_1$ then the magnitude of their difference (ignoring sign) that is, $|q_1 - q_2|$, is at least 1. Hence in this case (2.11) shows that $|r_1 - r_2|$ must be greater than or equal to $b$. However, since both $r_1$ and $r_2$ are less than $b$, and are not negative, the magnitude of their difference $|r_1 - r_2|$ must be *less* than $b$. It is therefore impossible to have $q_2 \neq q_1$. Hence $q_1 = q_2$ and (2.11) then shows that $r_1 = r_2$. In other words, the integers $q$ and $r$ satisfying (2.10) are unique.

In fact the division theorem has already been assumed in the method given in Section 2.2.5.3 for conversion from decimal form to an arbitrary base. For example, in equations (2.4) to (2.7) it was implicitly assumed that in each case the quotient and remainder were unique. Hence the representation of a positive integer with respect to a given base is unique.

If $d$ is an integer which divides each of two other integers $a$ and $b$, then $d$ is called a **common divisor** of $a$ and $b$. If $d$ is the largest such common divisor it is called the **greatest common divisor** (**g.c.d.**) of $a$ and $b$, often written as $(a,b)$. If $(a,b) = 1$, then $a$ and $b$ are said to be **relatively prime**. This means that there is no integer larger than 1 which divides both $a$ and $b$. You've probably come across the term **prime number**, which means an integer which has no factors other than 1 and itself, that is, a prime number cannot be expressed as the product of two integers each smaller than itself. The rather quaintly-named **Fundamental Theorem of Arithmetic** states that any positive integer which is not a prime can be expressed in a unique way as a product of primes (apart from writing the factors in a different order). For example, 77 can only be written as $7 \times 11$, and similarly

$$5733 = 3 \times 3 \times 7 \times 7 \times 13 = 3^2 \times 7^2 \times 13$$

The definitions involving two integers can be extended to three or more in an obvious and natural way. Thus $d = (a_1, a_2, \ldots, a_m)$ is the g.c.d. of a set of integers $a_1, a_2, \ldots, a_m$ if it is the largest integer which divides each of them; and if $d = 1$ then these integers are said to be relatively prime.

**Example 2.23**  **Common divisors**

Let $a = 180$, $b = 72$. Dividing $b$ into $a$ gives

$$180 = 72 \times 2 + 36$$

so in (2.10) $q = 2$, $r = 36$. Clearly $2|180$ and $2|72$ so 2 is a common divisor of $a$ and $b$. Since $180 = 5 \times 3^2 \times 2^2$ and $72 = 3^2 \times 2^3$, you can see that other common divisors of $a$ and $b$ are 4, 6, 9, 12, 18, 24 and 36. The largest of these is 36, so $(180,72) = 36$. Similarly if a third integer is 48, then $(180,72,48) = 24$ since $36 \nmid 48$ Notice that the g.c.d. is always unique.

If instead $b = 77$, then $a$ and $b$ have no factors in common since $77 = 7 \times 11$, so $(a,b) = 1$.

Nothing in this section so far depends upon which number base is being used. Recall from Section 2.2.5.1 that $x = (x_n x_{n-1} \ldots x_2 x_1)_b$ is divisible by $b^k$ if and only if the last $k$ digits $x_k, x_{k-1}, \ldots, x_1$ are zero. In fact this result can be extended to any divisor of the base $b$: if $d|b$ then $x$ is divisible by $d^k$ if and only if $(x_k \ldots x_2 x_1)_b$ is divisible by $d^k$. Some illustrations of this result are now given, followed by a proof.

**Example 2.24**     **Divisibility by powers of a base**

(a)   First consider decimal numbers, so $b = 10$. Since $2|10$, it follows that a decimal number $x = x_n \ldots x_2 x_1$ is divisible by 2 if $x_1$ is divisible by 2; is divisible by 4 if $x_2 x_1$ is divisible by 4; and is divisible by 8 if $x_3 x_2 x_1$ is divisible by 8. These results were quoted in Section 1.3.

Similarly, since $5|10$ then $x$ is divisible by 5 if $x_1$ is divisible by 5; and $x$ is divisible by $5^2 = 25$ if $x_2 x_1$ is divisible by 25.

(b)   Consider the octal number $x = (2714)_8$, and take $d = 2$, which is a divisor of 8. The last digit 4 of $x$ is divisible by 2, so $x$ is divisible by 2. The last two digits of $x$ are $(14)_8 = 12$, which is divisible by $2^2$, so $x$ is divisible by 4. However, the last three digits of $x$ are $(714)_8 = 460$ which is not divisible by $2^3$, so $x$ is not divisible by 8. The decimal equivalent of $x$ is 1484.

It is easy to see why the result works. Since $d|b$, there is a positive integer $q$ such that $b = qd$. Hence

$$x = (x_n x_{n-1} \ldots x_2 x_1)_b = x_n b^{n-1} + x_{n-1} b^{n-2} + \ldots + x_2 b + x_1$$
$$= x_n q^{n-1} d^{n-1} + \ldots + x_{k+1} q^k d^k$$
$$+ x_k q^{k-1} d^{k-1} + \ldots + x_2 q d + x_1$$

and the terms involving $x_n$ down to $x_{k+1}$ are certainly divisible by $d^k$. Hence $x$ is divisible by $d^k$ if and only if the remaining terms are divisible by $d^k$, and these terms constitute $(x_k \ldots x_2 x_1)_b$.

The theorems in Section 2.2.6 on divisibility by $b + 1$ and $b - 1$ can be modified in a similar way. For example, the first theorem becomes:

If $d|(b - 1)$ then $a = (a_n \ldots a_2 a_1)_b$ is divisible by $d$ if and only if $S = a_1 + a_2 + \ldots + a_n$ is divisible by $d$.

The proof requires only the following slight modifications from that of the original theorem. Since $b - 1 = dc$ for some positive integer $c$, it follows from (2.8) that

$$b^k = 1 + dct_k = 1 + dt'_k$$

for all positive integers $k$, with each $t'_k$ being a positive integer. The expansion of $a$ now gives

$$a = a_n(1 + dt'_{n-1}) + a_{n-1}(1 + dt'_{n-2}) + \ldots + a_2(1 + dt'_1) + a_1$$
$$= (a_n + a_{n-1} + \ldots + a_2 + a_1) + dP$$

where $P$ is also a positive integer. Since $d|dP$, it follows that $d|a$ if and only if $d|S$.

**Example 2.25**  **Divisibility by a factor of $b - 1$**

The hexadecimal number $(1F3763)_{16}$ has sum of hexits

$$(1 + F + 3 + 7 + 6 + 3)_{16} = 1 + 15 + 3 + 7 + 6 + 3 = 35$$

This is not divisible by 15, but is divisible by 5 which is a factor of 15, so that $5|(1F3763)_{16}$. The decimal equivalent is 2045795.

The theorem on divisibility by $b + 1$ is modified in an identical way, the only difference being that $d|(b + 1)$ (see Exercise 2.84). It is interesting that by taking $b = 1000$ in this case, the tests given in Section 1.3.5 for divisibility of decimal numbers by 7, 11 or 13 can be derived. The key fact is that $1001 = 7 \times 11 \times 13$ – see Problem 2.9.

**Exercise**  **2.76**  Find the quotient and remainder in the division theorem when $a = 2581$ and $b = 19$.

**Exercise**  **2.77**  Determine all the common divisors and the g.c.d. for

(a)  60, 72 and 108

(b)  315 and 5670.

**Exercise**  **2.78**  Test whether $(11853)_{12}$ is divisible by 3, 9, 27 or 81. Check by finding the decimal equivalent.

**Exercise**  **2.79**  If $a$ and $b$ are positive integers such that $a|b$ and $b|a$, what can you conclude?

**Exercise**  **2.80**  If $a|b$ and $b|c$, prove that $a|c$.

**Exercise**  **2.81**  If $a$, $b$ and $c$ are positive integers show that $a|b$ if and only if $ac|bc$.

**Exercise**  **2.82**  Prove that if $d = (a,b)$ then $d|(a + b)$ and $d|(a - b)$.

**Exercise**  **2.83**  Show that $3|(a^3 - a)$ for any positive integer $a$.

**Exercise**  **2.84**  If $d|(b + 1)$, state and prove the modified version of the theorem on divisibility by $b + 1$ in Section 2.2.6. Hence test $(14BCB)_{13}$ for divisibility by 7.

**Exercise**  **2.85**  Find three relatively prime numbers from the set 66, 78, 110, 165, 195.

**Exercise**  **2.86**  The **least common multiple (l.c.m.)** of two positive integers $a$ and $b$ is the smallest positive integer divisible by both $a$ and $b$, and is denoted by $[a,b]$. Prove:

(a)  $[ca,cb] = c[a,b]$ for any positive integer $c$

(b)  $[a,b] = ab/(a,b)$.

### 2.3.2 Euclid's algorithm

In Example 2.23 the g.c.d. of two integers was obtained by finding all the divisors of each number, and then selecting those possessed in common. This is a very inefficient process, since divisors not held in common are needlessly calculated.

A procedure which computes the g.c.d. directly is attributed to Euclid, and is the oldest known algorithm. It appeared around BC 300 in Book VII of Euclid's *Elements*, perhaps the most successful mathematics textbook ever written. It is based on the division theorem given in the previous section, and before giving a formal statement of Euclid's scheme it's helpful to look at a simple example.

**Example 2.26**  **Repeated divisions**

Consider the integers $a = 252$ and $b = 54$. Applying the division algorithm (2.10) gives

$$252 = 54 \times 4 + 36 \tag{2.12}$$
$$\phantom{252 = }a \phantom{= }b \phantom{\times }q \phantom{\times }r_2$$

Now apply the division process again, but with $a$ replaced by $b$, and $b$ by $r_2$:

$$54 = 36 \times 1 + 18 \tag{2.13}$$
$$\phantom{5}b \phantom{= }r_2 \phantom{\times 1 + }r_3$$

Next, divide $r_3$ into $r_2$:

$$36 = 18 \times 2 + 0 \tag{2.14}$$
$$r_2 \phantom{= }r_3 \phantom{\times 2 + }r_4$$

The remainder $r_4$ is zero, so the process stops. You can easily check that the last non-zero remainder $r_3 = 18$ is the g.c.d. of $a$ and $b$. To see why this works first notice that, from (2.14), 18 divides 36. Substituting $r_2 = 18 \times 2$ into (2.13) gives

$$b = 54 = 18 \times 2 + 18 = 3 \times 18$$

showing that 18 divides 54. Finally, substituting for $b$ and $r_2$ into (2.12) gives

$$252 = 3 \times 18 \times 4 + 18 \times 2 = 14 \times 18$$

showing that 18 is a common divisor of 252 and 54. To show it is the greatest common divisor, suppose $d$ is some other divisor of 252 and 54. From (2.12) it follows that $d$ must divide 36; from (2.13) it then follows that $d$ must divide 18, since it divides both 54 and 36. In other words $d \le 18$, so 18 is indeed the *greatest* common divisor.

## Euclid's algorithm

Successively apply the division theorem as follows:

$$r_i = r_{i+1}q_{i+1} + r_{i+2} \qquad (2.15)$$

for $i = 0, 1, 2, 3, \ldots$, where $r_0 = a$ and $r_1 = b$. The last non-zero remainder $r_n$ is the g.c.d. of the integers $a$ and $b$ (where $b < a$).

You are asked in Problem 2.19 to develop a general proof. Notice that the condition on the remainder in the division theorem gives $b > r_2 \geq 0$ at the first step, $r_2 > r_3 \geq 0$ at the second step, and so on. Hence the sequence of remainders in (2.15) satisfies $b > r_2 > r_3 > r_4 > \ldots \geq 0$. Since the remainders continually decrease, eventually a zero remainder will be obtained. Indeed, the most extreme case would be where each remainder is just 1 less than the preceding one in the sequence, when there would be a total of $b$ remainders. Thus in general it will not be necessary to carry out the division (2.15) more than $b$ times. In practice far fewer iterations will be needed, as the following example illustrates.

---

**Example 2.27**  **Euclid's algorithm**

Taking $a = 364$ and $b = 154$ in the expression (2.15) gives

$$(i = 0) \quad 364 = 154 \times 2 + 56$$
$$(r_2)$$
$$(i = 1) \quad 154 = \phantom{0}56 \times 2 + 42$$
$$(r_3)$$
$$(i = 2) \quad \phantom{0}56 = \phantom{0}42 \times 1 + 14$$
$$(r_4)$$
$$(i = 3) \quad \phantom{0}42 = \phantom{0}14 \times 3 + 0$$

The last non-zero remainder is $r_4 = 14$, so $(364, 154) = 14$.

---

It's interesting to quote a result known as **Lamé's theorem**. This states that the number of divisions required to find the g.c.d. of two positive integers never exceeds five times the number of decimal digits in the smaller of the two integers. This is a very conservative estimate – for instance, in Example 2.27 only four divisions are needed, whereas Lamé's result predicts that at most $5 \times 3 = 15$ divisions may be required. Surprisingly, the proof of the theorem involves **Fibonacci numbers**, to be discussed in Section 5.3.2. These are defined by

$$F_1 = 1, \quad F_2 = 1, \quad F_3 = F_2 + F_1 = 2, \quad F_4 = F_3 + F_2 = 3$$

and so on, each number in the sequence being the sum of the preceding two, so in general

$$F_{n+2} = F_{n+1} + F_n, \qquad n = 1, 2, 3, 4, \ldots \qquad (2.16)$$

The properties of these numbers are investigated in some detail in Chapter 5, but it is of interest here to notice that *any* two consecutive Fibonacci numbers are relatively prime, for example $F_4 = 3$, $F_5 = 5$. To prove that this is true in general, suppose that $(F_{n+2}, F_{n+1}) = d$ so that $d|F_{n+1}$ and $d|F_{n+2}$. From (2.16), $F_n = F_{n+2} - F_{n+1}$ so that $d$ must also divide $F_n$. Similarly $F_{n-1} = F_{n+1} - F_n$, so $d|F_{n-1}$, and continuing in the same way ends up with $d|F_2$. Since $F_2 = 1$, this establishes that $d = 1$, as required. You are asked in Problem 2.20 to show that Euclid's algorithm applied to $F_{n+1}$ and $F_{n+2}$ requires exactly $n$ divisions to establish that $(F_{n+1}, F_{n+2}) = 1$.

Euclid's algorithm can also be used to express the g.c.d. $d = (a,b)$ in the form

$$d = ax + by \qquad (2.17)$$

where $x$ and $y$ are integers.

---

**Example 2.28**    **Solving equation (2.17)**

Return to $a = 252$ and $b = 54$ in Example 2.26. Start with equation (2.13) containing $r_3 = d = 18$, and rewrite it as

$$\begin{array}{ccc} 18 = & 54 - & 36 \times 1 \\ d & b & r_2 \end{array}$$

Next, substitute into this for $r_2$ from (2.12) to obtain

$$\begin{aligned} 18 &= 54 - (252 - 54 \times 4) \times 1 \\ &= -252 + 54 \times 5 \\ &= 252 \times (-1) + 54 \times 5 \end{aligned}$$

showing that in (2.17) $x = -1$ and $y = 5$. However, the solution of (2.17) for $x$ and $y$ is not unique, for instance you can also take $x = 2$, $y = -9$ in this example (see Exercise 2.93).

The solutions of (2.17) can be interpreted geometrically. The points marked in Figure 2.4 are those having integer co-ordinates. The straight line represented by (2.17) only passes through the points of this integer lattice for which $x$ and $y$ are solutions.

In the same way, using the steps of Euclid's algorithm in Example 2.27, starting at $i = 2$, you get

$$\begin{aligned} d = 14 &= 56 - 42 \times 1 \\ &= 56 - (154 - 56 \times 2) \times 1 \qquad \text{(using } i = 1) \\ &= -154 + 56 \times 3 \\ &= -154 + (364 - 154 \times 2) \times 3 \qquad \text{(using } i = 0) \\ &= 364 \times 3 + 154 \times (-7) \end{aligned}$$

This is the required expression (2.17) with $a = 364$, $b = 154$, $x = 3$, $y = -7$, $d = 14$.

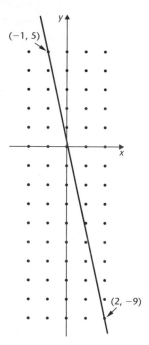

*Figure 2.4* $252x + 54y = 18$

When there are a large number of divisions in the implementation of Euclid's algorithm, tracing through the repeated substitutions as in Example 2.28 can prove rather tedious. A more direct way of using the quotients in Euclid's algorithm to solve (2.17) without having to do any substitutions is set out in Exercise 2.90.

**Exercise**   **2.87**   Use Euclid's algorithm to find the g.c.d. of each of the following pairs of integers:

(a) 945, 81   (b) 9361, 473   (c) 21593, 725.

**Exercise**   **2.88**   It can be shown (see Problem 2.17) that $(a,b,c) = (a,(b,c))$. That is, the g.c.d. of three integers $a$, $b$, $c$ is the g.c.d. of one of them taken with the g.c.d. of the other two. Use this with Euclid's algorithm to find the g.c.d. of 13530, 9192, 1128.

**Exercise**   **2.89**   For each of the pairs of integers in Exercise 2.87(a) and (b), find integers $x$ and $y$ such that $ax + by = d$, where $d = (a,b)$.

**Exercise**   **2.90**   A solution of equation (2.17) is $x = s_n$, $y = t_n$, where

$$s_i = s_{i-2} - q_{i-1}s_{i-1}$$
$$t_i = t_{i-2} - q_{i-1}t_{i-1} \qquad (2.18)$$

for $i = 2, 3, 4, \ldots, n$ with $s_0 = 1$, $s_1 = 0$, $t_0 = 0$, $t_1 = 1$ and the quotients $q_i$ are given by Euclid's algorithm (2.15) (for a proof, see Problem 2.21). Apply this method when $a = 513$, $b = 133$.

**Exercise**    **2.91**    Use (2.18) to obtain a solution of $ax + by = 1$, where $a$ and $b$ are the integers in Exercise 2.87(c).

**Exercise**    **2.92**    Prove that the equation $ax + by = c$ has a solution with $x$ and $y$ integers if and only if $d \mid c$, where $d = (a,b)$.

**Exercise**    **2.93**    Show that if $x = x_0$, $y = y_0$ is an integer solution of $ax + by = d$ where $d = (a,b)$, then

$$x = x_0 - \frac{b}{d}p, \quad y = y_0 + \frac{a}{d}p$$

is also an integer solution for any integer $p$.

**Exercise**    **2.94**    Use the result in the preceding exercise to find the most general form of the solution of the equation

$$23x + 39y = 200$$

Find, if possible, a solution with both $x$ and $y$ positive.

**Exercise**    **2.95**    You wish to spend exactly £10 on stamps costing 28p and 23p. How many stamps of each kind must you buy?

**Exercise**    **2.96**    An area of wall measuring 3.5 m square is to be completely covered with tiles. There are two sizes of tiles, the first measuring $19 \times 10$ cm and the second $17 \times 5$ cm. How many of each size of tile will be needed?

**Exercise**    **2.97**    Use equation (2.16) to write down the Fibonacci numbers from $F_5$ up to $F_{15}$. Show that $(F_9,F_6) = F_{(9,6)}$, $(F_{10},F_5) = F_{(10,5)}$, $(F_{12},F_6) = F_{(12,6)}$, $(F_{15},F_{10}) = F_{(15,10)}$.

**Exercise**    **2.98**    Continue from the preceding exercise and write down $F_{16}$ up to $F_{24}$. Use Euclid's algorithm to find $(F_{16},F_{24})$. Can you spot anything about your result?

**Exercise**    **2.99**    Check for all the Fibonacci numbers you computed in the two preceding exercises that if $r \mid s$ then $F_r \mid F_s$.

<br>

## 2.4    Modular arithmetic

At the beginning of this chapter, in Section 2.1.1, you were introduced to the idea of 'modular arithmetic' through reading the time from a clock face (see Example 2.1). Recall that the position of the hour hand is unaffected by 12-hour periods, so what matters is the number of hours which have elapsed after multiples of 12 have been removed. Similarly, the position of the minute hand is unaffected by 60-minute periods. In each case, what does affect the position of the hour or minute hand is the net result after multiples of 12 or 60 respectively have been discounted. In other words, what matters is the *remainder* after division by 12 or 60 respectively. The arithmetic is said to be 'modulo' 12 or 60 respectively. Other examples discussed in Section 2.1 were a 24-hour digital display, where arithmetic is modulo 24; days of the week, where arithmetic is modulo 7; and using a compass where angles are measured modulo 360. Car odometers usually work modulo 100,000, so if you are offered an old used car showing only 25,000 miles, then beware! The ideas of this section will be used extensively in the theory of codes in Chapter 4.

### 2.4.1 Definitions and properties

The Division theorem in Section 2.3.1 (page 87) showed that the remainder obtained when an integer $a$ is divided by some positive integer $m$ is unique. This leads to the following definition:

(D1) If two integers $a$ and $b$ have the *same* remainder after division by $m$, then $a$ and $b$ are said to be **congruent modulo** $m$, written

$$a \equiv b (\mathrm{mod}\ m) \tag{2.19}$$

and $m$ is called the **modulus**. The relationship (2.19) is called a **congruence**.

The definition (D1) is equivalent to each of the following alternative definitions:

(D2) $m$ divides $a - b$.

(D3) $a = b + mt$ for some integer $t$.

To prove (D2) is equivalent to (D1), note that if (D1) holds then the division theorem implies

$$a = mq_1 + r, \quad b = mq_2 + r$$

Therefore $a - b = m(q_1 - q_2)$, showing that $m \mid (a - b)$. Conversely, if (D2) holds then $a - b = mk$ for some integer $k$. If $b = mq + r$ then

$$a = mk + b = mk + mq + r$$
$$= m(k + q) + r$$

This shows that $a$ has the same remainder $r$ as $b$ when divided by $m$, so (D1) is established.

The equivalence of (D3) to (D1) is established similarly. Notice that in particular if $a = 0 (\mathrm{mod}\ m)$ then this means that $m \mid a$.

If $a \neq b (\mathrm{mod}\ m)$ then $a$ and $b$ are said to be **incongruent** modulo $m$.

When $a$ is divided by $m$ to give the unique remainder $r$ with $0 \leq r < m$, then $a \equiv r (\mathrm{mod}\ m)$ and $r$ is called the **least (non-negative) residue** of $a$ modulo $m$. To compute this using a calculator, say for example 2791 (mod 43), divide 2791 by 43 to get 64.90698 and retain the integer part 64. This is the quotient $q$ in $2791 = 43q + r$, so the least residue is therefore $r = 2791 - 43 \times 64 = 39$.

Example 2.29 **Congruences**

Suppose that the modulus is $m = 12$, as in the 'clock arithmetic' in Section 2.1.1. Then as you saw in Example 2.1,

$$19 \equiv 7(\text{mod } 12) \quad \text{and} \quad 100 \equiv 4(\text{mod } 12)$$

These congruences satisfy the first form of definition (D1), since after division by 12, 19 has remainder 7 and 100 has remainder 4. Thus 7 is the least residue of 19 modulo 12, and 4 is the least residue of 100 modulo 12. Alternatively, the second form of definition (D2) gives $12|(19 - 7)$ and $12|(100 - 4)$; and (D3) holds because $19 = 7 + 12 \times 1$, and $100 = 4 + 12 \times 8$.

As another example, $43 \equiv 19(\text{mod } 12)$ since $12|(43 - 19)$, thereby satisfying (D2). However, 19 is not the least residue of 43 modulo 12, since $19 > 12$; the least residue is 7. On the other hand, 43 and 8 are incongruent modulo 12 since $12 \nmid (43 - 8)$.

There is no requirement for $a$ and/or $b$ in (2.19) to be positive. For example,

$$23 \equiv -1(\text{mod } 12) \quad \text{and} \quad -4 \equiv 8(\text{mod } 12)$$

since in the first case $12|[23 - (-1)]$, and in the second $12|(-4 - 8)$.

---

The arithmetic of congruences is in many ways similar to that of equalities, and this justifies the use of the symbol $\equiv$ in place of $=$. For example, you can add (or subtract) the same integer to both sides of a congruence, or multiply both sides by an integer. Stated formally, these conditions are:

If $a \equiv b(\text{mod } m)$ and $c$ is any integer then the following properties hold:

(P1) $a + c \equiv (b + c)(\text{mod } m)$

(P2) $a - c \equiv (b - c)(\text{mod } m)$

(P3) $ac \equiv bc(\text{mod } m)$

To prove (P1), simply notice that

$$(a + c) - (b + c) = a - b$$

so $m|(a + c) - (b + c)$ because of (D2). The properties (P2) and (P3) are proved similarly. Another useful property involving multiplication is:

(P4) If $a_1 \equiv b_1(\text{mod } m)$ and $a_2 \equiv b_2(\text{mod } m)$ then $a_1 a_2 \equiv b_1 b_2(\text{mod } m)$.

To prove (P4), use (D3) to write

$$a_1 = b_1 + mt_1, \quad a_2 = b_2 + mt_2$$

for some integers $t_1$ and $t_2$. Multiplying gives

$$\begin{aligned} a_1 a_2 &= b_1 b_2 + b_1 mt_2 + b_2 mt_1 + m^2 t_1 t_2 \\ &= b_1 b_2 + m(b_1 t_2 + b_2 t_1 + mt_1 t_2) \end{aligned}$$

so that $a_1a_2 - b_1b_2$ has a factor $m$, and the definition (D2) establishes the required result.

The property (P4) holds for the product of more than two congruences with the same modulus – see Exercise 2.111.

You have to be more careful when considering division of both sides of a congruence by $c$. For example $20 \equiv 8(\text{mod } 12)$, but dividing both sides by 2 in an 'obvious' way gives $10 \equiv 4(\text{mod } 12)$ which is not correct, since 12 does not divide $10 - 4$. However, division of both sides of a congruence by $c$ *is* valid provided $(c,m) = 1$. This condition is not satisfied in this example, since $c = 2$, $m = 12$ and $(c,m) = 2$. In fact the most general result on division is as follows:

(P5)  If $a \equiv b(\text{mod } m)$, and $c$ is a non-zero integer such that $c|a$ and $c|b$, then

$$\frac{a}{c} \equiv \frac{b}{c} \left( \text{mod } \frac{m}{d} \right) \text{ where } d = (c,m).$$

To prove property (P5), let $a = a_1c$ and $b = b_1c$ where $a_1$ and $b_1$ are integers. Since by (D3) $a - b = mt$ for some integer $t$ it follows that $(a_1 - b_1)c = mt$. However $c = c_1d$, $m = m_1d$ where $(c_1,m_1) = 1$, so substituting gives $(a_1 - b_1)c_1d = m_1dt$. This reduces to

$$t = (a_1 - b_1)c_1/m_1$$

Since $c_1$ and $m_1$ are relatively prime, and $t$ is an integer, it must be the case that $m_1|(a_1 - b_1)$. Definition (D2) shows that $a_1 \equiv b_1(\text{mod } m_1)$, which is the required result.

Notice that in particular when $(c,m) = 1$ then $a/c \equiv (b/c)(\text{mod } m)$.

---

**Example 2.30**  **Properties of congruences**

Consider the congruence $36 \equiv 8(\text{mod } 14)$ which by (D2) means that $14|(36 - 8)$. By property (P1), adding 3 to both sides of the congruence gives $39 \equiv 11(\text{mod } 14)$, which is correct since $14|(39 - 11)$. Similarly by property (P2), subtracting 3 from both sides gives $33 \equiv 5(\text{mod } 14)$; and by property (P3) multiplying both sides by 3 gives $108 \equiv 24(\text{mod } 14)$.

If a second congruence is $23 \equiv 9(\text{mod } 14)$ then the property (P4) shows that

$$36 \times 23 \equiv 9 \times 8(\text{mod } 14)$$

or $828 \equiv 72(\text{mod } 14)$. The correctness of this is again seen from (D2), since $14|(828 - 72)$.

For division, property (P5) shows that dividing both sides of $36 \equiv 8(\text{mod } 14)$ by 4 gives $9 \equiv 2(\text{mod } 7)$ since $d = (4,14) = 2$.

Similarly, for the case $20 \equiv 8(\text{mod } 12)$ mentioned above, the correct result on dividing both sides by 2 is $10 \equiv 4(\text{mod } 6)$ since $d = (2,12) = 2$.

It's interesting to return to the results in Section 2.2.6 on divisibility of a base-$b$ number by $b - 1$ and $b + 1$. In the proof of the first case, the crucial fact in (2.8) was

$$b = 1 + (b - 1), \quad b^k = 1 + (b - 1)t_k$$

where $k = 2, 3, 4, \ldots$ and the $t_k$ are positive integers. Using the congruence notation, these expressions can be written as

$$b \equiv 1(\mathrm{mod}(b - 1)), \quad b^k \equiv 1(\mathrm{mod}(b - 1)) \tag{2.20}$$

The theorem itself can now be restated:

The number $(a_n \ldots a_1)_b \equiv 0(\mathrm{mod}(b - 1))$ if and only if $a_1 + a_2 + \ldots + a_n \equiv 0(\mathrm{mod}(b - 1))$

Similarly, the theorem on divisibility by $b + 1$ states that $(a_n \ldots a_1)_b \equiv 0(\mathrm{mod}(b + 1))$ if and only if $a_1 - a_2 + a_3 - \ldots + (-1)^{n-1}a_n \equiv 0(\mathrm{mod}(b + 1))$. Its proof involved results which can now be written as

$$b \equiv -1(\mathrm{mod}(b + 1)), \quad b^k \equiv (-1)^k(\mathrm{mod}(b + 1)) \tag{2.21}$$

In fact both (2.20) and (2.21) are special cases of the following result:

$$\left.\begin{array}{l} \text{If } a \equiv b(\mathrm{mod}\ m) \quad \text{then} \quad a^k \equiv b^k(\mathrm{mod}\ m) \\ \text{for any positive integer } k \end{array}\right\} \tag{2.22}$$

An easy way to prove (2.22) is to apply (P4) repeatedly with $a_1 = a_2 = a$, $b_1 = b_2 = b$, so that $a^2 = b^2(\mathrm{mod}\ m)$, $a^3 = b^3(\mathrm{mod}\ m)$, and so on.

| Example 2.31 | **Least residue for a power of 2** |

The result in (2.22) can be used to obtain congruences involving very large powers of integers. For example, the congruence $2 \equiv 2(\mathrm{mod}\ 11)$ can be used to find the least residue modulo 11 of (say) $2^{573}$. Repeatedly square the congruence (that is, repeatedly take $k = 2$ in (2.22)) and reduce modulo 11 at each step to obtain the sequence:

$$
\begin{aligned}
2 &\equiv 2(\mathrm{mod}\ 11) \\
2^2 &\equiv 2^2(\mathrm{mod}\ 11) \equiv 4(\mathrm{mod}\ 11) \\
(2^2)^2 &\equiv 4^2(\mathrm{mod}\ 11) \\
\text{i.e. } 2^4 &\equiv 16(\mathrm{mod}\ 11) \equiv 5(\mathrm{mod}\ 11) \\
2^8 &\equiv 25(\mathrm{mod}\ 11) \equiv 3(\mathrm{mod}\ 11) \\
2^{16} &\equiv 9(\mathrm{mod}\ 11) \\
2^{32} &\equiv 81(\mathrm{mod}\ 11) \equiv 4(\mathrm{mod}\ 11) \\
2^{64} &\equiv 16(\mathrm{mod}\ 11) \equiv 5(\mathrm{mod}\ 11) \\
2^{128} &\equiv 25(\mathrm{mod}\ 11) \equiv 3(\mathrm{mod}\ 11) \\
2^{256} &\equiv 9(\mathrm{mod}\ 11) \\
2^{512} &\equiv 81(\mathrm{mod}\ 11) \equiv 4(\mathrm{mod}\ 11)
\end{aligned}
$$

Note that the residues occur in a repeating pattern 4, 5, 3, 9, $\ldots$.

To find $2^{573} \pmod{11}$ it is first necessary to express $2^{573}$ in terms of powers in the above list. The simplest way to do this is to convert 573 to binary form using the scheme given in Section 2.2.2.1, which produces $573 = (1000111101)_2$. Expanding the binary expression gives

$$573 = 2^9 + 2^5 + 2^4 + 2^3 + 2^2 + 1$$
$$= 512 + 32 + 16 + 8 + 4 + 1$$

so that

$$2^{573} = 2^{512 + 32 + 16 + 8 + 4 + 1}$$
$$= 2^{512} 2^{32} 2^{16} 2^8 2^4 2 \tag{2.23}$$

Now multiply together the congruences for each of the powers of 2 in (2.23) – property (P4) shows that the result is

$$2^{573} \equiv 4 \times 4 \times 9 \times 3 \times 5 \times 2 \pmod{11} \tag{2.24}$$
$$\equiv 4320 \pmod{11}$$

Dividing 4320 by 11 produces a remainder 8, so finally $2^{573} \equiv 8 \pmod{11}$. Since $2^{573} \approx 3.1 \times 10^{172}$, which is an extremely large number, the relative ease with which the least residue modulo 11 can be found is impressive.

The method of the preceding example can be used to compute the least residue of $b^N$ modulo $m$ for arbitrary positive integers $b$, $N$ and $m$. First express $N$ in binary form $(a_k a_{k-1} \ldots a_1)_2$. Then find the smallest positive remainder modulo $m$ for $b$, $b^2$, $b^4$, ..., $b^r$ where $r = 2^{k-1}$, by repeatedly squaring and reducing modulo $m$ if necessary. Finally, multiply together the remainders as in (2.24) and reduce the result modulo $m$.

Another useful application of (2.22) is the rule called 'casting out 9's' for testing whether there is an error in an addition or multiplication of two (or more) decimal numbers. Since $10 \equiv 1 \pmod{9}$, the expression (2.22) gives $10^k \equiv 1 \pmod{9}$ for any positive integer $k$. Thus a decimal number $X = x_n \ldots x_2 x_1$ satisfies the condition

$$X = x_n 10^{n-1} + x_{n-1} 10^{n-2} + \ldots + x_2 10 + x_1$$
$$\equiv x_n (1 \pmod{9}) + x_{n-1} (1 \pmod{9}) + \ldots + x_2 (1 \pmod{9}) + x_1$$
$$\equiv (x_n + x_{n-1} + \ldots + x_1) \pmod{9}$$

where property (P3) has been used to obtain $x_n (1 \pmod{9}) \equiv x_n \pmod{9}$, and so on. This establishes yet again the result first given in Section 1.3.7, that $X$ is divisible by 9 if and only if the sum of its digits is divisible by 9. However, more can be achieved. For convenience write $s(X)$ for the sum of digits $x_1 + x_2 + \ldots + x_n$. Suppose $Y = y_m \ldots y_1$ is a second decimal number, so that $Y \equiv s(Y) \pmod{9}$. From the result in Exercise 2.110 it follows that

$$X + Y = (s(X) + s(Y)) \pmod{9}$$

Hence working with arithmetic modulo 9, the sum of the digits of $X + Y$ must be equal to the sum of the digits of $X$ plus the sum of the digits of $Y$, that is,

$$s(X + Y) = (s(X) + s(Y))(\text{mod } 9)$$

If this does *not* hold then an error has occurred in computing $X + Y$.

**Example 2.32**  **Casting out 9's**

Consider $3719 + 842$, set out as follows:

$$
\begin{array}{ll}
& \textit{Sum of digits} \\
3719 & 3 + 7 + 1 + 9 = 20 \equiv 2(\text{mod } 9) \\
\underline{842} & \underline{8 + 4 + 2 \quad\;\; = 14 \equiv 5(\text{mod } 9)} \\
4551 & 4 + 5 + 5 + 1 = 15 \equiv 6(\text{mod } 9)
\end{array}
$$

Since $6 \neq 2 + 5$, an error in the addition has been detected – it should in fact be 4561, for which the sum of digits is $16 \equiv 7(\text{mod } 9)$, and 7 does equal $2 + 5$.

Similarly, using the property (P4) it follows that a corresponding result holds for the *product XY*, namely

$$s(XY) \equiv s(X)\,s(Y)(\text{mod } 9)$$

For example, using the same two numbers, the correct product is

$$3719 \times 842 = 3131398$$

and the digit sum of the product is $28 \equiv 1(\text{mod } 9)$. For the individual numbers

$$s(3719) \times s(842) \equiv 2 \times 5(\text{mod } 9) \equiv 10(\text{mod } 9) \equiv 1(\text{mod } 9)$$

so there is agreement. Notice, however, that in either case agreement does *not* guarantee correctness, as there might be an error $0(\text{mod } 9)$ in $s(X + Y)$ or $s(XY)$. For example, if the sum of 3719 and 842 was recorded as 4516 due to an erroneous interchange of the last two digits, the digit sum would remain at $16 \equiv 7(\text{mod } 9)$. Thus the digit sums would agree even though the addition result is incorrect.

**Exercise**  **2.100**  State which of the following are valid congruences:

(a) $1528 \equiv 10(\text{mod } 11)$ (b) $2131 \equiv 1903(\text{mod } 19)$

(c) $147 \equiv -2(\text{mod } 75)$ (d) $401 \equiv -418(\text{mod } 39)$.

**Exercise**  **2.101**  Find for what positive integers $m$ the following congruences hold:

(a) $37 \equiv 30(\text{mod } m)$ (b) $93 \equiv -27(\text{mod } m)$ (c) $2197 \equiv 0(\text{mod } m)$.

**Exercise** **2.102** Find the least residue for each of the following:

(a) 2184(mod 5)   (b) 2964(mod 23)   (c) 2590(mod 35)

(d) 7793(mod 124).

**Exercise** **2.103** Verify that $365 \equiv 1(\bmod\ 7)$. Use this to determine on which day of the week 8 April 1996 fell, given that 1 February 1992 was a Saturday.

**Exercise** **2.104** The International Standard Book Number (ISBN) consists of a 10-digit number $x_1 x_2 x_3 \ldots x_{10}$ where the digits $x_1$, $x_2$, ..., $x_9$ can take any of the values 0, 1, 2, ..., 9 but the so-called **check digit** $x_{10}$ can also take the value ten, denoted by the Roman numeral X. The check digit is chosen so that

$$x_1 + 2x_2 + 3x_3 + \ldots + 9x_9 + 10x_{10} \equiv 0(\bmod\ 11)$$

(a) Find the check digit if the first nine digits of an ISBN are 013834094.

(b) An ISBN is recorded as 089-74590X, where the fourth digit has been lost. Determine the missing digit.

The ISBN will be studied in detail in Section 4.4.1.

**Exercise** **2.105** Show that the definition (D3) is equivalent to the definition (D1).

**Exercise** **2.106** Prove the properties (P2) and (P3).

**Exercise** **2.107** Simplify the following congruences using property (P5):

(a) $130 \equiv 80(\bmod\ 25)$   (b) $186 \equiv 6(\bmod\ 36)$   (c) $215 \equiv 30(\bmod\ 37)$.

**Exercise** **2.108** Prove directly that if $a \equiv b(\bmod\ m)$ and $c|a$, $c|b$, $c|m$ then $a/c = (b/c)(\bmod\ m/c)$.

**Exercise** **2.109** Prove directly that if $a \equiv b(\bmod\ m)$ and $c|a$, $c|b$ and $(c,m) = 1$ then $a/c = (b/c)(\bmod\ m)$.

**Exercise** **2.110** Prove that if $a_1 \equiv b_1(\bmod\ m)$ and $a_2 \equiv b_2(\bmod\ m)$ then $a_1 + a_2 \equiv (b_1 + b_2)(\bmod\ m)$.

**Exercise** **2.111** Prove that the property (P4) holds for three or more congruences with the same modulus.

**Exercise** **2.112** For each of the following, find the least positive residue:

(a) $2^{179}(\bmod\ 7)$   (b) $5^{20}(\bmod\ 16)$   (c) $3^{283}(\bmod\ 13)$.

**Exercise** **2.113** Prove that if $a \equiv b(\bmod\ m)$ and $n|m$ then $a \equiv b(\bmod\ n)$.

**Exercise** **2.114** Show that if $a \equiv b(\bmod\ m)$ and $c$ is a positive integer then $ac \equiv bc(\bmod\ mc)$.

**Exercise** | **2.115** Use the method of casting out 9's to detect whether any of the following contains an error:

(a) $5412 + 2837 = 8329$  (b) $5412 \times 2837 = 15363844$

(c) $273 \times 314 \times 1126 = 96252972$.

**Exercise** | **2.116** What is the last digit of $7^{98}$? (*Hint*: use the method of Example 2.31 with modulo 10 arithmetic.)

### 2.4.2 Solving linear congruences

In view of the similarities found in the previous section between equations and congruences, you might expect that solving a congruence containing an unknown should be manageable. This is certainly true provided the congruence is 'linear', that is, there are no squared or higher powers of the unknown.

**Example 2.33** **Unknown modulus**

In Exercise 2.101 you were asked to solve congruences of the form $a \equiv b(\text{mod } x)$ where $a$ and $b$ are given integers. From the second form of definition (D2) in Section 2.4.1 it follows that $x$ is any integer which divides $a - b$. For example, the solution of $11 \equiv 2(\text{mod } x)$ is $x = 3$ or $x = 9$ since these are the only two positive integers ($> 1$) which are factors of $11 - 2 = 9$. The solution of such a congruence is unique only when $a - b$ is a prime number (that is, has no factors other than itself and 1). For example, the congruence $20 \equiv 9(\text{mod } x)$ has the unique solution $x = 11$.

A more general type of congruence has the form

$$ax + b \equiv c(\text{mod } m) \tag{2.25}$$

where $a$, $b$, $c$ and $m$ are given integers.

**Example 2.34** **Equation (2.25) with $a = 1$**

If $a = 1$ the solution of (2.25) is simply

$$x \equiv c(\text{mod } m) - b$$
$$\equiv (c - b)(\text{mod } m), \qquad \text{by property (P2)}$$

Using definition (D3) in Section 2.4.1 this can be written as

$$x = c - b + mt$$

where $t$ is an arbitrary integer. For example, the solution of

$$x + 2 \equiv 4(\text{mod } 5)$$

is $x \equiv 2(\text{mod } 5)$, or $x = 2 + 5t$ for any integer $t$. Thus any number congruent to $2(\text{mod } 5)$ solves the congruence, that is, $2, 7, 12, \ldots, -3, -8, -13, \ldots$ obtained by taking $t = 0, 1, 2, \ldots, -1, -2, -3, \ldots$ .

If $b = 0$ then (2.25) reduces to

$$ax \equiv c(\text{mod } m) \tag{2.26}$$

Again using the definition (D3), this can be written as

$$ax = c + mt$$

which is valid for any integer $t$. Assuming $a \neq 1$, the problem is to find a value of $t$ such that $x = (c + mt)/a$ is an integer. In simple cases this can be done by trying out values of $t$ until you find one which works. In general, however, you need to solve the equation

$$ax - mt = c \tag{2.27}$$

for integers $x$ and $t$. According to Exercise 2.92, this can only be done if the g.c.d. $e$ of $a$ and $m$ divides $c$. Otherwise, (2.27) has no solution, which means that the original congruence (2.26) also has no solution in this case.

If $e$ does divide $c$, suppose $a_0 = a/e$, $m_0 = m/e$ and $c_0 = c/e$, with $a_0$ and $m_0$ relatively prime. Dividing both sides of (2.27) by $e$ produces

$$a_0x - m_0t = c_0 \tag{2.28}$$

You can find a solution $x = x_0$, $t = t_0$ of (2.28) either by trying out values in simple cases, or by using the procedure set out in Exercise 2.90. The general solution of (2.28) is

$$x = x_0 + m_0p, \quad t = t_0 + a_0p \tag{2.29}$$

for arbitrary integer values of $p$ (see Exercise 2.93). Hence $x = x_0 + m_0p \equiv x_0(\text{mod } m_0)$ is the solution of the congruence (2.26).

Notice that in particular if $(a, m) = 1$ then $e = 1$ and there is a *unique* solution modulo $m$ of (2.26).

| Example 2.35 | Solving (2.26) |

(a)  To solve the congruence

$$5x \equiv 2(\text{mod } 7)$$

identify $a = 5$, $c = 2$, $m = 7$. Since $e = (a, m) = 1$, equation (2.28) is just

$$5x - 7t = 2$$

You can easily spot that $x_0 = 6$, $t_0 = 4$ is a solution of this equation. Hence the solution of the congruence is $x \equiv 6(\text{mod } 7)$.

(b)  The congruence

$$5x \equiv 8(\text{mod } 10)$$

has no solution because the g.c.d. of $a = 5$ and $m = 10$ is $e = 5$, which does not divide $c = 8$.

(c) Consider the congruence

$$5x \equiv 25(\mathrm{mod}\ 75) \tag{2.30}$$

Here $a = 5$, $c = 25$, $m = 75$ and $e = (5, 75) = 5$, so that $a_0 = 1$, $c_0 = 5$, $m_0 = 15$. Hence equation (2.28) is

$$x - 15t = 5$$

which you can see at once has a solution $x_0 = 5$, $t_0 = 0$. Therefore $x \equiv 5(\mathrm{mod}\ 15)$ is the solution of (2.30), that is, $x = 5,\ 20,\ 35,\ \ldots$ all satisfy (2.30) – for example $5 \times 35 = 175 \equiv 25(\mathrm{mod}\ 75)$.

---

If $b \neq 0$ in the congruence (2.25), simply subtract $b$ from both sides using property (P2) in Section 2.4.1 to produce

$$ax \equiv (c - b)(\mathrm{mod}\ m)$$

which has the form of the congruence in (2.26). The solution then proceeds as illustrated in Example 2.35.

**Exercise**  **2.117** Find the solution of each of the following congruences:

(a) $x - 3 \equiv 5(\mathrm{mod}\ 11)$  (b) $5x \equiv 20(\mathrm{mod}\ 8)$  (c) $4x \equiv 8(\mathrm{mod}\ 24)$.

**Exercise**  **2.118** Find where possible the solution of each of the following congruences:

(a) $15x \equiv 11(\mathrm{mod}\ 21)$  (b) $15x \equiv 11(\mathrm{mod}\ 165)$  (c) $15x \equiv 11(\mathrm{mod}\ 29)$

(d) $15x \equiv 11(\mathrm{mod}\ 39)$  (e) $15x \equiv 12(\mathrm{mod}\ 39)$.

**Exercise**  **2.119** Find the solution of the congruences:

(a) $18x + 10 \equiv 11(\mathrm{mod}\ 35)$

(b) $157x \equiv 1(\mathrm{mod}\ 725)$ (use the result in Exercise 2.91)

(c) $125x \equiv 35(\mathrm{mod}\ 70)$.

**Exercise**  **2.120** If $x$ satisfies the congruence $ax \equiv 1(\mathrm{mod}\ m)$ then $x$ is called an **inverse** $a^{-1}$ of $a$ (modulo $m$). This is by analogy with ordinary arithmetic, where $ax = 1$ means that $x = 1/a = a^{-1}$. For example, $3 \times 4 \equiv 1(\mathrm{mod}\ 11)$ so $3^{-1} = 4(\mathrm{mod}\ 11)$. Find the smallest positive inverse modulo 11 of 5, 6 and 7.

### 2.4.3  Chinese remainder theorem

After you learned how to solve ordinary linear equations, you probably then moved on to consider a pair of equations to be solved simultaneously. In the same way it's now appropriate to look at solving two (or more) linear congruences. The problem is attributed to Chinese mathematicians (whose names do not seem to be known) of the first century, and can be stated very simply: find an integer $x$ which produces specified remainders $a_1, a_2, a_3, \ldots$ when divided by given integers $m_1, m_2, m_3, \ldots$ respectively.

| Example 2.36 | **A counting problem** |
|---|---|

A child playing with some toy bricks divides them into three equal piles (a 'pile' consists of two or more bricks). However, when the child tries to make them into five equal piles, there are four left over. How many bricks are there? You can easily see that if there are 24 bricks, then this would give three equal piles of eight each, or five piles of four each with four left over. However, other numbers of bricks will do. For example:

39 bricks: 3 piles of 13; or 5 piles of 7, with 4 left over
54 bricks: 3 piles of 18; or 5 piles of 10, with 4 left over

There is no unique answer, but these numbers satisfying the given conditions differ by multiples of 15.

In mathematical terms, you are looking for an integer $x$ which is exactly divisible by 3, and which leaves a remainder of 4 when divided by 5, that is,

$$x \equiv 0 (\text{mod } 3) \quad \text{and} \quad x \equiv 4 (\text{mod } 5)$$

In view of the solutions $x = 24, 39$ or $54$ above, it's easy to see that the general solution of this pair of congruences is $x \equiv 24 (\text{mod } 15)$, and hence is unique modulo 15. Notice that the solution $x = 9$ is rejected because the child has at least $5 \times 2 = 10$ bricks.

---

The preceding example illustrates the following general result.

## Chinese remainder theorem

Let $m_1, m_2, \ldots, m_r$ be $r$ integers each greater than 1, such that every pair is relatively prime, that is, $(m_i, m_j) = 1$ for all $i \neq j$. Then the system of congruences

$$x \equiv a_1 (\text{mod } m_1), \quad x \equiv a_2 (\text{mod } m_2), \ldots, \quad x \equiv a_r (\text{mod } m_r) \tag{2.31}$$

where the $a_i$ are given integers, has a unique solution modulo $M = m_1 m_2 \ldots m_r$.

**Proof** Let $M_i = M/m_i$, for $i = 1, 2, \ldots, r$. It follows (see Exercise 2.125) that because the $m_i$ are pairwise prime then $(M_i, m_i) = 1$ for $i = 1, 2, \ldots, r$. Consider the congruence

$$M_1 x \equiv 1 (\text{mod } m_1) \tag{2.32}$$

This congruence has precisely the form of (2.26) in Section 2.4.2, and from the theory developed there it follows that because $(M_1, m_1) = 1$ then (2.32) has a unique solution modulo $m_1$, which can be denoted by $x = b_1 \pmod{m_1}$. Substituting this into (2.32) produces

$$M_1 b_1 \equiv 1 \pmod{m_1}$$
$$= 1 + m_1 t_1 \tag{2.33}$$

for some integer $t_1$ (using definition (D3) in Section 2.4.1). Now define $X_1$ as

$$X_1 \equiv a_1 M_1 b_1 \pmod{M}$$
$$= a_1 M_1 b_1 + M t_2 \tag{2.34}$$

for some integer $t_2$, and see what happens when $X_1$ is substituted into each of the congruences (2.31). Firstly, from (2.33) and (2.34)

$$X_1 = a_1 + a_1 m_1 t_1 + a_1 M t_2$$

and since $m_1 | M$ it follows that $X_1 \equiv a_1 \pmod{m_1}$, that is, $X_1$ satisfies the *first* congruence in (2.31). However, since $m_2 | M_1$ and $m_2 | M$ it follows from (2.34) that $X_1 \equiv 0 \pmod{m_2}$, and similarly $X_1 \equiv 0 \pmod{m_k}$ for $k = 3, 4, \ldots, r$.

Now repeat the process by replacing $M_1$ and $m_1$ in (2.32) with $M_2$ and $m_2$. Using an identical argument it follows that the congruence

$$M_2 x \equiv 1 \pmod{m_2}$$

has a unique solution $x \equiv b_2 \pmod{m_2}$, and that $X_2 \equiv a_2 M_2 b_2 \pmod{M}$ satisfies

$$X_2 \equiv a_2 \pmod{m_2}, \quad X_2 \equiv 0 \pmod{m_k}, \qquad k \neq 2$$

Hence $X_2$ satisfies the *second* congruence in (2.31).

You should now be able to see the pattern in general: each congruence

$$M_i x \equiv 1 \pmod{m_i}, \quad i = 1, 2, \ldots r \tag{2.35}$$

has a unique solution $x \equiv b_i \pmod{m_i}$, and

$$X_i = a_i M_i b_i \pmod{M}, \qquad i = 1, 2, \ldots, r \tag{2.36}$$

satisfies the congruences

$$X_i \equiv a_i \pmod{m_i}, \quad X_i \equiv 0 \pmod{m_k}, \qquad k \neq i \tag{2.37}$$

The way to combine these solutions of the individual congruences so as to produce a solution of the simultaneous congruences (2.31) is to define

$$X \equiv (X_1 + X_2 + \ldots + X_r) \pmod{M} \tag{2.38}$$

Because of (2.37) it follows that $X$ in (2.38) satisfies each of the congruences in (2.31), as required.

**Example 2.37** **Solving (2.31)**

The proof of the theorem is useful because it actually includes a constructive method for obtaining the solution. To solve the system

$$x \equiv 1(\text{mod } 5), \quad x \equiv 3(\text{mod } 7), \quad x \equiv 4(\text{mod } 11)$$

first identify $a_1 = 1$, $a_2 = 3$, $a_3 = 4$, $m_1 = 5$, $m_2 = 7$, $m_3 = 11$. It is clear that $(m_1, m_2) = (5, 7) = 1$, and similarly $(m_1, m_3) = 1$, $(m_2, m_3) = 1$ so the condition of the theorem is satisfied. Next, compute

$$M = 5 \times 7 \times 11 = 385, M_1 = M/5 = 77, M_2 = M/7 = 55, M_3 = M/11 = 35$$

Then solve the three congruences (2.35), which are

$$77x \equiv 1(\text{mod } 5), \quad 55x \equiv 1(\text{mod } 7), \quad 35x \equiv 1(\text{mod } 11)$$

Using the method of Section 2.4.2 you can confirm that the respective solutions are

$$3(\text{mod } 5), \quad 6(\text{mod } 7), \quad 6(\text{mod } 11)$$

so that $b_1 = 3$, $b_2 = 6$, $b_3 = 6$. The required solution of the original congruences is given by (2.36) and (2.38) as

$$
\begin{aligned}
X &\equiv (a_1 M_1 b_1 + a_2 M_2 b_2 + a_3 M_3 b_3)(\text{mod } 385) \\
&\equiv (1 \times 77 \times 3 + 3 \times 55 \times 6 + 4 \times 35 \times 6)(\text{mod } 385) \\
&\equiv 2061(\text{mod } 385) \\
&\equiv 136(\text{mod } 385)
\end{aligned}
$$

The smallest positive solution which satisfies the three congruences is therefore $x = 136$.

**Example 2.38** **Computer arithmetic with large numbers**

The Chinese remainder theorem can be used to carry out arithmetic with extremely large integers, exceeding those which a computer can handle (however big a computer's memory may be, there will always be a limit to the largest integer it can store).

As a very simple illustration of the ideas involved, suppose the computer memory is limited to numbers up to 100, but it is required to do arithmetic with integers up to $10^5$. The trick is to reduce these numbers modulo certain integers (pairwise relatively prime) whose product $M$ is greater than $10^5$. For example, taking $m_1 = 99$, $m_2 = 98$, $m_3 = 97$ gives $M = 941094 > 10^5$. Do the arithmetic with the least residues, and then recover the required result via the Chinese remainder algorithm.

For example, to find $z = a + b$ where $a = 2173$ and $b = 3548$, first compute the least residues:

$$
\begin{aligned}
a &\equiv 94(\text{mod } 99) \equiv 17(\text{mod } 98) \equiv 39(\text{mod } 97) \\
b &\equiv 83(\text{mod } 99) \equiv 20(\text{mod } 98) \equiv 56(\text{mod } 97)
\end{aligned}
$$

Using the addition property (P1) in Section 2.4.1 gives

$$z \equiv (94 + 83)(\mathrm{mod}\ 99) \equiv 177(\mathrm{mod}\ 99) \equiv 78(\mathrm{mod}\ 99) \left.\vphantom{\begin{array}{c}1\\1\\1\end{array}}\right\}$$
$$z \equiv (20 + 17)(\mathrm{mod}\ 98) \equiv 37(\mathrm{mod}\ 98) \qquad\qquad \textbf{(2.39)}$$
$$z \equiv (39 + 56)(\mathrm{mod}\ 97) \equiv 95(\mathrm{mod}\ 97)$$

In order to determine the value of $z$, it is now required to solve the congruences (2.39) which have the form (2.31) with $a_1 = 78$, $a_2 = 37$, $a_3 = 95$. Since $M_1 = M/m_1 = 9506$, $M_2 = M/m_2 = 9603$ and $M_3 = M/m_3 = 9702$, the three congruences in (2.35) to be solved are

$$9506x \equiv 1(\mathrm{mod}\ 99), \quad 9603x \equiv 1(\mathrm{mod}\ 98), \quad 9702x \equiv 1(\mathrm{mod}\ 97) \qquad \textbf{(2.40)}$$

Using the method of Section 2.4.2 you can check that the solutions are respectively

$$50(\mathrm{mod}\ 99), \quad 97(\mathrm{mod}\ 98), \quad 49(\mathrm{mod}\ 97)$$

so that $b_1 = 50$, $b_2 = 97$, $b_3 = 49$ (in the notation of the theorem). The required solution of (2.39) is given by (2.36) and (2.38) as

$$\begin{aligned} z &\equiv (a_1 M_1 b_1 + a_2 M_2 b_2 + a_3 M_3 b_3)(\mathrm{mod}\ M) \\ &\equiv (78 \times 9506 \times 50 + 37 \times 9603 \times 97 + 95 \times 9702 \times 49)(\mathrm{mod}\ 941094) \\ &\equiv 116701377(\mathrm{mod}\ 941094) \\ &\equiv 5721(\mathrm{mod}\ 941094) \end{aligned}$$

Since $a + b < 941094$ it follows that $z = 5721$.

Notice that the congruences (2.40) are independent of the values of $a$ and $b$, so the integers $b_1$, $b_2$, $b_3$ do not have to be recalculated for each problem.

---

**Exercise**  **2.121**  For what positive integers $c < 12$ does the congruence $8x \equiv c(\mathrm{mod}\ 12)$ possess solutions? Find the solution in each case.

**Exercise**  **2.122**  Solve the following systems of congruences:
(a)  $x \equiv 5(\mathrm{mod}\ 7)$,  $x \equiv 6(\mathrm{mod}\ 8)$
(b)  $x \equiv 2(\mathrm{mod}\ 3)$,  $x \equiv 1(\mathrm{mod}\ 5)$,  $x \equiv 4(\mathrm{mod}\ 7)$
(c)  $x \equiv 0(\mathrm{mod}\ 2)$,  $x \equiv 2(\mathrm{mod}\ 5)$,  $x \equiv 3(\mathrm{mod}\ 7)$,  $x \equiv 4(\mathrm{mod}\ 9)$.

**Exercise**  **2.123**  Find the smallest multiple of 9 which leaves a remainder of 1 when divided by each of the integers 2, 5 and 11.

**Exercise**  **2.124**  Whilst on a touring holiday with British Spaceways you need to change some money. You have a little less than 1000 Earth Dollars to exchange. The rates per dollar offered by an Interplanetary Currency Bureau are 5 Jovian Francs, 13 Venusian Marks or 18 Martian Pounds. The bureau will not handle small change, that is, amounts less than one monetary unit. If you change your money into francs you would be left with a single dollar; changing into marks would leave you with 6 dollars; and changing into pounds would leave you with 9 dollars. How many Earth Dollars do you have?

**Exercise** | **2.125** Prove that if $m_1$, $m_2$ and $m_3$ are pairwise relatively prime, and $M_1 = m_2 m_3$, $M_2 = m_1 m_3$, $M_3 = m_1 m_2$ then $(M_1, m_1) = 1$, $(M_2, m_2) = 1$, $(M_3, m_3) = 1$. This result extends to $r$ integers $m_1, m_2, \ldots, m_r$.

**Exercise** | **2.126** Use the procedure of Example 2.38 to find:

(a) the sum of 3421 and 5679

(b) the product of 321 and 403.

### 2.4.4  $\mathbb{Z}_m$

**Example 2.39**  **Odd and even numbers**

If you add two even positive integers you get another even integer, the sum of two odd numbers also gives an even number, but the sum of an odd number and an even number is odd. This can be summarized as:

$$\text{even} + \text{even} = \text{even}, \quad \text{odd} + \text{odd} = \text{even}, \quad \text{odd} + \text{even} = \text{odd} = \text{even} + \text{odd}$$

To write this more concisely, denote 'even' by 0 and 'odd' by 1, to obtain

$$0 + 0 = 0, \quad 1 + 1 = 0, \quad 1 + 0 = 1 = 0 + 1 \tag{2.41}$$

You should recognize these rules as simply those for addition of 0 and 1, modulo 2, which can be written formally as

$$0 + 0 \equiv 0(\text{mod } 2), \quad 1 + 1 \equiv 0(\text{mod } 2), \quad 1 + 0 \equiv 1(\text{mod } 2) \equiv 0 + 1 \tag{2.42}$$

In this section it will be convenient to dispense with the congruence notation using $\equiv$ and (mod ), so that (2.41) will be used instead of (2.42). Just remember that the arithmetic being used is modular.

The rules for multiplication modulo 2 are similarly

$$0 \times 0 = 0, \quad 1 \times 0 = 0 = 0 \times 1, \quad 1 \times 1 = 1 \tag{2.43}$$

and (2.41) and (2.43) can be expressed in the following tabular forms:

| + | 0 | 1 |   | × | 0 | 1 |
|---|---|---|---|---|---|---|
| 0 | 0 | 1 |   | 0 | 0 | 0 |
| 1 | 1 | 0 |   | 1 | 0 | 1 |

The set of integers $\{0,1\}$ subject to the above rules is written as $\mathbb{Z}_2$. It is sometimes called the **congruence class** or **residue class** modulo 2, since *any* integer has a least residue of either 0 or 1 on division by 2.

In general $\mathbb{Z}_m$ denotes the set of integers $0, 1, 2, 3, \ldots, m-1$ subject to modulo $m$ arithmetic.

**Example 2.40** $\mathbb{Z}_3$ and $\mathbb{Z}_4$

$\mathbb{Z}_3$ consists of 0,1,2 subject to modulo 3 arithmetic, so the addition and multiplication tables are

*Table 2.7* Addition and multiplication for $\mathbb{Z}_3$

| + | 0 | 1 | 2 |
|---|---|---|---|
| 0 | 0 | 1 | 2 |
| 1 | 1 | 2 | 0 |
| 2 | 2 | 0 | 1 |

| × | 0 | 1 | 2 |
|---|---|---|---|
| 0 | 0 | 0 | 0 |
| 1 | 0 | 1 | 2 |
| 2 | 0 | 2 | 1 |

The tables are to be interpreted as meaning, for example

$$2 + 1 \equiv 3(\text{mod } 3) \equiv 0(\text{mod } 3)$$
$$2 \times 2 \equiv 4(\text{mod } 3) \equiv 1(\text{mod } 3)$$

which can be written in the short form $2 + 1 = 0, 2 \times 2 = 1$.
Similarly $\mathbb{Z}_4$ is 0,1,2,3 subject to

*Table 2.8* Addition and multiplication for $\mathbb{Z}_4$

| + | 0 | 1 | 2 | 3 |
|---|---|---|---|---|
| 0 | 0 | 1 | 2 | 3 |
| 1 | 1 | 2 | 3 | 0 |
| 2 | 2 | 3 | 0 | 1 |
| 3 | 3 | 0 | 1 | 2 |

| × | 0 | 1 | 2 | 3 |
|---|---|---|---|---|
| 0 | 0 | 0 | 0 | 0 |
| 1 | 0 | 1 | 2 | 3 |
| 2 | 0 | 2 | 0 | 2 |
| 3 | 0 | 3 | 2 | 1 |

where, for example,

$$3 + 2 \equiv 5(\text{mod } 4) \equiv 1(\text{mod } 4), \quad \text{i.e. } 3 + 2 = 1$$
$$3 \times 3 \equiv 9(\text{mod } 4) \equiv 1(\text{mod } 4), \quad \text{i.e. } 3 \times 3 = 1$$

The operations of addition and multiplication for $\mathbb{Z}_m$ are the same as those for ordinary numbers in many respects. Thus if $a$, $b$, $c$ are three members of $\mathbb{Z}_m$ then

(1) $a + b, a \times b$ are in $\mathbb{Z}_m$;

(2) $a + b = b + a, a \times b = b \times a$ (commutative laws);

(3) $(a + b) + c = a + (b + c), (a \times b) \times c = a \times (b \times c)$ (associative laws);

(4) $a \times (b + c) = (a \times b) + (a \times c)$ (distributive law);

(5) $a + 0 = a, a \times 1 = a$;

(6) for every $a$ in $\mathbb{Z}_m$ there is a unique element denoted by $-a$ such that $a + (-a) = 0$.

The rules (1)–(5) are all familiar, and simply say you can manipulate integers in $\mathbb{Z}_m$ in the usual ways you are familiar with, provided you remember that the arithmetic is modulo $m$. Only (6) needs a little explanation It means that for every member of $\mathbb{Z}_m$ there is another member which when added to $a$ gives zero. This number is denoted by $-a$, and is called the **additive inverse** of $a$ (because it is a member of $\mathbb{Z}_m$, $-a$ is a non-negative integer). For example, look at the addition array for $\mathbb{Z}_3$ in Table 2.7. In each row there is a zero entry, corresponding to

$$0 + 0 = 0, \quad 1 + 2 = 0, \quad 2 + 1 = 0$$

What rule (6) says is that

$$-0 = 0, \quad -1 = 2, \quad -2 = 1$$

that is, the additive inverse of 0 is 0 (not surprising!), the additive inverse of 1 is 2, and the additive inverse of 2 is 1.

Similarly, using the addition array for $\mathbb{Z}_4$ in Table 2.8 shows that the additive inverses of 0,1,2,3 are respectively 0,3,2,1. To determine these, simply read across each row in the addition table until you reach 0, then read upwards to the top row to obtain the additive inverse. For example, the bottom row has 0 in the second position, reading upwards to the top row gives 1, that is, $3 + 1 = 0$, so $-3 = 1$.

Technically, $\mathbb{Z}_m$ is an example of what is called a **commutative ring** with a unit element 1. Such a ring is defined as a set of elements which can be added, subtracted and multiplied according to the rules (1) to (6). Rings need not just consist of sets of numbers – another example is the set of polynomials having real coefficients and finite degree (see Problem 2.31).

You will have noticed that the question of 'division' of elements belonging to $\mathbb{Z}_m$ has not yet been mentioned. For a set of integers subject to the normal rules of arithmetic, division is not in general possible since division of one integer by another does not always produce an integer (for example, $4 \div 2 = 2$ but $5 \div 2 \neq$ integer). However, if you consider the set of rational numbers (see Chapter 1) then division is always possible, except by zero. A number like $2 \div 3 = 2/3$ can be regarded as $2 \times 3^{-1}$, where $3^{-1}$ is the number such that $3^{-1} \times 3 = 1$, and $3^{-1} (= 1/3)$ is called the **multiplicative inverse** of 3. To determine how this idea can be extended to $\mathbb{Z}_m$, first look at the multiplication array in Table 2.7 for $\mathbb{Z}_3$. You can see that

$$1 \times 1 = 1, \quad 2 \times 2 = 1$$

so $1^{-1} = 1$ and $2^{-1} = 2$, that is, both non-zero elements of $\mathbb{Z}_3$ have a multiplicative inverse. However, moving on to Table 2.8, you can see that in $\mathbb{Z}_4$

$$2 \times 1 = 2, \quad 2 \times 2 = 0, \quad 2 \times 3 = 2$$

so not only does 2 not have a multiplicative inverse, but also the product $2 \times 2$ is zero. This is not something peculiar to $\mathbb{Z}_4$. As another example, in $\mathbb{Z}_6$ where the arithmetic is modulo 6 then

$$3 \times 1 = 3, \quad 3 \times 2 = 0, \quad 3 \times 3 = 3, \quad 3 \times 4 = 0, \quad 3 \times 5 = 3$$

Thus you cannot take for granted in $\mathbb{Z}_m$ two properties which you are familiar with from the arithmetic of rational numbers. Firstly, it's possible to have a number without a multiplicative inverse, for example 2 in $\mathbb{Z}_4$ or 3 in $\mathbb{Z}_6$. Secondly, the product of two *non-zero* integers can be *zero*, for example $2 \times 2$ in $\mathbb{Z}_4$ or $3 \times 2$ and $3 \times 4$ in $\mathbb{Z}_6$. This latter fact has a further implication: if $a$, $b$ and $c$ are in $\mathbb{Z}_m$ and $ab = ac$ (with $a \neq 0$) then it does *not* necessarily follow that $b = c$, because factorizing gives $a(b - c) = 0$ and although $a \neq 0$ you need not have $b - c = 0$. Again returning to $\mathbb{Z}_6$, $3 \times 2 = 3 \times 4$ but $2 \neq 4$. Furthermore, the method you're familiar with for solving quadratic equations using ordinary arithmetic also is no longer valid. For example, to solve the equation

$$x^2 - 3x + 2 = 0 \tag{2.44}$$

you factorize as

$$(x - 2)(x - 1) = 0$$

and then conclude that since the product is zero, one of the factors must be zero, that is, either $x - 2 = 0$ or $x - 1 = 0$, so the solution of the equation is $x = 2$ or $x = 1$. Suppose now that the equation is to be solved in $\mathbb{Z}_6$, that is, using modulo 6 arithmetic. Substituting $x = 1$ or $x = 2$ into (2.44) still gives zero, but in fact there are two more solutions $x = 4$ and $x = 5$. Substituting these values into (2.44) gives respectively

$$16 - 12 + 2 = 0, \quad 25 - 15 + 2 = 0 \quad (\text{mod } 6)$$

It's natural to ask when an element in $\mathbb{Z}_m$ does have a multiplicative inverse and this is answered by the following:

## Theorem: Multiplicative inverses in $\mathbb{Z}_m$

An integer $a \neq 0$ in $\mathbb{Z}_m$ has a multiplicative inverse if and only if $(a,m) = 1$. In particular, if $m$ is a prime number then every non-zero member of $\mathbb{Z}_m$ has a multiplicative inverse.

**Proof** If $a$ has a multiplicative inverse then this means there is an integer $b$ such that $ab \equiv 1(\text{mod } m)$. Hence by the definition (D3) in Section 2.4.1, there is an integer $t$ such that $ab = 1 + mt$, that is,

$$ab - mt = 1$$

(notice that this is not a modulo $m$ equation). If $d$ is any common divisor of $a$ and $m$ then it is also a divisor of $ab - mt$, but since this is equal to 1 it follows that $d = 1$, that is $(a,m) = 1$.

Conversely, suppose $(a,m) = 1$. Then from Exercise 2.92 it follows that there exist integers $x$ and $y$ such that

$$ax + my = 1$$

Rearranging this equation gives $ax = 1 - my$, that is, $ax \equiv 1 \pmod{m}$, showing that $x$ is a multiplicative inverse of $a$ in $\mathbb{Z}_m$.

If $m$ is a prime number then it has no factors other than $m$ and 1, so $(a,m) = 1$ for all non-zero integers $a$ in $\mathbb{Z}_m$.

Notice that when the modulus $m$ is a prime number, the theorem removes the difficulty discussed earlier associated with a zero product. For suppose $ab = 0$ and $a \neq 0$. Then $a^{-1}$ exists and multiplying both sides by $a^{-1}$ gives

$$a^{-1}(ab) = (a^{-1}a)b = 1b = b$$
$$= a^{-1}(0) = 0$$

so $b = 0$. Hence when $m$ is prime, $ab = 0$ implies at least one of $a$ or $b$ is zero.

### Example 2.41  Multiplicative inverses

The theorem explains why 2 has no inverse in $\mathbb{Z}_4$, since $(2,4) = 2$. However, $1^{-1}$ and $3^{-1}$ do exist in $\mathbb{Z}_4$ because $(1,4) = 1$ and $(3,4) = 1$.

Now move on to $\mathbb{Z}_5$. You should check the following addition and multiplication tables for $\mathbb{Z}_5$:

Table 2.9 Addition and multiplication for $\mathbb{Z}_5$

| + | 0 | 1 | 2 | 3 | 4 | | × | 0 | 1 | 2 | 3 | 4 |
|---|---|---|---|---|---|---|---|---|---|---|---|---|
| 0 | 0 | 1 | 2 | 3 | 4 | | 0 | 0 | 0 | 0 | 0 | 0 |
| 1 | 1 | 2 | 3 | 4 | 0 | | 1 | 0 | 1 | 2 | 3 | 4 |
| 2 | 2 | 3 | 4 | 0 | 1 | | 2 | 0 | 2 | 4 | 1 | 3 |
| 3 | 3 | 4 | 0 | 1 | 2 | | 3 | 0 | 3 | 1 | 4 | 2 |
| 4 | 4 | 0 | 1 | 2 | 3 | | 4 | 0 | 4 | 3 | 2 | 1 |

The additive inverses are found as before using the zeros in the addition table, giving $-1 = 4$, $-2 = 3$, $-3 = 2$ and $-4 = 1$.

Because 5 is a prime number the theorem ensures that all the non-zero members of $\mathbb{Z}_5$ have multiplicative inverses. These are obtained by tracing the 1's in the multiplication table, for example in the penultimate row, reading from 1 up to 2 in the top row shows that $3^{-1} = 2$. Similarly $1^{-1} = 1$, $2^{-1} = 3$ and $4^{-1} = 4$.

Armed with these inverses, you can now solve linear equations over $\mathbb{Z}_5$. For example, the equation $2x = 3$ is solved by multiplying both sides by $2^{-1}$, giving

$$2^{-1} \times 2x = 2^{-1} \times 3$$

that is,

$$1x = 3 \times 3 = 4$$

so the solution is $x = 4$. Of course, in this simple example you could read off the answer from the multiplication table in Table 2.9.

If to the conditions (1) to (6) defining a commutative ring you add:

(7) for every $a \neq 0$ in $\mathbb{Z}_m$ there is a unique element denoted by $a^{-1}$ such that $a \times a^{-1} = 1$ then the resulting set of $m$ integers is an example of what is called a **finite field** of **order** $m$. The theorem on inverses in $\mathbb{Z}_m$ shows that $m$ must be a prime number for (7) to be satisfied.

**Example 2.42**   **Quadratic equations**

The simplest quadratic equation is $x^2 = a$, where $a$ is a positive integer, and the solution can be denoted by $\sqrt{a}$, the square root of $a$. Consider as an example the following list of squares in $\mathbb{Z}_5$, which can be extracted from Table 2.9:

$$\begin{array}{c c c c c} x & 1 & 2 & 3 & 4 \\ x^2 & 1 & 4 & 4 & 1 \end{array}$$

Reading upwards and writing $a$ for $x^2$ and $\sqrt{a}$ for $x$ gives

$$\begin{array}{c c c} a & 1 & 4 \\ \sqrt{a} & 1 \text{ or } 4 & 2 \text{ or } 3 \end{array}$$

This illustrates firstly that $\sqrt{a}$ need not be unique, and secondly that $\sqrt{a}$ may not even exist – in this example there are no members $x$ of $\mathbb{Z}_5$ satisfying $x^2 = 2$ or $x^2 = 3$. Notice also that there is nothing to be gained by selecting a 'negative' square root – for example, taking $\sqrt{4} = -2$ is the same as $\sqrt{4} = 3$ (see Example 2.41).

Incidentally, it is interesting to note that for $\mathbb{Z}_5$

$$2^1 = 2, \quad 2^2 = 4, \quad 2^3 = 8 = 3, \quad 2^4 = 16 = 1$$

so that every (non-zero) member of $\mathbb{Z}_5$ can be expressed as a power of 2. For this reason 2 is called a **primitive element** for $\mathbb{Z}_5$. It can be shown that every finite field has a primitive element.

As a more general quadratic equation, still working with $\mathbb{Z}_5$, consider

$$x^2 + 4x + 3 = 0 \tag{2.45}$$

This can be factorized as

$$(x + 3)(x + 1) = 0$$

so the solution is $x = -3 = 2$, or $x = -1 = 4$. Because the arithmetic is modulo the prime number 5, difficulties like that encountered in solving (2.44) over $\mathbb{Z}_6$ do not arise here.

It is instructive to use the formula given in Section 1.4.2, to solve (2.45). This gives the roots of the equation as

$$\begin{aligned} x &= \frac{-4 \pm \sqrt{(16 - 4 \times 3)}}{2} \\ &= (-4 \pm \sqrt{4})2^{-1} \\ &= (-4 + 2)2^{-1}, \text{ or } (-4 + 3)2^{-1}, \text{ or } (-4 - 2)2^{-1}, \text{ or } (-4 - 3)2^{-1} \\ &= -2 \times 2^{-1}, \text{ or } -1 \times 2^{-1}, \text{ or } -6 \times 2^{-1}, \text{ or } -7 \times 2^{-1} \end{aligned}$$

using the two values 2 and 3 found earlier for $\sqrt{4}$. There appear to be four solutions, but in fact these reduce to 2 (twice) and 4 (twice). To see this, use $2^{-1} = 3$ from Example 2.41 to get

$$-2 \times 2^{-1} = -2 \times 3 = -6 = 4$$
$$-1 \times 2^{-1} = -1 \times 3 = 2$$
$$-6 \times 2^{-1} = -6 \times 3 = -18 = 2$$
$$-7 \times 2^{-1} = -7 \times 3 = -21 = 4$$

using modulo 5 arithmetic.

**Exercise**    **2.127**   Write down the addition and multiplication tables for $\mathbb{Z}_6$ and $\mathbb{Z}_7$. In each case list the additive and multiplicative inverses, where they exist.

**Exercise**    **2.128**   Find a primitive element for $\mathbb{Z}_7$.

**Exercise**    **2.129**   Find the multiplicative inverses of

(a) 3 in $\mathbb{Z}_{11}$    (b) 11 in $\mathbb{Z}_{15}$   (c) 4 in $\mathbb{Z}_{17}$

(d) 27 in $\mathbb{Z}_{72}$   (e) 23 in $\mathbb{Z}_{65}$.

**Exercise**    **2.130**   Which integers in $\mathbb{Z}_{24}$ have a multiplicative inverse? Find the inverse in each case.

**Exercise**    **2.131**   Show that if $x$ and $y$ in $\mathbb{Z}_m$ have a multiplicative inverse then so does $xy$, and determine $(xy)^{-1}$. Use this to find $(15)^{-1}$ in $\mathbb{Z}_{23}$.

**Exercise**    **2.132**   Determine the squares of the members of $\mathbb{Z}_{11}$. Hence determine, where they exist, the square roots of the members of $\mathbb{Z}_{11}$.

**Exercise**    **2.133**   Solve in $\mathbb{Z}_{11}$ the quadratic equations:

(a) $7x^2 + 10x + 5 = 0$    (b) $8x^2 + 7x + 10 = 0$.

## Miscellaneous problems

**2.1**    Use an appropriate test from Section 1.3 to deduce that if a clock shows noon then it will also show noon after 165,607,344 hours have passed.

**2.2**    An inveterate gambler on horse races likes to choose the horse to bet on by counting in a special way. Suppose there are five horses in a race, and list them in order $A$, $B$, $C$, $D$, $E$. Count from left to right, then right to left, then left to right, and so on, as shown below:

| A | B | C | D | E | | A | B | C | D | E | | A | B | C | D | E | | A | B | C | D | E |
|---|---|---|---|---|---|---|---|---|---|---|---|---|---|---|---|---|---|---|---|---|---|---|
| 1 | 2 | 3 | 4 | 5 | | 9 | 8 | 7 | 6 | 5 | | 9 | 10 | 11 | 12 | 13 | | ........ | | | 14 | 13 |

The gambler stops when he reaches his 'lucky number' 43 (actually his present age) and bets on this horse. Which one will it be? Which horse would counting up to 1000 produce?

The next race has seven runners, $A$ to $G$. Which horse does the gambler bet on with the count to 43; or with the count to 1000?

**2.3** A widely used 5-bit code is called 'two-out-of-five' since in each byte $abcde$ two of the bits are 1 and the other three are 0. Show by listing all the possibilities that there are ten such 5-bit bytes.

Each 5-bit byte can be converted to a decimal digit by the rule $7a + 4b + 2c + d$. For example, 01010 becomes

$$7 \times 0 + 4 \times 1 + 2 \times 0 + 1 = 5$$

The exception is 11000 which corresponds to 0. Determine the decimal digits corresponding to the other 5-bit bytes.

This scheme forms the basis of the United States Zip postal code (see Example 4.7, Chapter 4).

**2.4** Using the notation in Exercise 2.11, show that if $b$ is a base greater than ten, and $Z$ is a symbol denoting any of the digits $1, 2, 3, \ldots, b - 1$ then

$$(ZZ.ZZ)_b = \frac{Z(b + 1)(b^2 + 1)}{b^2}$$

**2.5** Use the notation in Exercise 2.11 to evaluate:

(a) $213 \div 16$, using binary arithmetic

(b) $1754 \div 64$, using octal arithmetic.

**2.6** To convert a decimal *fraction*, say 0.8125, to binary form, the algorithm in Section 2.2.2.1 for converting decimal to binary is modified as follows.

*Step 1:* Multiply by 2, retain the integer part.
*Step 2:* If the remaining fractional part is 0, stop.
*Step 3:* If the remaining fractional part is not 0, repeat Step 1.

The required binary fraction is obtained by taking the integer parts in the order they are generated.

For the example 0.8215 this produces:

|  | Integer part |  |
|---|---|---|
| $0.8125 \times 2 = 1.625$ | 1 | |
| $0.625 \times 2 = 1.25$ | 1 | ↓ read down |
| $0.25 \times 2 = 0.5$ | 0 | |
| $0.5 \times 2 = 1.0$ | 1 | |

Hence $0.8215 = (0.1101)_2 = 1 \times 2^{-1} + 1 \times 2^{-2} + 0 \times 2^{-3} + 1 \times 2^{-4}$. The algorithm need not necessarily terminate – see part (c) below.

Use the algorithm to convert the following decimals to binary:

(a) 1.6875  (b) 2.46875  (c) 0.4.

**2.7**    It is possible to use a negative integer $b$ as a number base, so that

$$(a_n \ldots a_2 a_1)_b = a_n b^{n-1} + \ldots + a_2 b + a_1$$

as usual, with each integer $a_i$ satisfying $0 \le a_i < |b|$ and $a_n \ne 0$.

(a)  Compute the decimal equivalents of

$$(1101101)_{-2}, \quad (21012)_{-3}$$

(b)  Construct a table giving the base $-2$ equivalents of the integers $-4$, $-3$, $-2$, $-1$, $1$, $2$, $3$, $4$

(c)  Find the base $-2$ equivalents of $-7$, $-19$, and $75$.

**2.8**    Explain how to convert a number expressed in base $b$ into its equivalent in base $b^r$, where $b$ and $r$ are positive integers. (Exercise 2.66 has $b = 3$, $r = 2$; Exercises 2.72 and 2.73 have $b = 10$, $r = 3$.)

**2.9**    In Problem 1.10 it was pointed out that $1001 = 7 \times 11 \times 13$. Express a decimal number

$$\ldots a_9 a_8 a_7 a_6 a_5 a_4 a_3 a_2 a_1$$

to base 1000, as in Exercises 2.72 and 2.73. Use the theorem on divisibility by $b + 1$ in Section 2.2.6 to prove the test for divisibility of decimal numbers by 7, 11 or 13 given in Section 1.3.5.

**2.10**    Show that if $a = (a_n \ldots a_2 a_1)_b$, where $b$ is odd, then $a$ is even if and only if the sum of its digits $a_1 + \ldots + a_n$ is even. This generalizes the result in Exercise 2.74.

**2.11**    Prove that the ternary number $a = (a_n \ldots a_2 a_1)_3$ is divisible by 5 if and only if the sum

$$(a_1 + a_5 + a_9 + \ldots) + 3(a_2 + a_6 + a_{10} + \ldots) + 4(a_3 + a_7 + a_{11} + \ldots)$$
$$+2(a_4 + a_8 + a_{12} + \ldots)$$

is divisible by 5.

Hence test $(2120011210)_3$ for divisibility by 5. Use other divisibility tests to discover single-digit factors of this number.

**2.12**    Write a base-5 number $a = (a_n \ldots a_2 a_1)_5$ in terms of the base 125. Using the modified version of the theorem on divisibility by $b + 1$ in Exercise 2.84, show that $a$ is divisible by 2, or 7, or 9 if and only if

$$(a_3 a_2 a_1)_5 - (a_6 a_5 a_4)_5 + (a_9 a_8 a_7)_5 - \ldots$$

is divisible by 2, or 7, or 9.

Apply this result to $(304012204)_5$. Does this number have any other simple factors?

**2.13** Prove the following rule for squaring $x = (x_n x_{n-1} \ldots x_2 b)_{2b}$, which generalizes the case $b = 5$ in Exercise 2.13.

First define $y = (x_n x_{n-1} \ldots x_2)_{2b}$ and evaluate the product $y(y+1)$. Then append onto the right-hand end the digits $(\frac{1}{2}b)0$ if $b$ is even, and $[\frac{1}{2}(b-1)]0$ if $b$ is odd. For example, when $x = (124)_8$ then $y = (12)_8$ and $y(y+1) = (12)_8 \times (13)_8 = (156)_8$ so that $x^2 = (15620)_8$ since $b = 4$.

Use this method to find $[(238)_{16}]^2$.

**2.14** To weigh an object with a pair of scales and a set of weights, the object is placed in one pan and balancing weights are put on the other pan.

Show that any object weighing a whole number of grams up to $2^n - 1$ grams can be weighed using $n$ weights of $1, 2, 2^2, \ldots, 2^{n-1}$ grams.

**2.15** In the preceding problem suppose that weights can be placed on *both* pans. For example, with weights of 1, 3 and 9 grams you can weigh objects whose weight has any integer value from 1 to 13 grams. The first 8 values are set out as follows:

| Weight in left pan ($a$) | 1 | 3 | 3 | $1+3$ | 9 | 9 | $1+9$ | 9 |
|---|---|---|---|---|---|---|---|---|
| Weight in right pan ($b$) | 0 | 1 | 0 | 0 | $1+3$ | 3 | 3 | 1 |
| Weight of object ($c$) in right pan | 1 | 2 | 3 | 4 | 5 | 6 | 7 | 8 |

These three rows can be expressed conveniently by writing them as ternary numbers as follows (the suffix $(\ )_3$ is dropped for compactness):

| $a$ | 001 | 010 | 010 | 011 | 100 | 100 | 101 | 100 |
|---|---|---|---|---|---|---|---|---|
| $b$ | 000 | 001 | 000 | 000 | 011 | 010 | 010 | 001 |
| $c$ | 001 | 002 | 010 | 011 | 012 | 020 | 021 | 022 |

Notice that $a = b + c$, and that the values of $c$ are the ternary representations of the integers from 1 to 8.

(a) Complete the above tables for $c$ going from 9 to 13.

(b) Give the ternary number table which shows how with weights 1, 3, 9 and 27 you can weigh any object from 1 to 40 grams.

**2.16** Let $a$, $b$ and $c$ be positive integers with $d = (a,b)$. Prove:

(a) $(a/d, b/d) = 1$    (b) $(a + bc, b) = (a,b)$

**2.17** If $a$, $b$ and $c$ are positive integers prove that $(a,b,c) = (a,(b,c))$.

Explain what this result means, and use it to find the g.c.d. of 156, 240 and 540.

Extend the result for $n$ integers.

**2.18** Let $a$, $b$ and $c$ be positive integers with $(a,b) = 1$.

(a) If $(a,c) = 1$ show that $(a,bc) = 1$

(b) If $c|(a+b)$ show that $(a,c) = 1$, $(b,c) = 1$.

**2.19** A proof of Euclid's algorithm in Section 2.3.2 can be developed as follows.

(a) If $x$, $y$, $q$ and $r$ are integers and $x = yq + r$ show that the common divisors of $x$ and $y$ are the same as the common divisors of $y$ and $r$. Hence deduce that $(x, y) = (y, r)$.

(b) Show that the sequence of remainders generated by (2.15) satisfies

$$(a,b) = (r_1, r_2) = (r_2, r_3) = \ldots = (r_{n-1}, r_n) = r_n$$

**2.20** Use the relationship (2.16) which defines Fibonacci numbers to show that Euclid's algorithm requires $n$ divisions to establish that $(F_{n+1}, F_{n+2}) = 1$.

**2.21** Verify that the formulae (2.18) in Exercise 2.90 for solving $d = ax + by$ are correct by substituting into (2.15) when $n = 4$. Use the method of induction (see the Interlude) to prove that (2.18) is valid for all values of $n$.

**2.22** Prove that if $a$, $b$, $x$, $y$ are integers satisfying $d = ax + by$ with $d = (a,b)$ then $(x,y) = 1$.

**2.23** (a) Show that if $a$ is an odd integer then

$$a^2 \equiv 1 \pmod{4} \quad \text{and} \quad a^2 \equiv 1 \pmod{8}$$

(b) Show that if $a$ is an even integer then

$$a^2 \equiv 0 \pmod{4}$$

**2.24** A famous result called **Fermat's Little Theorem** states that if $p$ is a prime number and $a$ is an integer not divisible by $p$ then $a^{p-1} \equiv 1 \pmod{p}$.

Use this, together with the result in Exercise 2.111, to find the smallest positive residue for $2^{51} \pmod{17}$.

**2.25** Show that if $k = 2^p$, where $p$ is a positive integer, then $4^k$ is congruent either to $4 \pmod 9$ or to $7 \pmod 9$. Hence prove that $6 \times 4n \equiv 6 \pmod 9$ for all positive integers $n$.

**2.26** A number $a$ is expressed in base 34. Determine tests for divisibility of $a$ by 3, or 5, or 7, or 11.

Check your results by first converting 18557 to base 34, and then applying the tests.

**2.27** Find the solution of the congruence $105x \equiv 987 \pmod{1001}$.

**2.28** Find the solution of the system of congruences

$$x \equiv 1 \pmod 5, \quad x \equiv 3 \pmod{11}, \quad x \equiv 4 \pmod{12}, \quad x \equiv 5 \pmod{13}$$

**2.29** Show that the pair of congruences

$$x \equiv a_1 \pmod{m_1}, \quad x \equiv a_2 \pmod{m_2}$$

with $(m_1, m_2) = d > 1$ has a particular solution ($x^*$, say) if and only if $d \mid (a_1 - a_2)$.

Show also that the general solution is $x \equiv x^*(\bmod\ e)$ where $e = [a,b]$ is the l.c.m. of $a$ and $b$, defined in Exercise 2.86.

Solve the systems

(a) $x \equiv 5(\bmod\ 6)$, $\quad x \equiv 8(\bmod\ 9)$

(b) $x \equiv 8(\bmod\ 10)$, $\quad x \equiv 23(\bmod\ 25)$.

**2.30** Find all the multiplicative inverses which exist for members of $\mathbb{Z}_{34}$. Also, find a member $p$ of $\mathbb{Z}_{34}$ such that every element which has a multiplicative inverse is a power of $p$.

**2.31** Show that the set of all polynomials of the form

$$a_0 x^n + a_1 x^{n-1} + \ldots + a_{n-1}x + a_n$$

where $n$ is a positive integer and the coefficients $a_i$ are real numbers, constitutes a commutative ring (verify that the axioms (1) to (6) in Section 2.2.4 are satisfied).

---

**Student project**

**2.32** You encountered recurring decimals in Section 1.1.2 and recurring binary fractions in Problem 2.6. It is possible to express any rational number $x/y$ (here $x$ and $y$ are integers from 0 to 9, $y \neq 0$ and $x < y$) as a fraction in an arbitrary base $b$. The procedure is as follows.

Divide $xb$ by $y$ using decimal arithmetic to produce

$$xb = yq_1 + r_1$$

Repeat with

$$r_1b = yq_2 + r_2$$
$$r_2b = yq_3 + r_3$$

and so on. Then

$$\frac{x}{y} = \frac{q_1}{b} + \frac{q_2}{b^2} + \frac{q_3}{b^3} + \ldots$$

which can be written as

$$\frac{x}{y} = (0.q_1,q_2,q_3,\ \ldots)_b$$

The commas are necessary since some or all of the $q$'s may be bigger than 9.

For example, to express 4/5 in base 8 you get

$$4 \times 8 = 5 \times 6 + 2$$
$$2 \times 8 = 5 \times 3 + 1$$
$$1 \times 8 = 5 \times 1 + 3$$
$$3 \times 8 = 5 \times 4 + 4$$
$$4 \times 8 = 5 \times 6 + 2$$

and these divisions now repeat endlessly. Hence

$$\frac{4}{5} = \frac{6}{8} + \frac{3}{8^2} + \frac{1}{8^3} + \frac{4}{8^4} + \frac{6}{8^5} + \frac{3}{8^6} + \cdots$$
$$= (0.6,3,1,4,6,3,1,4,6, \ldots)_8$$

which is a recurring octal fraction.

(a) Verify that the above procedure is valid.

(b) Expand (i) $2/15$ in base 2; (ii) $1/7$ in base 100 (compare with Section 1.1.2); (iii) $79/135$ in base 60.

(c) Express the octal recurring fraction

$$(0.4,1,7,2,4,1,7,2,4,1,7, \ldots)_8$$

as a rational decimal number.

(d) When does a rational number have a terminating (that is, non-recurring) expansion in base 2; or in base 60?

## Student project

**2.33** Recall the special property of the number 6174, described in Section 1.1.1. Try to find a four-digit number to base 5 which has the corresponding property.

Try to develop a proof of this property (see Problem 1.19).

Show that there are no corresponding four-digit numbers for bases 4, 6, 7 or 8.

## Student project

**2.34** (a) Consider the pair of linear congruences

$$ax + by \equiv e(\bmod\ m)$$
$$cx + dy \equiv f(\bmod\ m)$$

where $a, b, c, d, e, f$ are integers and $m$ is a positive modulus. Show that if $D = ad - bc$ and $(D,m) = 1$ then there is a unique solution modulo $m$ given by

$$x \equiv D^{-1}(de - bf)(\bmod\ m)$$
$$y \equiv D^{-1}(af - ce)(\bmod\ m)$$

where $DD^{-1} \equiv 1(\bmod\ m)$.

Hence solve the system

$$5x + 4y \equiv 2(\bmod\ 11)$$
$$2x + 3y \equiv 9(\bmod\ 11)$$

(b) For readers familiar with the ideas of matrix algebra, you can define congruence of matrices as follows:

Let $A$ and $B$ be two $n \times n$ matrices with integer elements. The elements in row $i$ and column $j$ of $A$ and $B$ respectively are denoted by $a_{ij}$ and $b_{ij}$. Then $A \equiv B(\text{mod } m)$ means that $a_{ij} \equiv b_{ij}(\text{mod } m)$ for all possible values of $i$ and $j$.

For example

$$\begin{bmatrix} 23 & 7 \\ 9 & 13 \end{bmatrix} = \begin{bmatrix} 1 & 7 \\ -2 & 2 \end{bmatrix}(\text{mod } 11)$$

since $23 \equiv 1(\text{mod } 11)$, $7 \equiv 7(\text{mod } 11)$, $9 \equiv -2(\text{mod } 11)$ and $13 \equiv 2(\text{mod } 11)$.

Show that if $C$ and $D$ are two more $n \times n$ matrices with integer elements then

$$AC \equiv BC(\text{mod } m), \quad DA \equiv DB(\text{mod } m)$$

Define the inverse of $A$ modulo $m$ to be the matrix of integers $A^{-1}$ such that

$$A^{-1}A = AA^{-1} = I(\text{mod } m)$$

where $I$ is the unit matrix having 1's along the principal diagonal (northwest to southeast). For example

$$\underset{A}{\begin{bmatrix} 1 & 2 \\ 4 & 3 \end{bmatrix}} \begin{bmatrix} 6 & 7 \\ 3 & 2 \end{bmatrix} = \begin{bmatrix} 12 & 11 \\ 33 & 34 \end{bmatrix} = \begin{bmatrix} 1 & 0 \\ 0 & 1 \end{bmatrix}(\text{mod } 11)$$

$$\begin{bmatrix} 6 & 7 \\ 3 & 2 \end{bmatrix} \underset{A}{\begin{bmatrix} 1 & 2 \\ 4 & 3 \end{bmatrix}} = \begin{bmatrix} 34 & 33 \\ 11 & 12 \end{bmatrix} = \begin{bmatrix} 1 & 0 \\ 0 & 1 \end{bmatrix}(\text{mod } 11)$$

so

$$A^{-1} = \begin{bmatrix} 6 & 7 \\ 3 & 2 \end{bmatrix}(\text{mod } 11)$$

If

$$A = \begin{bmatrix} a & b \\ c & d \end{bmatrix}$$

and $D = ad - bc$ is non-zero and relatively prime to $m$, show that

$$A^{-1} = D^{-1}\begin{bmatrix} d & -b \\ -c & a \end{bmatrix}(\text{mod } m)$$

where $D^{-1}$ is defined in part (a). Use this approach to solve the pair of congruences in part (a).

(c) Try to obtain an expression for the inverse modulo $m$ of a $3 \times 3$ matrix of integers $A$ in terms of the adjoint of $A$. Hence solve the system of three congruences

$$2x + 5y + 6z \equiv 4(\text{mod } 7)$$
$$2x \quad\quad + z \equiv 2(\text{mod } 7)$$
$$x + 2y + 3z \equiv 3(\text{mod } 7)$$

## Further reading

Section 2.1   Blocksma M. (1989). *Reading the Numbers*. New York: Penguin, p. 41
Lines M.E. (1990). *Think of a Number*. Bristol, Adam Hilger: p. 60
McLeish J. (1992). *Number*. London: Flamingo
Molluzzo J.C. and Buckley F. (1986). *A First Course in Discrete Mathematics*. Belmont CA: Wadsworth, p. 63

Section 2.2   Barrow J.D. (1992). *Pi in the Sky*. London: Penguin, p. 26

Sections   Molluzzo J.C. and Buckley F. (1986). op cit., Chapter 1
2.2.1–2.2.4

Section   Grimaldi R.P. (1994). *Discrete and Combinatorial Mathematics, An Applied*
2.2.2.1   *Introduction*, 3rd edn. Reading MA: Addison-Wesley, p. 220

Section 2.2.5   Childs L. (1983). *A Concrete Introduction to Higher Algebra*. Berlin: Springer-Verlag, p. 37
Rosen K.H. (1993). *Elementary Number Theory and its Applications*, 3rd edn. Reading MA: Addison-Wesley, Section 1.3

Section   Dierker P. F. and Voxman W.L. (1986). *Discrete Mathematics*. San Diego: Harcourt
2.2.5.2   Brace Jovanovich, Section 2.2

Section 2.2.6   Rosen K.H. (1993). op. cit., Section 4.1

Section 2.3   Childs L. (1983). op. cit., Chapter 3
Rosen K.H. (1993). op. cit., Chapter 2

Section 2.4   Biggs N.L. (1989). *Discrete Mathematics*, Rev. edn. Oxford: Oxford University Press, Chapter 6
Childs L. (1983). op. cit., p. 47; p. 112; p. 65
Rosen K.H. (1993) op. cit., Chapter 3
Schroeder M.R. (1986). *Number Theory in Science and Communication*, 2nd enlarged edn. Berlin: Springer-Verlag, Part III

# 3 Combinatorics

As mentioned at the start of Chapter 1, numbers came into use because of the need to count objects – for example, the number of animals in a herd. Combinatorics deals with the more sophisticated problem of counting the number of ways of arranging objects in some particular fashion. To take a very simple example, suppose that a small sandwich bar offers a choice of three fillings: cheese ($C$), ham ($H$) or turkey ($T$), on either brown ($B$) or white ($W$) bread. How many different sandwiches with a single filling are on offer? In this case it's easy to list all the possibilities and count them. If, for example, $CB$ means cheese on brown bread, and so on, then there are *six* different sandwiches available, as follows:

$$CB, HB, TB, CW, HW, TW \tag{3.1}$$

If combinations of sandwich fillings are allowed, for example $CHB$ (cheese and ham on brown bread) then listing all the possibilities would be more complicated. Likewise, if many more kinds of fillings (for example, tuna, beef, egg mayonnaise, ...) and of bread (for example, soft rolls, French sticks, bagels, ...) were available then listing all the different combinations would be a lengthy and tedious chore. Clearly systematic methods are required. The techniques of counting combinations and other arrangements of objects form the subject matter of this chapter.

## 3.1 How many ways?

### 3.1.1 Multiplication principle

You can see that in the list of sandwiches set out in (3.1), each of the *three* kinds of filling can be combined with each of the *two* kinds of bread. It's obvious that

125

the total number of different sandwiches is equal to the product $3 \times 2 = 6$. This is an example of the **Multiplication Principle**:

> Suppose a process consists of $m$ stages, and there are $n_1$ choices at the first stage, $n_2$ at the second stage, up to $n_m$ at the last stage. If the choices are all independent of each other, then the total number of possible choices is $n_1 \times n_2 \times n_3 \times \ldots \times n_m$.

In the example of selecting a sandwich, the first stage is the choice of filling, and the second stage is the choice of bread.

---

**Example 3.1** **Applications of the principle**

(a) The ASCII code was introduced in Section 2.1.3. Symbols are represented by a string of seven bits, such as 1000001 which represents the letter A. There are *two* choices (0 or 1) for each of the *seven* bits, so the total number of different 7-bit bytes is

$$2 \times 2 \times 2 \times 2 \times 2 \times 2 \times 2 = 2^7 = 128$$

(b) Suppose that each member of a class of students is to be allocated a 3-digit decimal number, with the first (hundreds) digit non-zero. This first digit can be any one of $1, 2, 3, \ldots, 9$ so there are *nine* choices for it. The other two digits can also be zero, so there are *ten* choices for each of these. Altogether the total number of different identity numbers is therefore $9 \times 10 \times 10 = 900$.

(c) Suppose you have five different coins (5p, 10p, 20p, 50p and £1) which you want to distribute amongst your three children (Anne, Betty, Charles). In how many ways can this be done? The easiest way to answer this question is to realize that there are *three* ways of dealing with each coin – it can be given to Anne, or to Betty or to Charles. Since there are *five* different coins the total number of possible distributions of the coins is $3 \times 3 \times 3 \times 3 \times 3 = 3^5 = 243$ (notice that some of the children might not receive any coins).

In general, the number of ways of distributing $n$ different objects between $r$ different 'addresses' is $r^n$.

(d) A manufacturing company identifies its products with two letters and two non-zero digits, for example $AB12$. How many different products can be identified in this way, if there is a rule that the code must not include two vowels? To answer this, first suppose there is no restriction on the letters which can be used. Then since there are 26 letters in the alphabet, and each of the digits can be any integer from 1 to 9, the total number of different identity labels is $26 \times 26 \times 9 \times 9 = 54{,}756$. However, some of these contain two vowels. Since there are five vowels ($A, E, I, O, U$) the number of labels containing two vowels is $5 \times 5 \times 9 \times 9 = 2025$, because there are five choices for each of the two letters, and nine choices for each of the two

digits. Therefore the number of different identifiers satisfying the manufacturer's requirement is the total number, less those containing two
vowels, that is, $54{,}756 - 2025 = 52{,}731$.

---

**Exercise** | **3.1** A dinner menu offers three choices of starter, four choices of main course, and three desserts. How many different three-course meals can be served?

**Exercise** | **3.2** How many 4-digit decimal numbers are there, assuming the first digit is not zero? How many begin with 2 or 3?

**Exercise** | **3.3** (a) How many binary numbers are there with not more than six digits?

(b) How many ternary numbers (that is, to base 3) are there with not more than four digits?

**Exercise** | **3.4** How many odd numbers are there between 100 and 999? How many of these have distinct digits (that is, all different from each other)?

**Exercise** | **3.5** A warehouse contains 135,142 different items. Each item has an identifier consisting of two letters of the alphabet followed by several digits, each being any integer from 0 to 9.

(a) The letters $I$, $O$ and $X$ are excluded. How many digits must be used?

(b) In order to avoid possible confusion, it is decided not to use any identifier from (a) which has $Z$ as the second letter. How many identifiers are now available?

**Exercise** | **3.6** In how many different ways can a £1 coin, a £5 note, a £10 note, a £20 note and a £50 note be distributed among Dave, Elaine, Fred and Gina? What is the largest amount anyone can get?

**Exercise** | **3.7** Students are required to select one statistics module from a choice of four, and one computer science module from a choice of three. How many different ways are there of taking two courses? If the rules change so that students can take only one course either in statistics or in computer science, how many ways are there of doing this?

**Exercise** | **3.8** The second part of the preceding exercise is an example of the **Addition Principle**. This can be expressed as follows:

Suppose a collection of 'objects' is divided up into $m$ distinct non-overlapping parts, with $n_1$ objects in the first part, $n_2$ objects in the second part, up to $n_m$ in the last part. Then the total number of ways of selecting a single object is $n_1 + n_2 + \ldots + n_m$.

In the second part of Exercise 3.7, the available modules are divided up into a group of statistics modules and a group of computer science modules, so $n_1 = 4$ and $n_2 = 3$. A single module can be selected either from one group or the other in $4 + 3 = 7$ ways. You probably feel that the addition principle is obvious, since it is essentially just a formal way of stating that 'a whole is equal to the sum of its parts'.

In a supermarket freezer cabinet there are ten different kinds of pizza, five different curries and six different vegetarian meals. There are several packets of every type. How many different ways are there of choosing
(a) one meal of each type?     (b) one meal?

(c) three meals?     (d) three different meals?

**Exercise**  **3.9**  A standard pack of 52 playing cards is separated into the four suits of clubs, diamonds, hearts and spades. If one card is selected from each of the four suits, how many different hands of four cards can be obtained? How many of these contain two aces?

**Exercise**  **3.10**  The Eurocar is manufactured as a hatchback, saloon or estate. Each version can be purchased with a choice of equipment levels designated $L$, $SL$ or $GL$. The engine size (in litres), transmission and colour can be selected as follows:

| | Hatchback | Saloon | Estate |
|---|---|---|---|
| Engine | 1.4 or 1.6 | 1.4, 1.6 or 2.0 | 1.6 or 2.0 |
| Transmission | automatic or manual | automatic or manual | manual |
| Available colours | 6 | 4 | 3 |

How many different vehicles are available?

### 3.1.2  Permutations

A **permutation** is an **ordered** arrangement of $n$ different objects. For example, if the objects are the letters $X$, $Y$, $Z$ then you can write down all the possible permutations of these three letters as follows:

$XYZ, XZY, YXZ, YZX, ZXY, ZYX$

As indicated by the definition, the order in which the letters occur is all-important. In this simple example all the six possibilities can be written down. There are three ways of selecting the first letter (it can be $X$, $Y$ or $Z$); once this has been done there are two choices for the second letter, and then only one choice for the third letter. Hence, by the multiplication principle, the total number of possibilities is $3 \times 2 \times 1 = 6$. In general, with $n$ objects, the first object can be selected in $n$ different ways; once this has been done there are $n - 1$ choices for the second object, then $n - 2$ choices for the third object (since two have been used), and so on, until the last object which can be chosen in only one way. Again the multiplication principle shows that the total number of possibilities is

$$n \times (n - 1) \times (n - 2) \times \ldots \times 2 \times 1 \qquad (3.2)$$

This product of the consecutive integers from 1 to $n$ in (3.2) is called **$n$ factorial** and is written $n!$. Any pocket calculator having mathematical functions will have an $n!$ button. The factorial rapidly becomes large, for example

$$20! \approx 2.433 \times 10^{18}, \quad 60! \approx 8.321 \times 10^{81}, \quad 100! \approx 9.333 \times 10^{157}$$

A useful approximation when $n$ is large is provided by a formula attributed to the Scottish eighteenth-century mathematician **Stirling**. This states that

$$n! \approx \left(\frac{n}{e}\right)^2 \sqrt{2\pi n}\left(1 + \frac{1}{12n} + \frac{1}{288n^2}\right) \qquad (3.3)$$

where e $(= 2.71828...)$ is defined in Section 1.4.3.

| Example 3.2 | **Permutation of non-distinct objects** |
|---|---|

Suppose it is required to find how many permutations there are of the six letters *abbbcc*. Each permutation can be regarded as a six-letter 'word' where the order of the letters is significant, so for example *abbbcc* and *babbcc* are different words. Notice that words defined in this way need not have any meaning in a linguistic sense.

The difference from previous cases is that the letter *b* occurs three times, and the letter *c* twice. If the three *b*'s were distinct, say $b_1$, $b_2$, $b_3$, and the two *c*'s were distinct, say $c_1$, $c_2$ then the total number of different words constructed from the six different letters would be 6! However, the three letters $b_1$, $b_2$, $b_3$ can be arranged amongst themselves in 3! $(= 6)$ different ways. When $b_1 = b_2 = b_3 = b$, these different permutations all reduce to the same thing. For example, the six words

$$b_1 b_2 b_3 acc, \quad b_1 b_3 b_2 acc, \quad b_2 b_1 b_3 acc, \quad b_2 b_3 b_1 acc, \quad b_3 b_1 b_2 acc, \quad b_3 b_2 b_1 acc$$

all become *bbbacc*. The total number of different words is therefore reduced by dividing by 3!. An identical argument applies to the two *c*'s, so there is a further division, this time by 2! Hence the overall number of different words is

$$\frac{6!}{3!2!} = \frac{6 \times 5 \times 4 \times 3 \times 2 \times 1}{(3 \times 2 \times 1)(2 \times 1)} = 60$$

| Example 3.3 | **Distribution of objects amongst different addresses** |
|---|---|

A different way of looking at the problem in Example 3.2 produces an interesting result. Suppose six different objects labelled 1 to 6 are to be distributed amongst three people, so that Anne gets one object, Betty gets three objects and Charles gets two. Each permutation of the six letters *abbbcc* can be regarded as a distribution of these objects, with Anne receiving $a$, Betty $b$ and Charles $c$. For example, $babcbc$ means that Anne has object number 2 because $a$ is the second letter; similarly Betty has objects numbers 1, 3 and 5 and Charles has objects 4 and 6. The key point is that the *position* of each letter in the six-letter word identifies the numbered label of the object which the corresponding person receives. Thus the number of different ways in which the six objects can be distributed in the required fashion is precisely the same as the number of different six-letter words, namely $6!/(3!2!) = 60$.

The same argument applies in general:

The number of different ways of distributing $n$ different objects between $r$ different addresses, with $n_1$ at address 1, $n_2$ at address 2, ..., $n_r$ at address $r$ and $n_1 + n_2 + \ldots + n_r = n$, is

$$\frac{n!}{n_1! n_2! \ldots n_r!}$$

Recall that when there are no restrictions on the number at each address, the number of ways of distributing $n$ different objects is $r^n$ (see Example 3.1(c)). In this case one or more of the addresses may be empty.

---

**Example 3.4**    **Distribution into identical boxes**

Suppose that eight different objects are to be distributed into three boxes, with three objects in two of the boxes, and two in the remaining box. If the boxes were all different then, as seen in Example 3.3, the number of ways in which the distribution could be done is $8!/(3!)(3!)(2!)$. However, if the boxes are *identical*, then what is important is not which box an object is in, but which objects occupy a box together. For example, if the objects are labelled from 1 to 8 and objects within the same box are enclosed within square brackets, then the two distributions

$$[1,2,3],[4,5,6],[7,8] \quad \text{and} \quad [7,8],[1,2,3],[4,5,6]$$

are the same.

Since the two boxes containing three objects are indistinguishable, there are $2!$ ways of arranging these boxes, so the number of distributions must be divided by $2!$, giving an overall total

$$\frac{8!}{(3!)^2 2!(2!)} = \frac{8 \times 7 \times 6 \times 5 \times 4 \times 3 \times 2 \times 1}{(3 \times 2 \times 1)^2 (2 \times 1)^2} = 280$$

Again the argument can be generalized to obtain:

The number of different ways of distributing $n$ different objects into $r$ identical boxes, with $r_0$ empty boxes, $r_1$ boxes containing one object, $r_2$ boxes containing two objects, ..., $r_k$ boxes containing $k$ objects, and $r_0 + r_1 + r_2 + \ldots + r_n = r$, is

$$\frac{n!}{(2!)^{r_2}(3!)^{r_3}\ldots(k!)^{r_k}r_0!r_1!r_2!\ldots r_k!}$$

Notice the two types of factor in the denominator. The first kind comes from the argument in Example 3.3 with distinct boxes. Since there are $r_2$ boxes containing two objects, the factor 2! occurs $r_2$ times; the factor 3! occurs $r_3$ times, and so on. In the numerical part of this example $n = 8$, $r = 3$, $r_0 = 0$, $r_1 = 0, r_2 = 1$ and $r_3 = 2$.

The second type of factor in the denominator arises because the $r_0$ empty boxes are indistinguishable, so can be arranged amongst themselves in $r_0!$ ways; similarly the $r_1$ boxes holding one object can be arranged in $r_1!$ ways, and so on.

You are asked in Problem 3.45 to investigate further the distribution of different objects into identical containers. The situation when the boxes are different but the objects are identical will be discussed in Example 3.11.

Now suppose that ordered arrangements of $r$ objects selected from $n$ objects are to be counted. For example, all the possible permutations of two integers selected from 1,2,3,4 are

12, 21, 13, 31, 14, 41, 23, 32, 24, 42, 34, 43

To work out the total number of permutations in this example, begin with the fact that there are *four* ways of selecting the first integer. Once this has been done there are *three* choices for the second integer, so by the multiplication principle the total number of permutations is $4 \times 3 = 12$. In general, if $r$ objects are selected from $n$, there are:

$n$            ways of choosing the first object
$n-1$       ways of choosing the second object
$n-2$       ways of choosing the third object
$\ldots$
$n-r+1$   ways of choosing the $r$th object.

Hence by the multiplication principle, the total number of permutations of $r$ objects from $n$ is

$$n(n-1)(n-2)\ldots(n-r+1) \tag{3.4}$$

and the product in (3.4) is denoted by $P(n,r)$. In the example above, $P(4,2) = 4 \times 3 = 12$. Notice that if $r > n$ then $P(n,r) = 0$. Other notations used are $_nP_r$ or $^nP_r$, but these are less convenient in print or on a computer screen.

To evaluate $P(n,r)$ using (3.4), notice that this consists of the product of $r$ consecutive integers starting with $n$ and decreasing. For example, $P(13,4)$ is the product of four integers $13 \times 12 \times 11 \times 10 = 17{,}160$. An alternative expression is

$$P(n,r) = \frac{n!}{(n-r)!} \tag{3.5}$$

To see why (3.5) is correct, write the numerator as

$$n! = n(n-1)(n-2)\ldots(n-r+1)(n-r)(n-r-1)\ldots 3 \times 2 \times 1 \qquad (3.6)$$

$$\overleftarrow{\hspace{1.5cm}} P(n,r) \overrightarrow{\hspace{1.5cm}} \overleftarrow{\hspace{1.5cm}} (n-r)! \overrightarrow{\hspace{1.5cm}}$$

The first $r$ terms in (3.6) give $P(n,r)$ in (3.4), and the remaining $n-r$ terms constitute $(n-r)!$. Dividing both sides of (3.6) by $(n-r)!$ produces (3.5).

Notice that if $r = n$ in (3.4) then this expression becomes the same as that for $n!$, that is, $P(n,n) = n!$. A slight problem arises if $r = n$ is substituted into (3.5), since this gives $P(n,n) = n!/0!$. To avoid this difficulty (and for other reasons) it is necessary to *define* $0! = 1$. Also, substituting $r = 0$ into (3.5) gives $P(n,0) = n!/n! = 1$, so (3.5) is valid for $0 \leqslant r \leqslant n$.

| Example 3.5 | **Applications of $P(n,r)$** |
| --- | --- |

(a) Suppose there are eight contestants in a piano-playing competition. Then the number of different ways in which the winner, runner-up and third-place can be chosen is $P(8,3) = 8 \times 7 \times 6 = 336$. However, suppose that (unknown to the contestants) the judges have decided in advance that one of the competitors (Mr X) is guaranteed a place amongst the first three. To find how many different ways the first three places can be awarded in this case, first compute the number of placings which do *not* include Mr X. This is the number of ways of selecting the top three from the other seven competitors, that is, $P(7,3) = 7 \times 6 \times 5 = 210$. Hence the number of cases which *do* include Mr X amongst the first three is $336 - 210 = 126$.

(b) Suppose 5-digit numbers are constructed with distinct digits selected from 1, 2, 3, 4, 5, 6, 7, 8, 9. How many of these numbers do not contain both 3 and 4?

To solve this problem the 5-digit numbers which do not contain both 3 and 4 must be separated into three non-overlapping types:

(i) those containing neither 3 nor 4

(ii) those containing 3 but not 4

(iii) those containing 4 but not 3.

The permutations of type (i) consist of five digits selected from 1, 2, 5, 6, 7, 8, 9, so there are $P(7, 5) = 7 \times 6 \times 5 \times 4 \times 3 = 2520$. For the numbers of type (ii), the digit 3 can occupy any one of the five positions. The remaining four digits are selected from 1, 2, 5, 6, 7, 8, 9 in $P(7, 4) = 7 \times 6 \times 5 \times 4 = 840$ ways, so overall there are $5 \times 840 = 4200$ numbers of type (ii). By an identical argument there are also 4200 numbers of type (iii). Hence by the addition principle (see Exercise 3.8) the required total is the sum of the three separate parts, that is, $2520 + 4200 + 4200 = 10,920$.

**Exercise 3.11** A DJ on a student radio station has eight CD singles to play. How many different programmes of music can be selected?

**Exercise 3.12** Use Stirling's formula (3.3) to estimate 20! and 60!.

**Exercise 3.13** Evaluate the following:
(a) $P(7,3)$     (b) $P(11,4)$     (c) $P(19,2)$
(d) $1! + 4! + 5!$             (e) $4! + 0! + 5! + 8! + 5!$.

**Exercise 3.14** Express the following in terms of factorials:
(a) $13 \times 12 \times 11$     (b) $30 \times 29 \times 28 \times 27$.

**Exercise 3.15** Simplify the following:
(a) $\dfrac{10!}{7!}$        (b) $\dfrac{(n+3)!}{n!}$        (c) $\dfrac{(n+2)!}{(n-2)!}$.

**Exercise 3.16** How many different permutations are there of the letters in the words
(a) DISCRETE?           (b) MATHEMATICS?

**Exercise 3.17** Show that $P(n, n-1) = P(n,n)$.

**Exercise 3.18** (a) How many different 9-letter 'words' (in the sense defined in Example 3.2) can be constructed from the letters *aadddffff*?
(b) Diana buys nine different plants at a garden centre. She carries four plants to her car, her son carries three and her daughter carries two. In how many different ways can the plants be carried?

**Exercise 3.19** You buy 40 different items at a supermarket. You take 13 of them, and three friends with you take 9 items each. In how many different ways can the items be shared out?

**Exercise 3.20** On another visit to the supermarket by yourself you buy 43 different items. You pack them into five (identical) plastic bags, with two bags containing 8 items each, and the other three bags 9 items each. In how many different ways can the bags be packed?

**Exercise 3.21** How many different programmes of music can be constructed using 5 tracks from a CD containing 23 tracks?

**Exercise 3.22** How many 4-digit numbers can be constructed from the digits 1, 2, 3, 4, 5, 6 if repetition of digits is (a) not allowed? (b) allowed?

**Exercise 3.23** In the preceding exercise, how many numbers are there between 1000 and 4000 in each case?

**Exercise 3.24** How many odd 4-digit numbers with distinct digits can be constructed from the digits 1, 2, 3, 4, 5, 6, 7, 8, 9?

Exercise | **3.25** How many 4-digit numbers greater than 3000, and having distinct digits, can be constructed from the digits 1, 2, 3, 4, 5, 6, 7?

Exercise | **3.26** How many numbers greater than 3000, and having distinct digits, can be constructed from the digits 1, 2, 3, 4, 5, 6, 7?

Exercise | **3.27** How many of the numbers in each of Exercise 3.25 and Exercise 3.26 are even?

Exercise | **3.28** Six-digit numbers are constructed with distinct digits selected from 1, 2, 3, 4, 5, 6, 7, 8. How many of these numbers do not contain both 1 and 2?

### 3.1.3 Combinations

In many situations the order of the objects within an arrangement is not important. In this case a selection of $r$ objects from $n$ is called a **combination**, denoted by $C(n,r)$. The notations $_nC_r, {}^nC_r$ and $\binom{n}{r}$ are also used.

For example, return to the problem of choosing two integers from 1, 2, 3, 4. If now the order does not matter then all the possible combinations are

12, 13, 14, 23, 24, 34

This is because in the list of permutations each pair 12,21; 13,31; 14,41; and so on, is now counted only *once*. The total number of combinations is therefore half the number of permutations, that is,

$$C(4,2) = P(4,2)/2 = 12/2 = 6$$

Example 3.6 | **Choosing a committee**

Suppose a club committee consisting of three people is to be chosen from a total membership of 35. There are 35 ways of choosing the first committee person, then 34 ways of choosing the second from amongst the remaining members, and similarly 33 ways of choosing the third person on the committee. By the multiplication principle the total number of possible committees is $35 \times 34 \times 33$. However, the order in which the committee members are selected does not matter. The three members can be arranged amongst themselves in 3! ways. Therefore the number of *different* committees is

$$C(35,3) = \frac{35 \times 34 \times 33}{3!} = \frac{35 \times 34 \times 33}{3 \times 2 \times 1}$$

$$= 6545 \tag{3.7}$$

Notice that the numerator in (3.7) is just $P(35,3)$, so that $C(35,3) = P(35,3)/3!$.

The general case can be dealt with in the same way as the examples above. When selecting $r$ objects from $n$, the total number of permutations is $P(n,r)$. However, because the $r$ objects can be arranged amongst themselves in $r!$ ways, it follows that the number of combinations is

$$C(n,r) = \frac{P(n,r)}{r!}$$

$$= \frac{n!}{(n-r)!r!} \tag{3.8}$$

An alternative formula is obtained using the expression (3.4) for $P(n,r)$, so that

$$C(n,r) = \frac{n(n-1)(n-2)\ldots(n-r+1)}{r!} \tag{3.9}$$

Notice that $C(n,,0) = P(n,0) = 1$, $C(n,1) = P(n,1) = n$ and $C(n,r) = 0$ if $r > n$. When evaluating the formula (3.9), both numerator and denominator consist of a product of $r$ terms.

**Example 3.7**  **Applications of $C(n,r)$**

(a)  In many lotteries six numbers are selected at random from the integers 1 to 49. The winner of the main prize has to guess correctly these six numbers. Since the order in which the numbers are drawn does not matter, the total number of different combinations is $C(49,6)$. Evaluating this using (3.9) gives

$$C(49,6) = \frac{49 \times 48 \times 47 \times 46 \times 45 \times 44}{6 \times 5 \times 4 \times 3 \times 2 \times 1}$$

$$= 13,983,816$$

so the odds against a single ticket winning the main prize are almost 14 million to one. A discussion of the ideas of probability will be given in Section 3.5.

(b)  Suppose a student has to select five modules from a total of 19, of which 11 are mathematics modules and 8 are computer science modules. The rules require that every student must take at least one module in each of the two subject areas. The total number of possible choices of 5 modules from the 19 on offer is $C(19,5)$. However, some of these will not include a module from each subject area. Specifically, there are $C(11,5)$ choices which contain *only* mathematics, and $C(8,5)$ choices which contain *only* computer science. These cases do not satisfy the rules and so must be subtracted from the overall total. Hence the number of choices which a student has is

$$C(19,5) - C(11,5) - C(8,5) = \frac{19 \times 18 \times 17 \times 16 \times 15}{5!}$$

$$- \frac{11 \times 10 \times 9 \times 8 \times 7}{5!}$$

$$- \frac{8 \times 7 \times 6 \times 5 \times 4}{5!}$$

$$= 11{,}628 - 462 - 56$$

$$= 11{,}110$$

where again (3.9) has been used.

(c) This example is more complicated and involves the addition principle (see Exercise 3.8) as well as the multiplication principle (Section 3.1.1). Suppose a box contains seven white balls ($W$) and four black balls ($B$). Balls are taken out one at a time without replacement, until four balls have been obtained. The problem is to find in how many ways the selection can include (i) exactly two white balls, (ii) at least three white balls.

Notice that the order in which the balls are selected is not important – for example, $BBWW$ is the same as $WWBB$. Suppose therefore the two white balls in part (i) are selected first. This can be done in $C(7,2)$ ways since there are seven white balls to choose from. The remaining two balls are to be black, and can be selected from the four black balls in $C(4,2)$ ways. Hence by the multiplication principle the answer to part (i) is

$$C(7,2) \times C(4,2) = \frac{7 \times 6}{2!} \times \frac{4 \times 3}{2!}$$

$$= 21 \times 6 = 126$$

In part (ii), the selection must contain either three or four white balls. The number of ways of choosing three white balls and one black ball, by an argument similar to that used to solve (i), is $C(7,3) \times C(4,1) = 35 \times 4 = 140$. The number of ways of choosing four white balls is $C(7,4) = 35$. Since these two cases are non-overlapping, the total number of ways of getting at least three white balls is the sum $140 + 35 = 175$.

In the preceding example you may have noticed that $C(7,3) = C(7,4)$. This is not surprising, since each combination of three objects from seven leaves a combination of the remaining four objects from seven. In general

$$C(n,r) = C(n,n-r) \tag{3.10}$$

and this is easily proved by replacing $r$ by $n - r$ in (3.8) to give

$$C(n, n - r) = \frac{n!}{[n - (n - r)]!(n - r)!}$$

$$= \frac{n!}{r!(n - r)!} = C(n, r)$$

Another useful identity is

$$C(n + 1, r) = C(n, r - 1) + C(n, r) \tag{3.11}$$

and this is proved using the formula (3.8) together with some algebraic manipulation. First replace $n$ by $n + 1$ in (3.8) to get

$$C(n + 1, r) = \frac{(n + 1)!}{(n + 1 - r)!r!}. \tag{3.12}$$

Next replace $r$ by $r - 1$ in (3.8) to get

$$C(n, r - 1) = \frac{n!}{(n - r + 1)!(r - 1)!}$$

Notice that from the definition of factorial in (3.2)

$$(n - r + 1)! = (n - r + 1)(n - r)(n - r - 1) \ldots 2 \times 1$$

$$= (n - r + 1)(n - r)! \tag{3.13}$$

Similarly, in the denominator of $C(n, r)$ in (3.8),

$$r! = r(r - 1)! \tag{3.14}$$

Applying (3.13) and (3.14) to the right side of (3.11) produces

$$C(n, r - 1) + C(n, r) = \frac{n!}{(n - r + 1)(n - r)!(r - 1)!} + \frac{n!}{(n - r)!r(r - 1)!}$$

$$= \frac{n!}{(n - r)!(r - 1)!} \left( \frac{1}{n - r + 1} + \frac{1}{r} \right)$$

$$= \frac{n!}{(n - r)!(r - 1)!} \left( \frac{r + n - r + 1}{(n - r + 1)r} \right)$$

$$= \frac{n!(n + 1)}{(n - r)!(r - 1)!(n - r + 1)r}$$

$$= \frac{(n + 1)!}{(n - r + 1)!r!} \tag{3.15}$$

where the last step again uses (3.13) and (3.14) together with $(n + 1)! = (n + 1)n!$ Since (3.15) is the same as (3.12), this proves the identity (3.11).

For example, setting $n = 4$ and $r = 3$ in (3.11) gives

$$C(5,3) = C(4,2) + C(4,3) \tag{3.16}$$

Problems involving combinations are often quite tricky and require careful thought, as the following examples illustrate.

| Example 3.8 | **In or out** |

Suppose a driver of a minibus offers a ride to one or more people from a group of five. How many different sets of passengers can be taken? One way of solving the problem is to note that there are $C(5,1)$ ways of taking one passenger, $C(5,2)$ ways of taking two passengers, and so on, so by the addition principle the number of ways of taking one or more passengers is

$$C(5,1) + C(5,2) + C(5,3) + C(5,4) + C(5,5)$$
$$= 5 + 10 + 10 + 5 + 1 = 31$$

However, an easier way of solving the problem is to use the idea introduced in Example 3.1(c). For each person there are *two* alternatives, either in the bus or not. Since there are *five* people, the total number of possible ways of taking passengers is $2^5 = 32$. This includes the case when there are no passengers, so the required number of arrangements is $32 - 1 = 31$, as before.

For further applications of this idea, see Exercises 3.43, 3.44 and 3.45.

| Example 3.9 | **Combinations with repetition** |

Return to the problem of buying sandwiches, with which this chapter opened. There were six kinds of sandwiches available as set out in (3.1) (page 125), but suppose you don't like brown bread or ham. This leaves you with a choice of only two sandwiches, either cheese or turkey, both on white bread. If you want to buy three sandwiches, the four possibilities available are set out in Table 3.1.

The first row means you buy three cheese and no turkey sandwiches, in the second case you buy two cheese and one turkey, and so on. Another way of expressing the possibilities is shown in Table 3.2, where $x$ indicates a purchase. You can now see from Table 3.2 that the total number of possibilities is the number of arrangements of *four* symbols consisting of

*Table 3.1* Buying two sandwiches

|  | Cheese | Turkey |
|---|---|---|
| | 3 | 0 |
| *Number purchased* | 2 | 1 |
| | 1 | 2 |
| | 0 | 3 |

*Table 3.2*  Choosing 3 from 2

| Cheese | Turkey |
|--------|--------|
| x x x  |        |
| x x    | x      |
| x      | x      |
|        | x x x  |

*Table 3.3*  Choosing 3 from 4

| CB   | TB  | CW  | TW  |
|------|-----|-----|-----|
| x x x |     |     |     |
| x x  | x   |     |     |
| x    |     | x x |     |
|      | x   | x   | x   |

*three* x's and *one* vertical line. Using the argument developed in Example 3.2, this number is

$$\frac{4!}{3!} = C(4,3) = 4$$

since the three x's can be arranged amongst themselves in 3! ways.

Suppose your friend also dislikes ham, but is happy with brown as well as white bread. Your friend wishes to buy three sandwiches from the four available out of the list in (3.1), namely $CB$ (= cheese on brown), $TB$ (= turkey on brown), $CW$ (= cheese on white) and $TW$ (= turkey on white). Some of the possibilities are shown in Table 3.3, where again x denotes a purchase, so for example the second row means two $CB$ sandwiches and one $TB$ sandwich are bought.

You can see from Table 3.3 that the total number of possibilities is the number of different arrangements of *six* symbols consisting of *three* x's and *three* vertical lines. This number is

$$\frac{6!}{3!3!}$$

which by the definition (3.8) is equal to $C(6,3)$.

The argument extends to the general case of selecting $r$ objects, with repetition allowed, from $n$ different objects. Each row in an array like Tables 3.2 and 3.3 contains $r$ x's and $n-1$ vertical lines. The number of different possibilities is the total number of arrangements of the $r+n-1$ symbols, namely $(r+n-1)!$, divided by the number of ways in which the x's can be rearranged amongst themselves (that is, $r!$) and the number of ways in which the vertical lines can be rearranged (that is, $(n-1)!$). This gives

$$\frac{(r+n-1)!}{(n-1)!r!}$$

which by the definition (3.8) is equal to $C(r+n-1,r)$.

Two other applications of this result now follow.

**Example 3.10**   **Integer solutions of equations**

Go back to Table 3.3, and suppose your friend buys $x_1$ sandwiches of type $CB$, $x_2$ of $TB$, $x_3$ of $CW$ and $x_4$ of $TW$. Since a total of three sandwiches is purchased, then $x_1 + x_2 + x_3 + x_4 = 3$, where each $x_i \geqslant 0$. The number of different (non-negative) integer solutions of this equation is therefore the same as the number of different sandwich purchases, that is, $C(6,3)$. Similarly, in general the number of different non-negative integer solutions of the equation

$$x_1 + x_2 + \ldots + x_n = r, \qquad x_i \geqslant 0$$

is $C(r+n-1,r)$.

**Example 3.11**   **Distribution of *identical* objects into different boxes**

Yet another way of regarding Table 3.3 is that it represents the distribution of three *identical* objects (represented by the $x$'s) between four different columns. In general, the number of different ways of distributing $r$ identical objects amongst $n$ different boxes, with empty boxes allowed, is $C(r+n-1,r)$. For example, if you have nine identical apples to share amongst four people, the number of different ways in which this can be done is

$$C(9+4-1,9) = C(12,9)$$

$$= C(12,3) = 220$$

How does this change if each person is to get at least one apple? Simply give one apple to each person, leaving five apples. These can then be distributed in total of $C(5+4-1,5) = C(8,5) = 56$ ways. This idea can be extended to the general case when empty boxes are not allowed.

**Exercise**   **3.29**   How many different ways are there of buying three CDs from the top ten?

**Exercise**   **3.30**   Evaluate the following:

(a) $C(7,3)$     (b) $C(11,4)$     (c) $C(19,2)$     (d) $C(75,72)$.

**Exercise**   **3.31**   Express the following in terms of factorials:

(a) $C(16,5)$     (b) $17 \times 13 \times 7 \times 4$.

**Exercise** **3.32** How many different selections are there in a lottery where five numbers are chosen at random from the integers 1 to 39?

**Exercise** **3.33** Consider again the two-out-of-five code described in Problem 2.3. Show that the number of different 5-bit bytes is $C(5,2)$.

**Exercise** **3.34** In how many different ways can four playing cards be selected from a standard pack of 52? What is the answer if the four cards consist of

(a) one card from each suit?

(b) two clubs, one diamond, one heart?

(c) four cards all from the same suit, including an ace?

**Exercise** **3.35** Prove directly that

$$C(n+1, r+1) = C(n, r) + C(n, r+1)$$

Check that this identity is correct when $n = 19$, $r = 14$, by evaluating both sides.

**Exercise** **3.36** In an examination you are asked to answer three questions out of five in Section A, and two out of five in Section B. In how many ways can candidates submit solutions to five questions?

**Exercise** **3.37** Every Saturday night you have a party with four friends. What is the least number of friends you must have so that there is a different group each week for a year (52 weeks)?

**Exercise** **3.38** A group of six students is to be divided into three pairs for laboratory work. In how many different ways can this be done?

**Exercise** **3.39** A committee of four people is to be chosen from 18 men and 12 women. How many possible committees are there? How many contain

(a) three men?

(b) at least one woman?

(c) at least one man and at least one woman?

**Exercise** **3.40** Five balls are taken out of a box one at a time without replacement. The box contains ten black and six white balls. How many ways are there of choosing the five balls

(a) without restrictions?

(b) when two are black?

(c) when all have the same colour?

(d) when at least two are white?

**Exercise** **3.41** A box contains 12 balls, of which five are red, four are blue and three are white. How many different selections of five balls can you make so that there is at least one ball of each colour?

**Exercise** | **3.42** How many four-letter combinations can be constructed from the letters *AAABBCDE*? How many four-letter 'words' are there? (A 'word' is as defined in Example 3.2 – for example, six different 'words' *AABB, ABAB, ABBA, BAAB, BABA, BBAA* can be formed from the combination *AABB*.)

**Exercise** | **3.43** There are eight books in the library on a particular topic which you are studying. How many ways are there of borrowing one or more of these books?

**Exercise** | **3.44** In how many ways can eight books be split up into two piles (not necessarily equal)

(a) when the books are all different?　　(b) when the books are identical?

**Exercise** | **3.45** A pizza take-away advertises that it offers a choice of over 200 different pizzas. You can choose one or more different toppings. How many toppings must be available?

**Exercise** | **3.46** How many different ways are there of choosing two ice-cream cones, with repetitions allowed, from a choice of five different flavours? Check your result by listing all the possibilities.

**Exercise** | **3.47** A supermarket freezer cabinet is well stocked with 12 flavours of ice-cream. In how many ways can you buy

(a) seven differently flavoured cartons of ice-cream?

(b) seven cartons when repetition of flavours is allowed?

(c) seven cartons when more than half must be vanilla flavour?

**Exercise** | **3.48** In how many ways can seven identical balls be distributed into three different boxes

(a) without repetition?

(b) when no box is to be empty?

(c) when the first box must contain an even number of balls?

**Exercise** | **3.49** In the preceding exercise, what is the number of different distributions if each of the seven balls is a different colour?

**Exercise** | **3.50** In how many ways can ten apples and 13 oranges be shared amongst six children, if each child is to receive at least one of each kind of fruit?

**Exercise** | **3.51** How many different solutions are there of the equation

$$x_1 + x_2 + x_3 + x_4 = 17$$

where each $x_i$ is a non-negative integer?

## 3.2 Binomial coefficients

The numbers $C(n,r)$ for $r = 1, 2, 3, \ldots, n-1$ are called **binomial coefficients** because they occur in the expansion of a 'binomial' (which means a sum of *two* terms $a + b$) raised to an integer power $n$. For example, by direct expansion you can verify that

$$(a+b)^2 = a^2 + 2ab + b^2$$
$$= C(2,0)a^2 + C(2,1)ab + C(2,2)b^2$$
$$(a+b)^3 = a^3 + 3a^2b + 3ab^2 + b^3$$
$$= C(3,0)a^3 + C(3,1)a^2b + C(3,2)ab^2 + C(3,3)b^3$$
$$(a+b)^4 = a^4 + 4a^3b + 6a^2b^2 + 4ab^3 + b^4$$
$$= C(4,0)a^4 + C(4,1)a^3b + C(4,2)a^2b^2 + C(4,3)ab^3 + C(4,4)b^4$$

### 3.2.1 Binomial theorem

This states that for any positive integer $n$

$$(a+b)^n = a^n + C(n,1)a^{n-1}b + C(n,2)a^{n-2}b^2$$
$$+ \ldots + C(n,n-2)a^2b^{n-2} + C(n,n-1)ab^{n-1} + b^n$$
$$= \sum_{r=0}^{n} C(n,r)a^{n-r}b^r. \tag{3.17}$$

Recall that when $r = 0$ then $C(n,0) = 1$ and when $r = n$ then $C(n,n) = 1$. The first and last terms in (3.17) are therefore respectively $C(n,0)a^n = a^n$ and $C(n,n)b^n = b^n$. To see why the coefficients in $(a+b)^n$ turn out to be the combinations of $n$ objects taken $1, 2, 3, \ldots$ at a time, it's easiest to look at a specific case, say $n = 3$. To examine how the expansion of $(a+b)^3$ is built up, write it as $(a_1 + b_1)(a_2 + b_2)(a_3 + b_3)$. Introducing suffices in this way enables a record to be kept of what happens. First multiply together the first two brackets to obtain

$$(a_1 + b_1)(a_2 + b_2) = a_1a_2 + (a_1b_2 + a_2b_1) + b_1b_2$$

and then multiply by $(a_3 + b_3)$ to get

$$(a_1 + b_1)(a_2 + b_2)(a_3 + b_3) = a_1a_2a_3 + (a_1a_3b_2 + a_2a_3b_1 + a_1a_2b_3)$$
$$+ (a_1b_2b_3 + a_2b_1b_3 + a_3b_1b_2) + b_1b_2b_3 \tag{3.18}$$

You can see that every term in (3.18) contains exactly one variable from each of the original brackets, with different suffices. For example, the second term $a_1a_3b_2$ in (3.18) has $a_1$ from the first bracket, $b_2$ from the second and $a_3$ from the third. On setting $a_1 = a_2 = a_3$ and $b_1 = b_2 = b_3 = b$, then the first bracket on the right in (3.18) becomes $3a^2b$: the coefficient 3 is the number of ways of choosing one '$b$' term from the original three terms $(a_1 + b_1)$, $(a_2 + b_2)$, $(a_3 + b_3)$, which is precisely $C(3,1)$. Similarly, the second bracket on the right in (3.18) becomes $3ab^2$, and the coefficient 3 is the number of ways of choosing two '$b$' terms from the original product, that is $C(3,2)$. The argument can be extended to the general case in (3.17): the coefficient of the term $a^{n-r}b^r$ is the number of ways of choosing $r$ $b$'s from the $n$ factors in $(a+b)^n$, and this number is $C(n,r)$.

| Example 3.12 | **Applications of the theorem** |

(a)  Expanding $(a + b)^{20}$ using (3.17) gives the first few terms as

$$(a + b)^{20} = a^{20} + C(20,1)a^{19}b + C(20,2)a^{18}b^2 + C(20,3)a^{17}b^3 + \ldots$$

$$= a^{20} + 20a^{19}b + 190a^{18}b^2 + 1140a^{17}b^3 + \ldots$$

The coefficient of, for example, $a^{13}b^7$ is obtained using (3.9) as

$$C(20,7) = \frac{20 \times 19 \times 18 \times 17 \times 16 \times 15 \times 14}{1 \times 2 \times 3 \times 4 \times 5 \times 6 \times 7}$$

$$= 77,520$$

(b)  Recall that a 'word' was defined in Example 3.2 as a sequence of letters whose order matters, although the word need not make linguistic sense. Suppose four-letter words are to be constructed from the letters $ABCDEFG$, where repetition of letters is allowed. For example, $AABC$ and $CCDD$ are acceptable words. Now count how many words there are containing 0, 1, 2, 3 or 4 vowels, as follows.

*0 vowels*: Since there are five consonants, there are five choices for each letter in a word, so by the multiplication principle the total number is $5^4$.

*1 vowel*: There are $C(4,1)$ choices for the position of the vowel in the word. There are two vowels to choose from, and five consonants for the other three letters. By the multiplication principle the total number is $C(4,1) \times 2 \times 5^3$.

*2 vowels*: There are $C(4,2)$ choices for the positions of the vowels, and the overall total is $C(4,2) \times 2^2 \times 5^2$.

Similarly, the numbers of words with 3 and 4 vowels are respectively $C(4,3) \times 2^3 \times 5$ and $C(4,4) \times 2^4$.

These numbers of words are in fact found more easily by realizing that they are equal to the terms in the binomial expansion of $(5 + 2)^4$ using (3.17), which gives

$$5^4 + C(4,1)5^3 \times 2 + C(4,2)5^2 \times 2^2 + C(4,3)5 \times 2^3 + 2^4$$

A similar application for probability problems will be given in Example 3.27.

**Exercise**    **3.52**   Use the binomial theorem to determine the expansions of:

(a) $(a+b)^5$        (b) $(1+x)^7$        (c) $(a+2b)^6$.

**Exercise**    **3.53**   Determine the coefficient of:

(a) $a^5b^5$ in $(a+b)^{10}$        (b) $a^9b^3$ in $(a^3+b)^6$

(c) $x^{15}$ in $(1+x)^{19}$        (d) $a^5b^3$ in $(a-2b)^8$.

**Exercise**    **3.54**   By taking $a=1$, $b=1$ in (3.17) prove that $2^n = \sum_{r=0}^{n} C(n,r)$, for $n = 1,2,3,\ldots$ Similarly, show that

$$1 + C(n,2) + C(n,4) + C(n,6) + \ldots = C(n,1) + C(n,3) + C(n,5) + \ldots$$

**Exercise**    **3.55**   By writing $11 = 10 + 1$, use the binomial theorem to evaluate $(11)^7$.

**Exercise**    **3.56**   By writing $1.05 = 1 + 0.05$, use the binomial theorem to show that $(1.05)^{30} > 4$.

**Exercise**    **3.57**   Five-letter words (in the sense described in Example 3.2) are constructed from the letters of the word MATHEMATICS. Repetition of letters is allowed. How many words contain at most two vowels?

**Exercise**    **3.58**   Prove that

$$C(n,r) = \left(\frac{n}{r}\right)C(n-1, r-1).$$

**Exercise**    **3.59**   Prove that for any positive integer $s \leqslant r$ then

$$C(n,r)C(r,s) = C(n,s)C(n-s, r-s).$$

**Exercise**    **3.60**   Prove that

$$C(2n,2) = 2C(n,2) + n^2$$

## 3.2.2   Pascal's triangle

If the binomial coefficients which arise in the expansion of $(a+b)^n$ in (3.17) are set out as in Table 3.4, then the resulting array, which extends downwards indefinitely, is known as **Pascal's triangle**. This is named after the seventeenth-century French mathematician Blaise Pascal, who developed many applications of the triangle in the theory of probability. However, the triangle was known in China, India and Persia as far back as the eleventh century. It is also called the 'arithmetical triangle', which is the name used by Pascal himself in his treatise of 1665.

The expansions of $(a+b)^n$ are immediately available from the table, for example using the row $n=4$ with $a=1$, $b=x$, gives

$$(1+x)^4 = 1 + 4x + 6x^2 + 4x^3 + 1$$

*Table 3.4*   Pascal's triangle – first form

|   | C(n,0) | C(n,1) | C(n,2) | C(n,3) | C(n,4) | C(n,5) | C(n,6) | C(n,7) | ... |
|---|--------|--------|--------|--------|--------|--------|--------|--------|-----|
| 0 | 1 | | | | | | | | |
| 1 | 1 | 1 | | | | | | | |
| 2 | 1 | 2 | 1 | | | | | | |
| 3 | 1 | 3 | 3 | 1 | | | | | |
| 4 | 1 | 4 | 6 | 4 | 1 | | | | |
| 5 | 1 | 5 | 10 | 10 | 5 | 1 | | | |
| 6 | 1 | 6 | 15 | 20 | 15 | 6 | 1 | | |
| 7 | 1 | 7 | 21 | 35 | 35 | 21 | 7 | 1 | |
| 8 | 1 | 8 | 28 | 56 | 70 | 56 | 28 | 8 | ... |
| . | . | . | . | . | . | . | . | . | ... |
| . | . | . | . | . | . | . | . | . | ... |

(*n* labels the rows)

*Table 3.5*   Pascal's triangle – second form

An often-preferred alternative to Table 3.4 is to arrange the numbers in the shape of the triangle shown in Table 3.5.

Just looking at the triangle (in either form) reveals some obvious features, such as the lines of 1's, and the lines 1, 2, 3, 4, 5, .... In fact the triangle has an amazing number of interesting properties, some of which are now described.

### 3.2.2.1   Properties of Pascal's triangle

*1 Construction*

The key feature is that it is not necessary to use the formula for $C(n,r)$ in (3.8) or (3.9) in order to obtain the entries in the triangle.

Firstly, notice that the first and last entries in each row are 1, since $C(n,0) = 1 = C(n,n)$. Every other number in Table 3.5 is simply the sum of the two numbers in the row above, to the left and right. For example, the first two rows are

$$1$$
$$1 \qquad 1$$

and the third row is constructed as shown:

where $2 = 1 + 1$. The next row is then computed from the third row as follows:

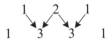

The numbers where the heads of the arrows meet are obtained by adding together the two entries at the start of the arrows, so that $1 + 2 = 3$, $2 + 1 = 3$. The next row is produced in exactly the same way:

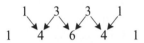

where $1 + 3 = 4$, $3 + 3 = 6$, $3 + 1 = 4$. Continuing this process you obtain:

$$\begin{array}{ccccccccc} & 1 & & 4 & & 6 & & 4 & & 1 \\ 1 & & 5 & & 10 & & 10 & & 5 & & 1 \end{array}$$ (3.19)

where $1 + 4 = 5$, $4 + 6 = 10$, $6 + 4 = 10$, $4 + 1 = 5$. The reason behind this easy way of building up the triangle is provided by the identity (3.11). Firstly, consider as an example the special case (3.16), which states that

$$C(5,3) = C(4,2) + C(4,3)$$

that is, $10 = 6 + 4$ which is one of the results in (3.19). In general, referring to Table 3.4 shows that two consecutive rows in Table 3.5 are

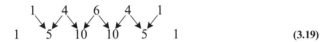

and the arrows indicate that

$$C(n + 1,r) = C(n,r - 1) + C(n,r)$$

which is precisely the identity (3.11).

## 2 Symmetry

You can see that each row in the triangle (either form) is the same whether read from left to right, or from right to left. In particular each row, and therefore the entire triangle in Table 3.5, is symmetrical relative to a vertical line through the middle. This is because of the identity (3.10), which states that $C(n,r) = C(n,n - r)$. Thus for example $C(6,2) = C(6,4) = 15$, as seen in the seventh row of the triangle in Table 3.5.

Notice that the first row of the triangle corresponds to $n = 0$, the second to $n = 1$ and so on, so the $(k + 1)$th row corresponds to $n = k$ and contains $k + 1$ terms which are the coefficients in $(a + b)^k$.

From $n = 3$ onwards, the second term and the penultimate term are $C(n,1) = n = C(n,n - 1)$.

### 3 Row sums

Adding up the terms in each row of the triangle after the first gives respectively

$$1 + 1 = 2, \quad 1 + 2 + 1 = 4 = 2^2, \quad 1 + 3 + 3 + 1 = 2^3,$$

$$1 + 4 + 6 + 4 + 1 = 2^4, \quad \text{and so on}$$

In general (see Exercise 3.54) the sum of the terms in the row corresponding to $n = k$ is $2^k$.

Another property you were asked to prove in Exercise 3.54 is that the sum of alternate terms in a row, starting from 1 on the left, is equal to the sum of the remaining terms. For example, in the seventh row

$$1 + 15 + 15 + 1 = 6 + 20 + 6$$

### 4 Column sums

Using the first form of the triangle in Table 3.4, the sum of the numbers in each column within a rectangular box shown below is equal to the number in the next row, indicated by the arrow. For example, in the third column $1 + 3 + 6 + 10 = 20$, which corresponds to

$$C(2,2) + C(3,2) + C(4,2) + C(5,2) = C(6,3) \tag{3.20}$$

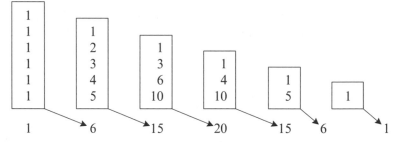

The result holds however many entries you take in a column, so that in general

$$C(r,r) + C(r + 1,r) + C(r + 2,r) + \ldots + C(n,r) = C(n + 1,r + 1) \tag{3.21}$$

for $r = 1, 2, 3, \ldots, n$. The example in (3.20) has $r = 2$, $n = 5$. You are asked in Problem 3.19 to develop a proof of (3.21).

### 5 Unimodality

Another property of Pascal's triangle which you can see by looking at Table 3.4 or Table 3.5 is that in each row the numbers steadily increase before reaching a maximum, and then steadily decrease again. A sequence of numbers having this property is called **unimodal**. Rows for which $n$ is even have a single largest number in the middle, whereas when $n$ is odd the two middle numbers have the largest value. Of course, you cannot assume that the unimodal property holds for every row of the triangle merely because it is true for the rows shown in the tables. A method for proving the result in general is outlined in Exercise 3.63.

*6 Link with Fibonacci numbers*
The sequence of Fibonacci numbers

$$F_1 = 1, F_2 = 1, \quad F_3 = F_2 + F_1 = 2, \quad F_4 = F_3 + F_2 = 3,$$

$$F_5 = 3 + 2 = 5, \quad F_6 = 3 + 5 = 8, \ldots$$

was introduced in Section 2.3.2 and will be discussed in detail in Section 5.3.2. In Problem 5.29 you are asked to prove that adding the numbers along diagonals in Table 3.4, as shown below, produces the Fibonacci numbers.

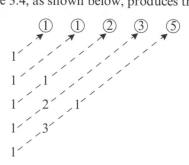

*7 Binary representation*
The following table lists the values of $k$, which is the number of 1's in the binary representation of a decimal number $n$ (see Table 2.2).

| $n$ | 1 | 2 | 3 | 4 | 5 | 6 | 7 | 8 |
|---|---|---|---|---|---|---|---|---|
| binary | $(01)_2$ | $(10)_2$ | $(11)_2$ | $(100)_2$ | $(101)_2$ | $(110)_2$ | $(111)_2$ | $(1000)_2$ |
| $k$ | 1 | 1 | 2 | 1 | 2 | 2 | 3 | 1 |
| $2^k$ | 2 | 2 | 4 | 2 | 4 | 4 | 8 | 2 |

The bottom row gives the values of $2^k$, and a surprising result is that these are equal to the number of *odd* coefficients in the corresponding row of Pascal's triangle. For example, when $n = 5$ then $2^k = 4$, and there are four odd numbers in the fifth row of the triangle (that is, 1,5,5,1). The proof that this result holds in general is beyond the scope of this book.

**Exercise**  **3.61**  Construct the next three rows in Tables 3.4 and 3.5. Verify that Property 7 holds in each case.

**Exercise**    **3.62**   Draw the diagram which arises when Property 4 is applied to Table 3.5.

**Exercise**    **3.63**   Express the quotient $C(n,r)/C(n,r-1)$ in its simplest possible form. Determine the conditions for which this ratio is either less than 1 or greater than 1. Hence show that the sequence of binomial coefficients in any row of Pascal's triangle is unimodal.

Finally, show that the maximum value of $C(n,r)$ occurs at $r = (1/2)n$ when $n$ is even, and at $r = (1/2)(n-1)$ and $r = (1/2)(n+1)$ when $n$ is odd.

**Exercise**    **3.64**   If you write the rows of Pascal's triangle as a single decimal number, then for $n = 2$ you get $121 = 11^2$, and for $n = 3$ similarly $1331 = 11^3$. Try to extend this way of obtaining powers of 11 from the triangle for $n = 3, 4, 5$. Can you explain what happens for a general value of $n$?

**Exercise**    **3.65**   Consider the third diagonal $1, 3, 6, 10, \ldots$ in Table 3.5. Notice that the sums of two adjacent terms are equal to perfect squares, for example $1 + 3 = 2^2$, $3 + 6 = 3^2$, $6 + 10 = 4^2$, $10 + 15 = 5^2$. Prove that this holds for any two adjacent terms however far you go along the diagonal (which extends indefinitely).

**Exercise**    **3.66**   What is the sum of all the numbers in the triangle above row $k$ (the rows being numbered $0,1,2,3,\ldots$ as in Table 3.4)?

**Exercise**    **3.67**   A third way of writing out Pascal's triangle is as follows:

$$
\begin{array}{ccccccc}
1 & 1 & 1 & 1 & 1 & 1 & 1 & \ldots \\
1 & 2 & 3 & 4 & 5 & 6 & \ldots \\
1 & 3 & 6 & 10 & 15 & \ldots \\
1 & 4 & 10 & 20 & \ldots \\
1 & 5 & 15 & \ldots \\
1 & 6 & \ldots \\
1 & \ldots
\end{array}
$$

Show that (after the first column) each element is equal to the difference of the two terms directly below it and below it to the left. For example, $4 = 10 - 6$.

The **harmonic triangle** shown below is constructed so that (after the first row) each element is equal to the difference of the two terms directly above it and above it to the right. For example, $1/6 = (1/2) - (1/3)$.

$$
\begin{array}{ccccc}
\frac{1}{1} & \frac{1}{2} & \frac{1}{3} & \frac{1}{4} & \frac{1}{5} & \ldots \\[4pt]
\frac{1}{2} & \frac{1}{6} & \frac{1}{12} & \frac{1}{20} & \ldots \\[4pt]
\frac{1}{3} & \frac{1}{12} & \frac{1}{30} & \ldots \\[4pt]
\frac{1}{4} & \frac{1}{20} & \ldots \\[4pt]
\frac{1}{5} & \ldots
\end{array}
$$

Construct the next diagonal of this triangle.

Some properties of the harmonic triangle are investigated in Problems 3.20 and 3.21.

## 3.3 Pigeonhole principle

On rare occasions mathematics allows a touch of frivolity to affect terminology. The so-called 'pigeonhole principle' is one such example. It arises from the obvious observation that if a number of pigeons enter some nesting boxes ('pigeonholes') and there are not enough to go round then some pigeons must share boxes. The precise origins of the principle are not known, but it was widely used by the nineteenth-century French mathematician Dirichlet. Other whimsical names sometimes used are the 'drawer' or 'shoebox' principle.

### 3.3.1 Basic version

A formal statement of the **pigeonhole principle (PP)** is: if $n + 1$ objects are put into $n$ boxes, then at least one box contains two or more objects. Clearly, the statement is also true if there are $m$ objects with $m > n$.

**Example 3.13** **Applications of the principle**

Despite the fact that the principle is self-evident, it can be used to obtain interesting and non-trivial results. The first two below are very straightforward.

(a) In a group of 13 (or more) people, there will be at least two who have a birthday in the same month of the year. Here the 'objects' are the 13 (or more) birthday months and the 'boxes' are the 12 months of the year. This is shown in Figure 3.1, where X indicates a birthday falling in a particular month, with two occurring in October. However, the PP does not make any other assertion about the way the objects are spread amongst the boxes. For example, the birthdays may occur in clusters in just some of the months, as shown in Figure 3.2. Indeed it's certainly possible (although admittedly very unlikely) that all 13 people have their birthday in the same month of the year, in which case all but one of the boxes would be empty.

| January | February | March | April |
|---------|----------|-------|-------|
| X | X | X | X |
| **May** | **June** | **July** | **August** |
| X | X | X | X |
| **September** | **October** | **November** | **December** |
| X | X X | X | X |

*Figure 3.1* Thirteen birthdays

| January | February | March | April |
|---|---|---|---|
| X X | X X |  | X X X |

| May | June | July | August |
|---|---|---|---|
| X |  | X |  |

| September | October | November | December |
|---|---|---|---|
| X X X |  |  | X |

*Figure 3.2* Clustering of birthdays

(b) A bag contains 12 socks, consisting of six pairs each having a different colour. You put your hand into the bag without looking inside, and take out a single sock. How many times must you do this before you can be sure of having a matching pair of socks? To solve this, identify the six colours as the 'boxes', and the socks as the 'objects'. To ensure that at least one box has two socks in it (that is there are two socks having the same colour), the PP shows that you must take out *seven* socks from the bag.

(c) This application is more subtle. Suppose there is a group of 12 people, some of whom are friends with each other. Then there will be two people who have the same number of friends in the group.

The verification of this statement proceeds as follows. To begin, notice that any one person has either $0, 1, 2, 3, \ldots$, or 11 friends. However, if someone has no friends in the group, then there cannot be anyone with 11 friends; conversely, if someone has 11 friends, then no one in the group is friendless. Therefore, if the 'boxes' are regarded as the numbers of friends a person has, both zero and 11 cannot be included simultaneously amongst the boxes. Hence there can only be a total of 11 boxes. These are to be occupied by 12 people, so by the PP at least two of them will be in the same box – in other words, at least two people will have the same number of friends.

(d) Perhaps surprisingly, results on properties of numbers can be derived using the PP. Suppose you select any eight numbers from the integers $1, 2, 3, \ldots$, 13. Then there will always be two integers amongst those you have chosen whose sum is 14. For example, if you select $1, 3, 5, 6, 7, 10, 11, 12$ then $3 + 11 = 14$. To prove that this always works, define the boxes as the following *seven* sets:

$$s_1 = \{1,13\}, \quad s_2 = \{2,12\}, \quad s_3 = \{3,11\}, \quad s_4 = \{4,10\},$$
$$s_5 = \{5,9\}, \quad s_6 = \{6,8\}, \quad s_7 = \{7\}$$

The 'objects' are the *eight* selected numbers, so by the PP at least one of the boxes will contain two numbers. This can't be $s_7$, so one of the two-number boxes $s_1$ to $s_6$ will be filled. The numbers occupying such a box add up to 14. For example, with the selection above,

$$1, 3, 5, 6, 7, 10, 11, 12 \longrightarrow \{1,13\}, \{2,12\}, \{3,11\}, \{4,10\}, \{5,9\}, \{6,8\}, \{7\}$$

where ✓ indicates the location of each selected number, with $s_3$ being filled. Notice that in this example the single-element box $s_7$ is also filled, but this is irrelevant since the PP states that at least one box contains *two* objects.

(e) As another example involving numbers, suppose you select any 51 numbers from the integers from 1 to 100. Then this selection will always contain two integers such that one divides the other (that is, without remainder). To prove this statement, take as the 'objects' the 51 chosen integers. To identify the 'boxes' needs a little care. If $a$ is a selected integer between 1 and 100, then by finding the highest power of 2 which divides $a$, it can be written as $a = 2^k b$, where $k$ is an integer such that $k = 0$ if $a$ is odd, and $k \geqslant 1$ if $a$ is even. The integer $b$ is odd, and so must be one of the 50 odd numbers $1, 3, 5, 7, \ldots, 99$. For example, if $a = 72 = 8 \times 9$ then $k = 3$ and $b = 9$. The boxes are now taken to be these 50 odd values of $b$. By the PP, two of the 51 selected numbers will occupy the same box – that is, have the *same* value of $b$. If these two particular integers are $a_1$ and $a_2$ then

$$a_1 = 2^k b, \quad a_2 = 2^p b, \quad k \neq p$$

so that $a_1$ divides $a_2$ if $k < p$, or $a_2$ divides $a_1$ if $k > p$.

(f) Another type of problem involves scheduling activities over a period of time. Suppose a personnel manager interviews at most 20 job applicants over a period of 14 consecutive days, with at least one candidate being interviewed each day. It is required to establish that no matter how the interviews are scheduled, there will be a period of consecutive days during which exactly seven candidates will be interviewed. To analyse this problem, let $x_i$ be the total number of interviews given by the end of the $i$th day, for $i = 1, 2, 3, \ldots, 14$. Since at least one interview takes place each day the running totals of interviews increase, that is,

$$x_1 < x_2 < x_3 < x_4 < \ldots < x_{13} < x_{14}$$

showing that all the $x_i$'s are distinct integers. Adding 7 throughout gives

$$x_1 + 7 < x_2 + 7 < x_3 + 7 < \ldots < x_{13} + 7 < x_{14} + 7$$

showing that $x_1 + 7, x_2 + 7, \ldots, x_{14} + 7$ are also distinct integers. Take as the 'objects' the 14 numbers $x_1, x_2, \ldots, x_{14}$ together with the 14 numbers $x_1 + 7, x_2 + 7, \ldots, x_{14} + 7$. However $x_1 \geqslant 1$, and since at most 20 interviews are held then $x_{14} \leqslant 20$, so that $x_{14} + 7 \leqslant 20 + 7 = 27$. In other words, the 28 integers $x_1, x_2, \ldots, x_{14} + 7$ can take (at most) the 27 integer

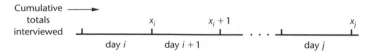

*Figure 3.3* Daily interviews

values between 1 and 27 inclusive. Identify as the 'boxes' these 27 integers. By the PP at least two of the objects must occupy the same box, that is, at least two of the 28 numbers must take the same value. Since the numbers $x_1, x_2, \ldots, x_{14}$ are all different, and the numbers $x_1 + 7, x_2 + 7, \ldots, x_{14} + 7$ are all different, the only way two of the 28 numbers can be equal is that there must be suffices $i, j$ such that $x_j = x_i + 7$. That is, $x_j - x_i = 7$ which means that from the start of day $i + 1$ to the end of day $j$, the personnel manager interviews exactly seven candidates, as shown in Figure 3.3

### 3.3.2 Extensions

In Problem 3.47 you are asked to investigate generalizations of the pigeonhole principle. In particular, suppose that $rn + 1$ (or more) objects are put into $n$ boxes. If no box contained more than $r$ objects then there would be at most $rn$ objects in total. Hence an *extended form* of the PP is that at least one box must contain $r + 1$ or more objects. The basic principle is the special case of this when $r = 1$.

Suppose $m_i (\geqslant 0)$ is the number of objects in box $i$, for $i = 1, 2, \ldots, n$, so $m_1 + m_2 + \ldots + m_n \geqslant rn + 1$. The *average* number of objects per box is the total number divided by $n$, that is,

$$\frac{m_1 + m_2 + \ldots + m_n}{n} \geqslant \frac{rn + 1}{n} = r + \frac{1}{n} > r$$

However, by the extended form of the PP, there is at least one integer $m_j$ such that $m_j \geqslant r + 1$. In other words:

Given any set of $n$ non-negative integers whose average is greater than some integer $r$, then at least one of the given integers is greater than or equal to $r + 1$.

This result is sometimes called the **averaging principle**.

**Example 3.14** **The case $n = 7$ and $r = 2$**

An illustration of these two generalizations is given in Figure 3.4. In (a) there are $15 = rn + 1$ objects (each represented by X) and one of the boxes contains 3 ($= r + 1$) objects, as stated by the extended form of the PP. Similarly, in Figure

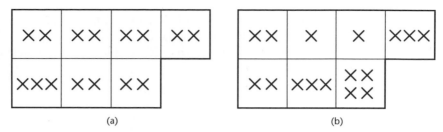

*Figure 3.4*   Distributions into seven boxes

3.4(b) there are 16 ($> rn + 1$) objects and three boxes contain three or more objects. Alternatively, think of an integer $p$ as being represented in Figure 3.4 by a box containing $p$ X's. In Figure 3.4(a) the seven integers are 2,2,2,2,3,2,2 and in Figure 3.4(b) they are 2,1,1,3,2,3,4. In each case the average of the integers is greater than 2 ($= r$), and there is at least one integer greater than or equal to 3 ($= r + 1$).

---

**Example 3.15**   **Applications of the averaging principle**

(a) Suppose the personnel manager in Example 3.13(f) interviews 26 applicants for another job over a period of five days. Let $m_i$ be the number of interviews held on day $i$, for $i = 1, 2, 3, 4, 5$. The average of these numbers is

$$\frac{m_1 + m_2 + m_3 + m_4 + m_5}{5} = \frac{26}{5} > 5$$

so that $r = 5$. By the averaging principle, there must be at least one day when at least six interviews are held.

In fact, it's also possible to deduce that there will be a period of two consecutive days when at least eight interviews are given. To verify this result, see what happens if you assume it does *not* hold. That is, assume that no two-day period contains more than *seven* interviews. Apply this to the first two days, then days three and four, and to the last day, to obtain

$$m_1 + m_2 \leqslant 7, \quad m_3 + m_4 \leqslant 7, \quad m_5 \leqslant 7$$

Adding together these inequalities gives

$$m_1 + m_2 + m_3 + m_4 + m_5 \leqslant 21$$

which is impossible since the sum of the $m_i$'s is 26. Therefore there must indeed be at least one two-day period containing at least eight interviews. However, there's no way of telling which period is involved.

(b) An easy application of the extended PP (see Exercise 3.80) produces the following alternative version of the averaging principle: if a set of non-negative integers has an average value $a$, then at least one of the integers will be greater than or equal to $a$. As an illustration of this, suppose that every member of a group of 460 first-year students must choose one mathematics module from six on offer. The average number of students per module is $460/6 = 76\frac{2}{3}$, so there will be at least one module taken by 77 or more students.

---

**Exercise**    **3.68**   A lecture theatre contains 30 students. Show that at least two of them have the same first initial to their name. How many students must there be to ensure that at least two have the same first two initials?

**Exercise**    **3.69**   A bag contains 24 socks, consisting of 12 pairs, each pair having a different pattern. How many socks must be taken out at random before getting a matched pair?

**Exercise**    **3.70**   A blindfolded person takes socks out of a drawer containing 12 black and 12 brown socks. How many must be taken out to be sure of getting

(a) at least one pair of matching socks?

(b) a pair of socks of each colour?

**Exercise**    **3.71**   If in the previous exercise the drawer contains 12 black, 12 brown and 12 grey socks, how many must be taken out to ensure

(a) a matching pair?

(b) a pair of socks of each colour?

**Exercise**    **3.72**   Suppose in Exercise 3.70 the drawer contains instead six pairs of black gloves and six pairs of brown gloves. How many gloves must be taken out to ensure a matching pair?

**Exercise**    **3.73**   Show that if any ten integers are selected from the set $\{1, 2, 3, 4, \ldots, 17\}$ then there will always be a pair of integers whose sum is 18.

**Exercise**    **3.74**   How many people must be selected at random from a telephone directory to ensure that at least two have

(a) a birthday on the same day and month of the year (excluding 29 February)?

(b) the same last two digits in their telephone number?

(c) the same year of birth? (assume age span is 18 to 120).

**Exercise**    **3.75**   Show that in any set of 17 positive integers, there will be two integers which have the same remainder after division by 16.

**Exercise**    **3.76**   An amateur photographer goes on a six-week holiday with 60 rolls of film, and is certain to use at least one roll every day. Show that there is a period of consecutive days in which the photographer shoots exactly 23 rolls of film.

**Exercise**   **3.77**   On a different one-week trip the amateur photographer uses 16 rolls of film. Show that there is at least one day on which the photographer uses at least three rolls of film. Show also that there is at least one three-day period in which at least six rolls are used.

**Exercise**   **3.78**   Show that in any group of 40 people there will be at least four of them whose birthday is in the same month.

**Exercise**   **3.79**   A compact disc has 11 tracks and a total playing time of 57 m 16 s. Show that there is at least one track which plays for longer than 5 m 13 s.

**Exercise**   **3.80**   Use the extended pigeonhole principle with $n$ integers whose sum is $rn + n$ to prove the alternative form of the averaging principle stated in Example 3.15(b). Also prove this result by a direct argument.

**Exercise**   **3.81**   A restaurant seats 110 people at 19 tables. Can you be sure that your party of six people (including yourself) can be accommodated at a single table?

## 3.4   Inclusion and exclusion

The idea developed in this section prevents 'double counting' in situations like the following.

**Example 3.16**   **University applicants**

Suppose that in a group of applicants for a degree course in business studies, 120 students have an A-level in mathematics, 135 have an A-level in statistics and 31 students have A-levels in both mathematics and statistics. An entry requirement for the course is that students must have an A-level in at least one of mathematics or statistics. How many of the applicants satisfy this requirement?

Notice that it is *not* correct simply to add together 120 and 135 to get 255, since those students holding A-levels in both subjects will be counted *twice*. A simple way of tackling this problem is to use a pictorial representation, as shown in Figure 3.5. The left-hand circle contains students having mathematics, and the right-hand circle those having statistics. The hatched area where the two circles intersect represents the 31 students who have both subjects. It is easy to see that the numbers in the three regions must be as shown in Figure 3.6. For example, there must be 89 students who have *only* mathematics, because together with the 31 who have mathematics and statistics this gives a total of 120 with mathematics. The total number who have at least one appropriate A-level is therefore $120 + 135 - 31 = 224$. The 31 students who have both subjects are *included* twice when the numbers in the two circles are added together, and so must be *excluded* from the overall total.

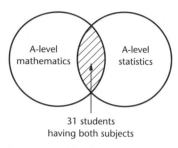

*Figure 3.5* A-levels

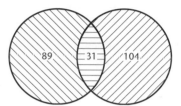

*Figure 3.6* Numbers per group

The method illustrated above is best described in general via the language of sets, and this is outlined first.

### 3.4.1 Sets and Venn diagrams

A **set** is a collection of objects which possess some well-defined property. For example, the set of all even positive numbers less than ten is $2, 4, 6, 8$. It's customary to denote a set by a capital letter, and to enclose the **elements** or **members** of the set within braces ('curly brackets'), so for this example

$$S_1 = \{2, 4, 6, 8\} \tag{3.22}$$

Another way to describe a set is to specify the defining properties of its members, so for example $S_1$ in (3.22) can be written as

$$S_1 = \{x | x \text{ is even and } 0 < x < 10\} \tag{3.23}$$

The vertical line means 'such that', so this characterization of $S_1$ states that it is the set of numbers $x$ such that $x$ is even, greater than zero and less than ten. The notations $a \in S_1$ and $b \notin S_1$ are used to indicate that $a$ is a member of $S_1$, but $b$ is not a member. If a set is not well defined, for example: 'the set of all fairly large even numbers', then it is called **fuzzy**. Such sets have proved very useful in developing 'computer intelligence', but are outside the scope of this book (for a non-technical introduction, see McNeill and Freiberger, 1994).

Some fairly standard notation is:

$\mathbb{Z}$ = set of all integers = $\{0, 1, -1, 2, -2, 3, -3, \ldots\}$
$\mathbb{Z}^+$ = set of positive integers = $\{1, 2, 3, 4, \ldots\}$
$\qquad\qquad\qquad = \{x | x \in \mathbb{Z}$ and $x > 0\}$
$\mathbb{N}$ = set of non-negative integers = $\{0, 1, 2, 3, 4, \ldots\}$
$\qquad\qquad\qquad = \{x | x \in \mathbb{Z}$ and $x \geqslant 0\}$
$\mathbb{R}$ = set of real numbers
$\mathbb{Q}$ = set of rational numbers = $\{x / y | x, y \in \mathbb{Z}, y \neq 0\}$
$\mathbb{Z}_m = \{0, 1, 2, 3, \ldots, m - 1\}$ subject to modulo $m$ arithmetic (see Section 2.4.4)
$\mathbb{C}$ = set of complex numbers = $\{x + iy | x, y \in \mathbb{R}, i^2 = -1\}$

For example, $S_1$ in (3.22) can be written as

$$S_1 = \{x | x \in \mathbb{Z}^+, x \text{ is even and } x < 10\} \qquad\qquad (3.24)$$

Notice that a set can have an infinite number of members (for example, $\mathbb{R}$). If every member of a set $A$ is also a member of a set $B$ then $A$ is said to be a **subset** of $B$, written as $A \subseteq B$. If however $B$ contains at least one element which is not in $A$ then $A$ is a **proper** subset of $B$, written as $A \subset B$. For example, the set $S_1$ in (3.22) and (3.23) satisfies $S_1 \subset \mathbb{Z}^+$; other examples are $\mathbb{Z}^+ \subset \mathbb{Z}$, $\mathbb{Q} \subset \mathbb{R}$ and $\mathbb{R} \subset \mathbb{C}$. Two sets are **equal** if they each have exactly the same members. For example, if

$$S_2 = \{x | x = 2m, m \in \mathbb{Z}^+ \text{ and } m < 5\}$$

then $S_2 = S_1$, where $S_1$ is the set in (3.22).

In Example 3.16 a useful pictorial representation of sets was introduced, whereby a set is regarded as the interior of a closed region in the plane. Circles are usually used, but any closed curve such as an ellipse, or even a rectangle, will do. Diagrams like Figures 3.5 and 3.6 are named after the nineteenth-century English mathematician **Venn**. Some further properties of sets are now conveniently defined using Venn diagrams. If $A$ and $B$ are any two sets, then their **intersection** $A \cap B$ is the set of elements belonging to both $A$ and $B$, as shown in Figure 3.7(a). The **union** $A \cup B$ is the set of elements belonging either to $A$ or to $B$ or to both, as shown in Figure 3.7(b).

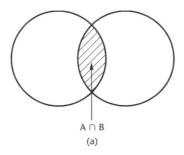

A ∩ B
(a)

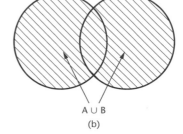

A ∪ B
(b)

*Figure 3.7* Intersection and union

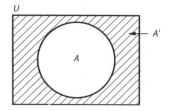

*Figure 3.8*  Complement of A

If $A$ and $B$ have no members in common they are said to be **disjoint**, and $A \cap B = \emptyset$, where $\emptyset$ is the **null** or **empty set** containing no elements. The **universal set** $U$ (or $E$) is the set of all elements relevant to the sets under consideration. The **complement** $A'$ or $A^c$ of a set $A$ is the set of elements which are in $U$ but **not** in $A$, as shown by the hatched region in Figure 3.8.

**Example 3.17**  **Properties of sets**

Consider the sets

$$A = \{x | x \in \mathbb{Z}^+, x < 7\}, \quad B = \{7, 9, 12\}, \quad C = \{6, 8, 10, 12\}$$

From the definition $A = \{1, 2, 3, 4, 5, 6\}$ and a suitable choice for the universal set is

$$U = \{x | x \in \mathbb{Z}^+, x < 13\} = \{1, 2, 3, 4, 5, 6, 7, 8, 9, 10, 11, 12\}$$

You can see that the following apply:

$$A \cap B = \emptyset, \quad A \cup B = \{1, 2, 3, 4, 5, 6, 7, 9, 12\}$$
$$A \cap C = \{6\}, \quad A \cup C = \{1, 2, 3, 4, 5, 6, 8, 10, 12\}$$
$$B \cap C = \{12\}, \quad B \cup C = \{6, 7, 8, 9, 10, 12\}$$
$$A' = \{7, 8, 9, 10, 11, 12\}, \quad B' = \{1, 2, 3, 4, 5, 6, 8, 10, 11\}$$
$$C' = \{1, 2, 3, 4, 5, 7, 9, 11\}$$

The definitions of union and intersection can be applied in an obvious way to more than two sets. For example,

$$A \cup B \cup C = (A \cup B) \cup C$$
$$= \{1, 2, 3, 4, 5, 6, 7, 8, 9, 10, 12\}$$

**Exercise**

**3.82**  List the elements of the following sets:

(a)  $A = \{x | x \in \mathbb{Z}^+, 5 < x < 12\}$

(b)  $B = \{x | x \in \mathbb{Z}, x^2 < 14\}$

(c)  $C = \{x | x \in \mathbb{N}, 2x + 1 < 0\}$.

**Exercise**    **3.83**    For each of the following, state whether $B$ is a subset of $A$
(a) $A = \{x|x \in \mathbb{Z}, |x| < 4\}$,   $B = \{1, 2, 3\}$
(b) $A = \{x|x \in \mathbb{Z}^+, x \text{ is even}\}$,   $B = \{2, 4, 6, 8, 9, 10\}$
(c) $A = \{x|x \in \mathbb{Z}_5\}$,   $B = \{1, 2\}$.

**Exercise**    **3.84**    Let $U = \{x|x \in \mathbb{Z}^+, x \leqslant 10\}$,   $A = \{1, 2, 3, 4\}$,   $B = \{5, 6, 7, 8, 9, 10\}$
$C = \{3, 4, 5\}$.
Determine
(a) $A \cap B$    (b) $A \cup B$    (c) $A'$    (d) $A \cap C$    (e) $B \cup C$    (f) $B' \cap (A \cup C)$.

**Exercise**    **3.85**    Is $\varnothing$ the same as $\{0\}$?

**Exercise**    **3.86**    If $A = \{2, 4, 6, 8\}$ and $B = \{1, 3, 5, 7\}$ give an example of two disjoint sets $C$ and $D$ for which $C \cup D = A \cup B$.

**Exercise**    **3.87**    Define 'tall' to mean over two metres in height, and 'old' to mean over 65 years. Let $A$ = set of all tall people, $B$ = set of all men, $C$ = set of all old people. Define a suitable universal set, and draw a Venn diagram identifying $A \cap C$, $B \cap C$, $A \cap B \cap C$.

**Exercise**    **3.88**    In a car rental fleet, 20 cars have a sunroof, 22 have air conditioning and five have both. Draw an appropriate Venn diagram and hence determine the total number of cars.

### 3.4.2    Inclusion–exclusion principle

A set $A$ is said to be **finite** if it contains a finite number of members, denoted by $n(A)$. If $A$ and $B$ are disjoint sets then it is clear from Figure 3.9 that

$$n(A \cup B) = n(A) + n(B) \tag{3.25}$$

where $A \cup B$ is the hatched region in the figure.

The result (3.25) is a statement in set notation of the **addition principle**, introduced in Exercise 3.8 where the term 'non-overlapping' was used instead of 'disjoint'. When $A$ and $B$ are not disjoint, then as in Example 3.16 you can see from Figure 3.7(b) that to count the number of elements in $A \cup B$ you add $n(A)$

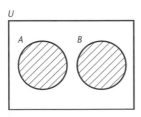

*Figure 3.9*   Union of disjoint sets

and $n(B)$, but then exclude from the total $n(A \cap B)$ because these elements have been counted twice. In formal terms,

$$n(A \cup B) = n(A) + n(B) - n(A \cap B) \tag{3.26}$$

and this is the **inclusion–exclusion (I–E) principle** when applied to two sets.

---

**Example 3.18**  **Applications of the principle**

(a)  Return to Example 3.16, where

$A$ = set of students having A-level mathematics
$B$ = set of students having A-level statistics

Then $n(A) = 120$, $n(B) = 135$ and $n(A \cap B) = 31$. The set $A \cup B$ consists of students having an A-level in at least one of the two subjects. Using (3.26), the required number of students is

$$n(A \cup B) = 120 + 135 - 31 = 224$$

(b)  In a survey of the newspaper-reading habits of 150 people it is found that 88 read the *Daily Record*, 60 read the *Daily Post* and 15 read neither paper. The problem is to find how many people read both newspapers. To solve this, identify

$U$ = set of 150 people surveyed
$A$ = set of *Daily Record* readers
$B$ = set of *Daily Post* readers.

Then $n(A) = 88$, $n(B) = 60$ and $n(U) = 150$. There are $150 - 15 = 135$ people who read at least one of the papers, that is, $n(A \cup B) = 135$. The number who read both papers is $n(A \cap B)$, and the I–E principle (3.26) gives

$$135 = 88 + 60 - n(A \cap B)$$

showing that $n(A \cap B) = 148 - 135 = 13$. The Venn diagram is given in Figure 3.10.

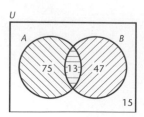

*Figure 3.10*  Newspaper readers

(c) Suppose it is required to count how many permutations of 123456 begin with 1 or end with 6 (or both). Define

$A$ = set of all permutations which begin with 1
$B$ = set of all permutations which end with 6.

Since the members of $A$ all begin with 1, the number of permutations of the remaining five digits is 5!, so $n(A) = 5!$ Using a similar argument, $n(B) = 5!$ The set $A \cap B$ consists of the permutations whose first digit is 1 *and* last digit is 6. Since there are four digits remaining, $n(A \cap B) = 4!$ The permutations to be counted are the members of the set $A \cup B$, and from (3.26)

$$n(A \cup B) = 5! + 5! - 4!$$
$$= 120 + 120 - 24 = 216$$

Another way of looking at the I–E principle for two sets is as follows. Suppose that in a collection of objects, each one may or may not possess one of two properties. Let $N(1)$ be the number of objects having property 1, and $N(2)$ the number having property 2. In the I–E principle (3.26) the first property is 'belonging to the set $A$', and the second property is 'belonging to the set $B$', so that $N(1) = n(A)$, $N(2) = n(B)$. If $N(1,2)$ denotes the number of objects possessing both properties, then an alternative way of regarding (3.26) is that the number of objects having at least one of the two properties is $N(1) + N(2) - N(1,2)$.

**Example 3.19   Divisibility of integers**

Suppose it is required to count how many positive integers between 1 and 50 inclusive are *not* divisible by 2 or 3. Take the first property to be divisibility by 2: the numbers satisfying this are the 25 even integers $2, 4, 6, \ldots, 48, 50$ so $N(1) = 25$. The second property is divisibility by 3, so the numbers in this case are the 16 integers $3, 6, 9, 12, \ldots, 45, 48$, and $N(2) = 16$. Finally, the numbers divisible by both 2 and 3, that is, by 6, are $6, 12, 18, \ldots, 48$, so that $N(1, 2) = 8$.

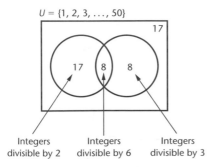

Integers divisible by 2    Integers divisible by 6    Integers divisible by 3

*Figure 3.11*   Divisibility by 2 and 3

Hence by the I–E principle the number of integers divisible by 2 or 3 (or both) is $25 + 16 - 8 = 33$. Since there are 50 numbers in total, the number of integers *not* divisible by either 2 or 3 is $50 - 33 = 17$. A Venn diagram representation is given in Figure 3.11.

Now turn to the situation where there are *three* sets which overlap as shown in Figure 3.12. When adding together the numbers of objects in $A$, $B$ and $C$, those in the dark region in Figure 3.12(a) will be counted three times, whereas those objects in the hatched regions will be counted twice. To demonstrate this, let the numbers of elements in the various regions be as shown in Figure 3.12(b). You can see that $n(A \cap B \cap C) = n_7$ and

$$n(A \cap B) = n_4 + n_7, \quad n(A \cap C) = n_5 + n_7, \quad n(B \cap C) = n_6 + n_7,$$
$$n(A) = n_1 + n_4 + n_5 + n_7, \quad n(B) = n_2 + n_4 + n_6 + n_7,$$
$$n(C) = n_3 + n_5 + n_6 + n_7$$

Adding together the last three expressions gives

$$n(A) + n(B) + n(C) = n_1 + n_2 + n_3 + 2(n_4 + n_5 + n_6) + 3n_7$$

The numbers $n_4$, $n_5$, $n_6$ occupy the hatched areas in Figure 3.12(a) and $n_7$ is the number in the dark area. You can also see from Figure 3.12(b) that

$$n(A \cup B \cup C) = n_1 + n_2 + n_3 + n_4 + n_5 + n_6 + n_7$$

By substituting the above expressions for the constituent parts you can verify that the following is correct:

$$n(A \cup B \cup C) = n(A) + n(B) + n(C)$$
$$- [n(A \cap B) + n(A \cap C) + n(B \cap C)] + n(A \cap B \cap C) \quad \text{(3.27)}$$

The expression (3.27) is the I–E principle for three sets: you *include* the numbers in the three individual sets, then *exclude* the numbers in the intersections of the pairs of sets, and finally *include* the number in the intersection of the three sets.

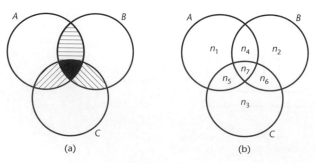

(a)                    (b)

*Figure 3.12*   Three overlapping sets

The alternative formulation of (3.27) now involves a collection of objects and *three* properties which each object may or may not possess. Define

$N(i)$ = number of objects having property $i$, for $i = 1, 2, 3$
$N(i, j)$ = number of objects possessing properties $i$ and $j$, for $i \neq j$
$N(1, 2, 3)$ = number of objects possessing all three properties

Then the number of objects having at least one of the three properties is

$$N(1) + N(2) + N(3) - [N(1, 2) + N(1, 3) + N(2, 3)] + N(1, 2, 3) \qquad \textbf{(3.28)}$$

**Example 3.20**  **The I–E principle for three sets**

(a) The annual MOT test in Britain for cars which are three or more years old includes checks on tyres, brakes and exhaust emissions. In a group of 66 cars tested by a garage, 15 had faulty tyres, 21 had faulty brakes and 18 exceeded the allowable emission limits. Also, 4 cars had faulty tyres and brakes, 6 failed on tyres and emissions, 9 failed on brakes and emissions, and 4 cars were unsatisfactory in all three respects. How many cars had no faults in these three checks?

To answer this question, let

$A$ = set of cars with faulty tyres
$B$ = set of cars with faulty brakes
$C$ = set of cars with unsatisfactory emissions.

The given information is

$n(A) = 15, \quad n(B) = 21, \quad n(C) = 18,$
$n(A \cap B) = 4, \quad n(A \cap C) = 6, \quad n(B \cap C) = 9,$
$n(A \cap B \cap C) = 4.$

From the I–E principle (3.27) the number of cars having at least one of the three faults is

$$n(A \cup B \cup C) = 15 + 21 + 18 - (4 + 6 + 9) + 4$$
$$= 39$$

Since $U$ is the set of all cars tested, $n(U) = 66$ so the number of cars with none of the three faults is $66 - 39 = 27$. The Venn diagram for this problem is given in Figure 3.13.

(b) Return to the problem of divisibility of integers in Example 3.19, and now compute how many positive integers between 1 and 50 (inclusive) are *not* divisible by 2, 3 or 5. It was found earlier that the numbers of integers divisible by 2 or 3 or both are respectively $N(1) = 25$, $N(2) = 16$, $N(1, 2) = 8$. The third property is divisibility by 5; this is satisfied by the numbers 5, 10, 15, ..., 45, 50 so $N(3) = 10$. The integers divisible by 2 and 5 are 10, 20, ..., 50 so $N(1, 3) = 5$; and the numbers divisible by 3 and 5 are

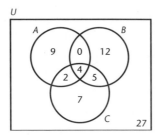

Figure 3.13 Car faults

15, 30, 45 so $N(2,3) = 3$. Finally, only 30 is divisible by 2, 3 and 5, so $N(1,2,3) = 1$. The I–E principle (3.28) states that the number of integers divisible by at least one of 2, 3 or 5 is

$$25 + 16 + 10 - (8 + 5 + 3) + 1 = 36$$

Since there are 50 integers altogether, the number of integers *not* divisible by 2, 3 or 5 is $50 - 36 = 14$.

The I–E principle can be extended to any number of sets $A_1, A_2, \ldots, A_m$ as follows:

$$\begin{aligned}
n(A_1 \cup A_2 \cup A_3 \cup \ldots \cup A_m) = & [n(A_1) + n(A_2) + \ldots + n(A_m)] \\
& - [n(A_1 \cap A_2) + n(A_1 \cap A_3) + \ldots + n(A_{m-1} \cap A_m)] \\
& + [n(A_1 \cap A_2 \cap A_3) + \ldots + n(A_{m-2} \cap A_{m-1} \cap A_m)] \\
& - \ldots + (-1)^{m-1} n(A_1 \cap A_2 \cap \ldots \cap A_m)
\end{aligned} \tag{3.29}$$

On the right-hand side in (3.29) the first square bracket contains the number of elements of sets taken one-at-a-time, the next contains sets two-at-a-time, then sets three-at-a-time, and so on, and the signs alternate. When $m = 2$ or 3, then (3.29) reduces to (3.26) and (3.27) respectively. The alternative form of the principle relating to $m$ properties states that the number of objects having at least one of the $m$ properties is

$$\begin{aligned}
& [N(1) + N(2) + \ldots + N(m)] - [N(1,2) + N(1,3) + \ldots + N(m-1,m)] \\
& + [N(1,2,3) + \ldots + N(m-2, m-1, m)] \\
& - \ldots + (-1)^{m-1} N(1,2,\ldots,m)
\end{aligned} \tag{3.30}$$

where $N(i_1, i_2, \ldots, i_p)$ is the number of objects possessing the $p$ properties $i_1, i_2, \ldots, i_p$. For example, when $m = 3$ then (3.30) reduces to (3.28).

**Example 3.21    Restricted integer solutions of equations**

Using the result given in Example 3.10 the number of solutions of the equation

$$x_1 + x_2 + x_3 + x_4 = 20 \tag{3.31}$$

where each $x_i$ is a non-negative integer, is $C(20 + 4 - 1, 20) = C(23, 20) = 1771$. Suppose it is required to determine how many of these solutions satisfy the additional restriction that each of the $x$'s is less than 9, that is, $x_i \leqslant 8$ for $i = 1, 2, 3, 4$. Without this condition, each of the $x$'s can be as large as 20 (when for example the solution of (3.31) has the form $x_1 = 20$, $x_2 = x_3 = x_4 = 0$, and so on).

To solve the problem, first use (3.30) with $m = 4$ to find the number of solutions which do *not* satisfy the required restriction. Define property 1 as the condition that a solution of (3.31) satisfies $x_1 > 8$, so that $N(1)$ is the number of non-negative integer solutions of (3.31) which have $x_1 > 8$. Rewrite (3.31) as

$$(x_1 - 9) + x_2 + x_3 + x_4 = 11$$

or

$$y + x_2 + x_3 + x_4 = 11 \tag{3.32}$$

where $y = x_1 - 9$. Then $N(1)$ is the number of non-negative solutions of (3.32), since $y \geqslant 0$ is equivalent to $x_1 \geqslant 9$ (that is, $x_1 > 8$). Again appealing to the result in Example 3.10, the number of such integer solutions of (3.32) is $C(11 + 4 - 1, 11) = 364$. An identical argument applies by defining property $i$ as the condition that a solution of (3.31) satisfies $x_i > 8$ for $i = 2, 3, 4$, so that $N(2) = N(3) = N(4) = 364$. The next step in using (3.30) is to compute the six numbers $N(1, 2)$, $N(1, 3)$, $N(1, 4)$, $N(2, 3)$, $N(2, 4)$, $N(3, 4)$. For example, $N(1, 2)$ is the number of solutions of (3.31) satisfying $x_1 > 8$, $x_2 > 8$. Rewrite (3.31) as

$$(x_1 - 9) + (x_2 - 9) + x_3 + x_4 = 2 \tag{3.33}$$

or equivalently $y_1 + y_2 + x_3 + x_4 = 2$. You can see from the same argument as before that the number of non-negative solutions of (3.33) is $N(1, 2) = C(2 + 4 - 1, 2) = 10$. Exactly the same result applies for the other five cases when the solution satisfies two of the properties, so $N(i, j) = 10$ in each case. Turning to the situation when three properties must be satisfied, you can see this is impossible: if any three of the $x$'s in (3.31) are greater than 8 then the left side is greater than 24, so (3.31) cannot be satisfied. Hence $N(i, j, k) = 0$ in all cases, and similarly $N(1, 2, 3, 4) = 0$ since if all $x_i > 8$ then the left side of (3.31) exceeds 32. It now follows from (3.30) that the number of solutions of (3.31) satisfying *at least one* of the conditions $x_i > 8$ is

$$N(1) + N(2) + N(3) + N(4) - [N(1, 2) + N(1, 3) + N(1, 4) + N(2, 3)$$
$$+ N(2, 4) + N(3, 4)]$$
$$= 4 \times 364 - 6 \times 10$$
$$= 1396$$

Finally, the problem as stated initially requires that *every* $x_i$ satisfies $x_i \leqslant 8$, which is equivalent to having *none* of conditions $x_i > 8$ satisfied (that is, *none* of the properties 1, 2, 3, 4 can hold). Since the total number of non-negative integer solutions of (3.31) is 1771, it follows that the number of integer solutions satisfying $0 \leqslant x_i \leqslant 8$ for each $i = 1, 2, 3, 4$ is $1771 - 1396 = 375$.

| Example 3.22 | Derangements |

A **derangement** is a permutation of $n$ ordered objects in which no object appears in its original position. For example, 4321 is a derangement of 1234 but 3241 is not.

To count how many derangements there are, define property $i$ as a permutation of the $n$ objects in which the $i$th object *does* appear in the $i$th position, for $i = 1, 2, \ldots, n$. The total number of permutations is $n!$, so the number of derangements of $n$ objects is

$$d_n = n! - \text{(number of permutations satisfying at least one of the } n \text{ properties)}$$

(3.34)

The number within brackets in (3.34) is computed using (3.30), as follows.

The number of permutations in which object $i$ remains in the $i$th position is

$$N(i) = (n - 1)!$$

since the remaining $n - 1$ objects can appear in any positions. If two objects $i_1$ and $i_2$ remain fixed in their original positions then the remaining $n - 2$ objects can be permuted to give $N(i_1, i_2) = (n - 2)!$. Similarly

$$N(i_1, i_2, i_3) = (n - 3)!, \quad N(i_1, i_2, i_3, i_4) = (n - 4)!$$

and so on, ending up with $N(1, 2, 3, \ldots, n) = 1$. The number of terms of type $N(i)$ in the first summation in (3.30) is the number of ways of choosing one fixed object, that is, $C(n, 1)$; the number of terms $N(i_1, i_2)$ in the second summation is the number of ways of choosing two fixed objects, that is, $C(n, 2)$; and so on. Hence the number of derangements is given by (3.30) as

$$d_n = n! - [C(n, 1)(n - 1)! - C(n, 2)(n - 2)! + C(n, 3)(n - 3)! + \ldots + (-1)^{n-1}]$$

(3.35)

This can be simplified to give the expression in Exercise 3.99.

| Exercise | **3.89** In a group of students, 73 have a personal computer, 125 have a personal stereo and 41 have both. How many own at least one of these devices? Draw an appropriate Venn diagram. |

| Exercise | **3.90** In a survey of the cola-drinking preferences of 200 people it is found that 121 like Brand $C$, 110 like Brand $P$ and 15 do not like either brand. How many like both brands? |

| Exercise | **3.91** How many permutations of 123456789 begin with 9 or end with 2 (or both)? |

**Exercise**  **3.92**  How many positive integers between 1 and 100 inclusive are not divisible by 3 or 5?

**Exercise**  **3.93**  A school wishes to attract science students. It proudly announces that in the previous academic year 115 students gained at least one A-level pass in Mathematics (*M*), Physics (*P*) or Chemistry (*C*), the specific numbers of passes being as follows:

| M | P | C | M + P | M + C | P + C | M + P + C |
|---|---|---|-------|-------|-------|-----------|
| 52 | 65 | 53 | 25 | 16 | 31 | 12 |

Has the school presented the results correctly?

**Exercise**  **3.94**  How many of the integers satisfying the conditions in Exercise 3.92 are divisible by 7?

**Exercise**  **3.95**  How many integers between 1 and 1000 inclusive are relatively prime to 1001?

**Exercise**  **3.96**  In how many ways can 50 identical chairs be distributed between three different conference rooms in a hotel? In how many of these distributions are there no more than half of the chairs in any one room?

**Exercise**  **3.97**  How many of the solutions of the equation in Exercise 3.51 satisfy the conditions $x_1 \leqslant 7$, $x_2 \leqslant 7$, $x_3 \leqslant 7$, $x_4 \leqslant 7$?

**Exercise**  **3.98**  Use an appropriate Venn diagram to prove (3.29) when $m = 4$.

**Exercise**  **3.99**  Show that the expression in (3.35) simplifies to

$$d_n = n! \left[ 1 - \frac{1}{1!} + \frac{1}{2!} - \frac{1}{3!} + \ldots + (-1)^n \frac{1}{n!} \right]$$

**Exercise**  **3.100**  A new managing director of a company wants to revitalize his team of four divisional managers. He decides that the managers will swap jobs every week, with no one doing their original job. For how many weeks can this scheme continue without repeating an allocation of jobs?

**Exercise**  **3.101**  Consider sequences $x_1, x_2, x_3, x_4, x_5$, where each element $x_i$ is a ternary number (that is, to base 3).

(a) How many different sequences are there?

(b) How many sequences contain no zeros?

(c) Use the inclusion–exclusion principle to find how many sequences contain at least one 0, at least one 1 and at least one 2.

**Exercise** | **3.102**  A standard expression is

$$e^{-1} = 1 - \frac{1}{1!} + \frac{1}{2!} - \frac{1}{3!} + \frac{1}{4!} - \dots$$

(compare with the formula for e in Exercise 1.25). Compute the sum of the first eight terms in this series, and compare it with the value of $e^{-1}$ correct to four decimal places. Hence deduce that $d_n \approx n!\,e^{-1}$ for $n \geqslant 7$, where $d_n$ is defined in Exercise 3.99.

**Exercise** | **3.103**  Research papers are accepted for publication by the *Journal of Interesting Results* if favourable reviews are obtained from two different expert referees. The editor receives six different articles (two copies of each) for possible publication. These are sent out to six reviewers who each get two different papers. In how many ways can this be done?

## 3.5  Probability

Expressions of uncertainty such as 'it's likely to rain later'; 'that horse has an even chance of winning'; 'smoking is likely to cause lung cancer'; are commonplace in everyday language. What does it mean when a weather forecaster says there is a 30% chance of rain tomorrow; or when you're told that the odds against winning the first prize in the National Lottery are 14 million to one?

Mathematical investigations of probability were begun in the sixteenth century by the Italian mathematician Cardano, and continued by Pascal, Fermat and others. The motivation was the age-old one of how to win in gambling games. A central idea is that of regularity in a random event. For example, if you toss a coin then the outcome is random in the sense that you cannot predict in advance whether it will come to rest with the 'head' or 'tail' side uppermost. However, if you toss an unbiased coin a large number of times (say more than 5000) then you would expect that the number of heads as a proportion of the total will be close to one-half, so there is a predictable long-term pattern. Instinctively you assume that the coin is 'very unlikely' to fall onto its edge. The aim of this section is to show how these concepts can be made more precise.

### 3.5.1  Basic ideas

One way of defining probability without going into a formal rigorous treatment using so-called 'sample spaces' is to think of a large number $N$ of repetitions of a random event (such as tossing an unbiased coin). Suppose that a certain selected event occurs $r$ times (for example, the coin coming up heads). Then if the ratio $r/N$ approaches a definite value $p$ as $N$ gets larger and larger, the

**probability** of the selected event is $p$. For the coin-tossing example, the probability of obtaining a head according to this definition is therefore $p = 1/2$. If the event never happens (that is, $r = 0$) then $p = 0$, and if the event always happens (that is, $r = n$) then $p = 1$. Thus $p$ is a number between 0 (representing impossibility) and 1 (representing certainty). Probabilities are also commonly expressed as percentages, ranging from 0% to 100%, by multiplying $p$ by 100.

An alternative definition of probability uses a theoretical rather than experimental line of reasoning. Suppose out of a total of $n$ equally likely events there are $m$ in which a selected event occurs; then the probability of this selected event is $m/n$. This approach can be used to compute probabilities, since the values of $m$ and $n$ can be found using the counting techniques developed in this chapter.

| Example 3.23 | **Evaluating probabilities** |
|---|---|

(a) If an unbiased coin is tossed twice, then there are four equally likely outcomes:

$$HH, HT, TH, TT$$

where $H$ denotes head, $T$ tail, and $HT$ denotes $H$ on the first toss and $T$ on the second, and so on. Out of these four possibilities there is only one which gives two heads. Hence the probability of getting two heads is $1/4$, or 25%.

(b) If an unbiased die is tossed, then each of the six faces bearing 1, 2, 3, 4, 5 or 6 dots is equally likely to come up. What is the probability of throwing a five? There is only *one* way in which the five comes up, so the probability of obtaining a five with a single throw of the die is $1/6$.

Suppose the die is tossed twice. What is the probability of getting a five on each occasion? To answer this, first count the total number of ways in which the die can come up. There are six possible outcomes on each throw, so by the multiplication principle (Section 3.1.1) there are a total of $6 \times 6 = 36$ ways altogether. There is only one way of getting five on each of the throws, so the probability is $1/36$.

(c) For the 3-digit student identity number described in Example 3.1(b), what is the probability that all the digits are the same? It was found that there are 900 different identity numbers. There are nine cases when the digits are identical ($111, 222, \ldots, 999$, since 000 is not allowed). Hence the required probability is $9/900 = 1/100$.

(d) Return to the selection of a committee of three people from 35, discussed in Example 3.6. Suppose there are 12 women and 23 men. What is the probability that the committee will contain one woman? It was found that the total number of committees is 6545. If the committee contains one woman, she can be chosen in $C(12, 1) = 12$ ways. The two men on the committee can be chosen in $C(23, 2) = 253$ ways, so by the multiplication principle the number of committees which include exactly one woman is $12 \times 253 = 3036$. Hence the required probability is $3036/6545 = 0.464$.

(e) Return to the problem of drawing four balls out of a box containing seven white and four black balls, described in Example 3.7(c). The total number of ways of choosing four balls is $C(11, 4) = 330$. From the results obtained in the earlier example it follows that (i) the probability that exactly two of the four balls selected will be white is $126/330 = 0.382$, (ii) the probability that at least three of the selected balls will be white is $175/330 = 0.531$.

(f) Two cards are chosen at random from a well-shuffled pack of standard playing cards. What is the probability that neither of the two cards selected is an ace? The total number of ways of selecting two cards is $C(52, 2)$. In order for neither card to be an ace, those selected must come from the 48 cards remaining after aces are excluded, so the number of outcomes in this case is $C(48, 2)$. Hence the probability that neither card is an ace is

$$\frac{C(48, 2)}{C(52, 2)} = \frac{48 \times 47}{52 \times 51} = 0.851$$

What is the probability that at least one of the two selected cards is an ace? There are two cases to consider: firstly, when one card is an ace, and the other is not. There are four aces to choose from, so this can happen in $C(4, 1) \times C(48, 1) = 192$ ways. Secondly, there are $C(4, 2) = 6$ ways in which both cards are an ace. By the addition principle, the total number of ways of having at least one ace is therefore $192 + 6 = 198$, and the probability of this is therefore $198/C(52, 2) = 0.149$. However, there is an easier way of obtaining this result, using the following argument.

If the probability that a certain event $E$ occurs is denoted by $P(E) = p$, then the probability that $E$ does not occur is $1 - p$. This follows from the second version of the probability definition: the number of occasions when $E$ occurs is $m$, so the number of occasions when $E$ does *not* occur is $n - m$. Hence the probability that $E$ does not occur is $(n - m)/n = 1 - m/n = 1 - p$. Borrowing from the language of sets in Section 3.4.1, it is convenient to denote the non-occurrence of $E$ by $E'$, the **complement** of $E$, so that

$$P(E') = 1 - P(E)$$

Applying this to the card problem, $E$ is the event: 'having no aces', and $P(E) = 0.851$. The event $E'$ is 'having either one or two aces', and $P(E') = 1 - 0.851 = 0.149$, as found above.

A common way of giving the probability of an event is to state the **odds** on or against the event. This means that if $p = m/n > 1/2$ is the probability of an event happening, then the **odds on** (or, in favour of) the event are $m$ to $n - m$ since it occurs $m$ times, and does not occur $n - m$ times. Similarly if $p < 1/2$, the **odds against** are $n - m$ to $m$; when $p = 1/2$, the odds are said to be **even**.

**Example 3.24** **Odds**

In part (a) of Example 3.23, the probability of getting two heads with two tosses of a coin is $1/4$, so the odds against this happening are 3 to 1.

In part (e) of Example 3.23, the probability that at least three of the four selected balls are white is $175/330 = 35/66$. Hence the odds that this occurs are 35 to 31 on.

---

If two events $E_1$ and $E_2$ cannot occur together they are called **mutually exclusive**. Let $P(E_1)$ be the probability that $E_1$ occurs, and $P(E_2)$ the probability of $E_2$. If $E_1$ occurs in $m_1$ ways, and $E_2$ in $m_2$ ways, from a total of $n$ outcomes, then $P(E_1) = m_1/n$, $P(E_2) = m_2/n$. Again borrowing from the language of sets, in this case the **intersection** of $E_1$ and $E_2$ satisfies $E_1 \cap E_2 = \varnothing$, where $\varnothing$ is the null event (that is, nothing occurs). The **union** $E_1 \cup E_2$ is the occurrence of either $E_1$ or $E_2$. The addition principle (first encountered in Exercise 3.8) shows that $E_1 \cup E_2$ occurs in $m_1 + m_2$ ways, so

$$P(E_1 \cup E_2) = (m_1 + m_2)/n = P(E_1) + P(E_2) = P(E_2 \cup E_1) \tag{3.36}$$

That is, the probability of *either* $E_1$ *or* $E_2$ occurring is the *sum* of the individual probabilities. The result extends to any number of events $E_1$, $E_2$, ..., $E_k$ which are pairwise mutually exclusive, that is, $E_i \cap E_j = \varnothing$ for all $i \neq j$: the **addition rule for probability** states that the probability of either $E_1$ or $E_2$ or ... or $E_k$ is

$$P(E_1 \cup E_2 \cup \ldots \cup E_k) = P(E_1) + P(E_2) + \ldots + P(E_k) \tag{3.37}$$

**Example 3.25** **Selecting cards**

Two cards are selected from a well-shuffled pack of playing cards. What is the probability of obtaining either two kings or two spades?

As in part (f) of Example 3.23, the total number of ways of choosing two cards is $C(52, 2)$. There are $C(4, 2)$ ways of choosing two kings, so the probability of this occurring is

$$\frac{C(4, 2)}{C(52, 2)} = \frac{4 \times 3}{52 \times 51} = \frac{1}{221}$$

There are 13 spades, so the probability of selecting two spades is

$$\frac{C(13, 2)}{C(52, 2)} = \frac{13 \times 12}{52 \times 51} = \frac{13}{221}$$

You cannot get two kings and two spades simultaneously (there is only one king of spades) so by the addition rule (3.36) the probability of either two kings or two spades is

$$\frac{1 + 13}{221} = 0.063$$

If the occurrence or non-occurrence of an event $E_1$ has no effect on the probability of an event $E_2$, and conversely, then $E_1$ and $E_2$ are said to be **independent** of each other. In this case if $E_1$ occurs in $m_1$ ways out of a total of $N_1$ outcomes, and $E_2$ in $m_2$ ways out of $N_2$, then by the multiplication principle in Section 3.1.1, the number of ways in which both $E_1$ and $E_2$ occur is $m_1 m_2$ out of a total of $N_1 N_2$ outcomes. Hence the probability that *both* $E_1$ and $E_2$ occur is

$$P(E_1 \cap E_2) = \frac{m_1 m_2}{N_1 N_2} = \left(\frac{m_1}{N_1}\right)\left(\frac{m_2}{N_2}\right) = P(E_1)\,P(E_2) = P(E_2 \cap E_1)$$

Similarly, if $E_1$, $E_2$, ..., $E_k$ are events independent of each other then the **multiplication rule for probability** states that the probability of $E_1$ and $E_2$ and ... $E_k$ occurring is

$$P(E_1 \cap E_2 \cap E_3 \cap \ldots \cap E_k) = P(E_1)\,P(E_2)\,P(E_3)\ldots P(E_k) \qquad \text{(3.38)}$$

**Example 3.26**  **Applications of (3.38)**

(a) One card is drawn from each of two packs of well-shuffled playing cards. What is the probability that both cards are aces? In each case the probability of drawing an ace is $4/52 = 1/13$ since there are 52 ways of drawing one card, and four ways of drawing an ace. Clearly, selecting an ace from one pack does not affect the selection from the second pack, so by (3.38) the probability that both cards are aces is $(1/13) \times (1/13) = 1/169$.

(b) Three unbiased coins are each tossed once. What is the probability that all three show tails? The probability that each coin shows a tail is $1/2$. Again, the occurrence of a tail for any one coin has no affect on what the other coins show, so by (3.38) the probability that all three come up tails is $(1/2) \times (1/2) \times (1/2) = 1/8$.

(c) A computer system consists of three independent subsystems, each of which has a one per cent probability of failing in a one-month period. The system fails if any one of the subsystems fails. What is the probability that the system breaks down during one month?

To answer this question, it's easier to first determine the probability that the system does *not* break down, that is, that none of the subsystems breaks down. The probability that any single subsystem fails is 0.01, so the probability that it does not fail is $1 - 0.01 = 0.99$. Since the subsystems are assumed to operate independently, by (3.38) the probability that all three do not fail is $0.99 \times 0.99 \times 0.99 = 0.9703$, and this is the probability that the complete system does *not* fail. Hence the probability of a system breakdown is $1 - 0.9703 = 0.0297$, or 2.97%.

To find the probability of system breakdown directly you have to consider the cases of one, two or three subsystem breakdowns (see Exercise 3.116).

**Example 3.27**    **Use of the binomial theorem**

Recall from part (f) of Example 3.23 that if the probability that an event occurs is $p$, then the probability that it does not occur is $q = 1 - p$. The probability that an event occurs $r$ times in $n$ independent trials is by the multiplication rule $p^r q^{n-r} C(n, r)$, since there are $C(n, r)$ ways of selecting $r$ successful outcomes from a total of $n$, the probability that the event occurs $r$ times is $p^r$, and the probability that the event does not occur $n - r$ times is $q^{n-r}$. By referring to (3.17) you can see that this is just the general term in the expansion of $(p + q)^n$ using the binomial theorem, which gives

$$(p + q)^n = p^n + p^{n-1} q C(n, 1) + p^{n-2} q^2 C(n, 2) + \ldots$$

The first term on the right-hand side gives the probability that the event occurs $n$ times, the second term that it occurs $n - 1$ times, the third that it occurs $n - 2$ times, and so on. For example, if an unbiased coin is tossed five times, then the probability of a head is $p = 1/2$, and of a tail is $q = 1 - p = 1/2$, so that

$$(p + q)^5 = \left(\tfrac{1}{2}\right)^5 + \left(\tfrac{1}{2}\right)^4 \left(\tfrac{1}{2}\right) C(5, 1) + \left(\tfrac{1}{2}\right)^3 \left(\tfrac{1}{2}\right)^2 C(5, 2)$$
$$+ \left(\tfrac{1}{2}\right)^2 \left(\tfrac{1}{2}\right)^3 C(5, 3) + \left(\tfrac{1}{2}\right) \left(\tfrac{1}{2}\right)^4 C(5, 4) + \left(\tfrac{1}{2}\right)^5$$

The probability of, say, two heads is the fourth term

$$\left(\tfrac{1}{2}\right)^2 \left(\tfrac{1}{2}\right)^3 C(5, 3) = \tfrac{5}{16}$$

Similarly, the probability of at least two heads is the sum of the first four terms in the series.

This application of the binomial theorem is an extension of that given in part (b) of Example 3.12 on page 144.

**Example 3.28**    **An apparent paradox**

Suppose that you play a game in which your friend tosses a six-sided die and you must guess what comes up. If your friend throws the die seven times, then on each occasion your probability of being correct is $1/6$. Applying the addition rule (3.37), your probability of guessing correctly on at least one throw seems to be $7 \times (1/6)$, which is greater than 1 – an impossibility, since a probability of 1 represents certainty! Clearly, it's by no means certain that you will make a correct guess during seven (or more) throws. What has gone wrong?

The reason an apparent paradox has arisen is because the seven events are not mutually exclusive: guessing correctly on one throw does not preclude guessing correctly on another occasion. Hence the addition rule (3.37) is not applicable.

The correct way to tackle this problem is to follow the argument introduced in part (c) of Example 3.26 and consider the probability of *not* making a correct guess. On any individual throw this is 5/6. Since the throws are independent of each other, the multiplication rule (3.38) applies, so that the probability of *not* guessing correctly for any of the seven throws is $(5/6)^7 = 0.279$. Hence the probability of guessing correctly at least once in the seven throws is $1 - 0.279 = 0.721$.

In view of Example 3.28 it's worthwhile reiterating when it's valid to add or to multiply probabilities. If two events cannot happen together (that is, are mutually exclusive) then the probability that *either* one *or* the other occurs is the *sum* of the individual probabilities. For example, with a six-sided die the probability of getting either a five or a six on a single throw is $(1/6) + (1/6) = 1/3$.

On the other hand, if two events occur independently of each other, then the probability that they *both* occur is the *product* of the individual probabilities. For example, the probability of getting a six with each of two throws of a die is $(1/6) \times (1/6) = 1/36$.

However, other situations can arise. If two events $E_1$ and $E_2$ are *not* mutually exclusive, this means that they can occur together. In this case the union $E_1 \cup E_2$ means that at least one of $E_1$ or $E_2$ occurs (that is, $E_1$, or $E_2$, or both). The number of ways in which this happens is given by the inclusion–exclusion formula (3.26). Translated into probabilities, this becomes

$$P(E_1 \cup E_2) = P(E_1) + P(E_2) - P(E_1 \cap E_2) \qquad\qquad (3.39)$$

When $E_1$ and $E_2$ *are* mutually exclusive, then $P(E_1 \cap E_2) = 0$ and (3.39) reduces to (3.36).

**Example 3.29**  **Newspaper readers**

Return to the survey of 150 people described in Example 3.18(b), in which 88 people read the *Daily Record*, 60 read the *Daily Post* and 13 read both papers. What is the probability that a person selected at random from the group reads at least one of the newspapers?

The probability of reading the *Record* is 88/150, the probability of reading the *Post* is 60/150, and the probability of reading both papers is 13/150. Hence by (3.39) the probability that a person reads at least one of the newspapers is

$$\tfrac{88}{150} + \tfrac{60}{150} - \tfrac{13}{150} = \tfrac{135}{150} = 0.9$$

A Venn diagram for this problem was given in Figure 3.10.

Extension of (3.39) to more than two events has the same form as (3.27) and (3.28) for three events, and (3.29) and (3.30) for the general case. For example, with three events

$$P(E_1 \cup E_2 \cup E_3) = P(E_1) + P(E_2) + P(E_3)$$
$$- [P(E_1 \cap E_2) + P(E_1 \cap E_3) + P(E_2 \cap E_3)]$$
$$+ P(E_1 \cap E_2 \cap E_3)$$

Many students find the notion of probability quite difficult to grasp initially. One widely believed misconception is the mythical 'Law of Averages'. The usual reasoning goes like this: suppose you toss an unbiased coin ten times, and somewhat to your surprise it comes up heads every time. Then because you expect 'on average' to get an equal number of heads and tails the 'law' states that on the next few tosses you should get more tails than heads so as to even out the totals. This way of thinking about probability is completely WRONG! Each toss of a coin is independent of all previous ones, so the probability of getting a tail on the eleventh (and subsequent tosses) remains one-half. What matters is the *proportion* of heads after an extremely *large number* of tosses, and this is very close to one-half. The concept of probability is that randomness has a regularity in the long run, but *not* necessarily in the short run. However, if you toss a coin 200 times and get heads every time, then you may well suspect that the coin is biased!

Finally, return to a question posed in the introduction in Section 3.5. What does it mean when a weather forecast predicts a 30% chance of rain tomorrow? Very little! Since the day cannot be repeated, the forecast cannot mean that out of 100 'tomorrows', 30 of them will be wet. Alternatively, perhaps it means that in 30% of the region covered by the forecast it will be wet; then again, it might mean that it will rain throughout the region for 30% of the day. This vague and imprecise use of the language of probability is best avoided.

Exercise **3.104** An unbiased coin is tossed four times. Write down all the possible outcomes. What is the probability of

(a) four heads?

(b) three heads and one tail?

(c) at least two heads?

Exercise **3.105** Two six-sided dice are tossed. What is the probability of obtaining

(a) two 2's?

(b) 2 and 4?

(c) a total of 9?

Exercise **3.106** In a lottery the first prize is given for correctly selecting three numbers from the integers 1 to 25. What are the odds against winning with a single ticket?

**Exercise** | **3.107** What is the probability of an event happening when the odds are
(a) 3 to 1 on?
(b) 4 to 1 against?
(c) 9 to 8 against?

**Exercise** | **3.108** An unbiased coin is tossed six times. What are the odds
(a) of getting three heads and three tails?
(b) that the number of heads is different from the number of tails?

**Exercise** | **3.109** Return to the problem in Exercise 3.40 of selecting five balls from a box of ten black and six white balls. What is the probability of obtaining
(a) two black balls?
(b) five balls all of the same colour?
(c) at least two white balls?

**Exercise** | **3.110** Return to the problem in Exercise 3.39 of selecting a committee of four people at random from 18 men and 12 women. What is the probability that the committee contains
(a) at least one woman?
(b) no women?
(c) two women?

**Exercise** | **3.111** Four cards are selected at random from a well-shuffled pack of standard playing cards (see Exercise 3.34). What is the probability of obtaining
(a) one card from each suit?
(b) four cards all from the same suit, including an ace?
(c) at least two cards from the same suit?

**Exercise** | **3.112** Two dice are tossed. What is the probability that the sum of the numbers showing is either 7 or 10?

**Exercise** | **3.113** A game consists of your friend drawing one playing card at random from each of two well-shuffled packs; you have to guess correctly the suit of each card in order to win. What is the probability of your winning a single game? If the game is played three times, what is your probability of winning at least once?

**Exercise** | **3.114** For the game described in the previous exercise, use your calculator to find how many times you must play in order to have a better than 50 per cent chance of winning once.

**Exercise** | **3.115** The ASCII code described in Section 2.1.3 uses 7-bit bytes. Suppose that due to interference or equipment fault, the probability of an error when transmitting a single bit is 0.01. All such errors are independent of each other. What is the probability of a byte containing

(a) no errors?

(b) at least one error?

(c) exactly one error?

**Exercise** | **3.116** Solve the problem in part (c) of Example 3.26 by finding the probabilities of one, two or three subsystem breakdowns.

**Exercise** | **3.117** Three unbiased coins are tossed once. It was seen in part (b) of Example 3.26 that the probability of all three showing tails is 1/8. Similarly, the probability that all three show heads is 1/8. Since these events are mutually exclusive, then by the addition rule the probability that all three coins show the same is $(1/8) + (1/8) = 1/4$.

However, consider the following argument. Since three coins are tossed, then two of them must come up alike – both heads, or both tails (this follows from the pigeonhole principle). The probability that the third coin shows the same as the other two is 1/2, since it either shows heads or tails. Hence the probability that all three are the same is 1/2. What is wrong with this reasoning?

**Exercise** | **3.118** The odds against a horse called 'Discrete' winning in a race are 5 to 1. The odds against a horse called 'Mathematics' winning in another race are 11 to 2. Each horse runs in only one race. What are the odds against both horses winning?

**Exercise** | **3.119** A communications satellite is constructed with five subsystems. To ensure operation, redundancy is built in so that the satellite works successfully provided at least two of the subsystems continue operating. What is the probability that the satellite functions after one year, assuming that each subsystem has a 50% chance of failing during the year?

**Exercise** | **3.120** It can be shown that if the thickness of a coin is 35% of its diameter, then the probability that it will land on its edge when tossed is 1/3. Such a coin is tossed five times. What is the probability of

(a) two heads?　　(b) at most two heads?

How does it compare with an unbiased thin coin?

**Exercise** | **3.121** In the survey of cola-drinking preferences described in Exercise 3.90, what is the probability that one person selected at random from the group likes

(a) at least one of the brands?

(b) both brands?

Exercise   **3.122**  In a group of 149 wine drinkers, 94 like red wine and 77 like white wine. What is the probability that one person selected at random from the group likes both red and white wine?

Exercise   **3.123**  In the preceding exercise, how many ways are there of selecting two people who like both red and white wine? What is the probability that two people selected at random from the group

(a)  like both red and white wine?

(b)  do not like the same kind of wine?

Exercise   **3.124**  In Exercise 3.122, the probability that one person selected at random from the group likes both red and white wine is 0.148. Using the multiplication rule, it would seem that when two people are selected at random, the answer to part (a) in Exercise 3.123 is $0.148 \times 0.148 = 0.022$. Why is this argument wrong?

Exercise   **3.125**  A patient suffering from a rare disease is offered a choice of two independent treatments, each of which has a 50% chance of success. The patient requests both treatments, arguing that this will give a total percentage chance of $50 + 50 = 100$ (that is, certainty) of recovery. What is the flaw in this reasoning? What is the patient's true probability of recovery if both treatments are applied?

Exercise   **3.126**  A machine produces electronic components. It is estimated on the basis of past production that 2% of the output is defective. Ten components are selected at random from a day's production. What is the probability that this sample contains

(a)  no defectives?

(b)  exactly two defectives?

(c)  at least two defectives?

Exercise   **3.127**  An old joke tells of a traveller who is afraid of being blown up on a flight. The airline assures the passenger that the chance of a terrorist bomb being on board the chosen flight is only one in a million (that is, a probability of $10^{-6}$). The passenger decides to try to smuggle a bomb onto the plane, arguing that the probability of there being two bombs on the flight is $10^{-6} \times 10^{-6} = 10^{-12}$, so the trip will be much safer. Explain the flaw in the passenger's argument.

### 3.5.2   Conditional probability

If two events $E_1$ and $E_2$ are not independent then the occurrence of $E_1$ depends upon whether $E_2$ has happened or not. In other words, the probability that $E_1$ occurs is **conditional** upon the occurrence of $E_2$, and this is written as $P(E_1|E_2)$, which is read as 'the probability of $E_1$ given $E_2$'.

**Example 3.30**  **Tossing a coin three times**

If an unbiased coin is tossed three times, then using the notation of part (a) of Example 3.23 the eight possible outcomes are

$$TTT, TTH, THT, THH, HTT, HTH, HHT, HHH \tag{3.40}$$

If $E_1$ is the event that exactly two tails occur, then $E_1 = \{TTH, THT, HTT\}$ so that $P(E_1) = 3/8$.

Suppose that it is known that at least one tail occurs: what is the probability of getting exactly two tails in this case? Let $E_2$ be the event that at least one tail occurs, so that $E_2$ consists of all the outcomes in (3.40) except the last, that is,

$$E_2 = \{TTT, TTH, THT, THH, HTT, HTH, HHT\}$$

You can see that $E_2$ consists of *seven* equally likely outcomes, of which *three* contain exactly two tails. Hence the required probability that exactly two tails occur, given that at least one tail occurs, is $P(E_1|E_2) = 3/7$.

In general the **conditional probability** is defined by

$$P(E_1|E_2) = \frac{P(E_1 \cap E_2)}{P(E_2)} \tag{3.41}$$

where it is assumed that $E_2$ is not impossible, that is, $P(E_2) > 0$, so conditional probability is a measure of the probability of both $E_1$ and $E_2$ occurring, relative to the probability of $E_2$. In Example 3.30 the intersection is

$$E_1 \cap E_2 = \{TTH, THT, HTT\}$$

so that $P(E_1 \cap E_2) = 3/8$. Since $P(E_2) = 7/8$, (3.41) gives

$$P(E_1|E_2) = \frac{3/8}{7/8} = \frac{3}{7}$$

as before.

Notice that interchanging $E_1$ and $E_2$ in (3.41) produces

$$P(E_2|E_1) = \frac{P(E_2 \cap E_1)}{P(E_1)}$$

$$= \frac{P(E_1 \cap E_2)}{P(E_1)} \tag{3.42}$$

since $E_1 \cap E_2$ and $E_2 \cap E_1$ each represent the occurrence of both $E_1$ and $E_2$.

**Example 3.31**  **Applications of (3.41)**

(a) In a group of students, 20% failed a mathematics exam, 10% failed a physics exam and 7% failed both mathematics and physics. What is the probability that a student (selected at random) who fails physics also fails mathematics? Let $E_1$ be the event that a student fails mathematics, and $E_2$ that of failing physics. Then $P(E_1) = 0.2$, $P(E_2) = 0.1$ and $P(E_1 \cap E_2) = 0.07$. Hence from (3.41) the required probability is

$$P(E_1|E_2) = \frac{0.07}{0.1} = 0.7$$

(b) A box contains 13 white balls and 7 black balls. Two balls are taken out one after the other without replacement. What is the probability that both balls are black? Let $E_1$ be the event that the first ball is black, and $E_2$ that the second ball is black. It is required to determine $P(E_1 \cap E_2)$, which from (3.42) is given by

$$P(E_1 \cap E_2) = P(E_1) \, P(E_2|E_1) \tag{3.43}$$

Because there are 20 balls in total, the probability that the first ball is black is $P(E_1) = 7/20$. If it is assumed that the first ball is black, then on the second selection there are 6 possibilities for choosing a black ball from the remaining 19 balls. Therefore the conditional probability that the second ball is black, given that the first ball is black, is $P(E_2|E_1) = 6/19$. Hence by (3.43) the required probability that both balls are black is

$$P(E_1 \cap E_2) = \tfrac{7}{20} \times \tfrac{6}{19} = \tfrac{21}{190}$$

Notice that this problem can be solved without using conditional probability. The number of ways of choosing two black balls is $C(7, 2)$ and the total number of ways of choosing two balls is $C(20, 2)$. Hence the probability that both balls will be black is

$$\frac{C(7, 2)}{C(20, 2)} = \frac{7 \times 6}{20 \times 19} = \frac{21}{190}$$

as before.

If $E_1$ and $E_2$ are independent then the probability that $E_2$ occurs is *not* conditional upon whether $E_1$ occurs, so in this case $P(E_2|E_1) = P(E_2)$. Notice that (3.43) then reduces to $P(E_1 \cap E_2) = P(E_1) \, P(E_2)$ which is the multiplication rule (3.38) for two independent events.

| Example 3.32 | **Testing for independence** |
| --- | --- |

A single card is taken out of a well-shuffled pack of 52 playing cards. Let $E_1$ be the event that a heart is chosen, and $E_2$ the event that a red king is chosen. Are $E_1$ and $E_2$ independent? Since there are 13 hearts, $P(E_1) = 13/52 = 1/4$. There are two red kings, so $P(E_2) = 2/52 = 1/26$. The intersection $E_1 \cap E_2$ is the selection of the king of hearts, so $P(E_1 \cap E_2) = 1/52$. From (3.42)

$$P(E_2|E_1) = \frac{(1/52)}{(1/4)}$$

$$= \frac{1}{3} \neq P(E_2)$$

showing that $E_1$ and $E_2$ are *not* independent.

**Example 3.33**  **Exit poll**

Suppose that in a local election there are two candidates, of whom $A$ receives 45% of the vote and $B$ receives 55%, with no spoiled ballot papers. A local TV station takes an exit poll by asking voters, as they leave polling stations, which candidate they voted for. Forty per cent of supporters of candidate $A$ and 30% of supporters of $B$ reply (truthfully) to this questioning. The TV interviewer wants to predict the result of the election on the basis of the exit poll, before the votes are counted.

Let $E_1$ be the event that a person votes for candidate $A$, so that $P(E_1) = 0.45$. The complement $E'$ is the event that a person does *not* vote for $A$. Because there are no wasted votes $E'_1$ is therefore a vote for candidate $B$, so that $P(E'_1) = 0.55$. Let $E_2$ be the event that a voter responds to the exit poll. The proportion of those who responded to the exit poll, given that they voted for $A$, is 0.4, that is, $P(E_2|E_1) = 0.4$. Similarly, since 30% of $B$'s supporters responded to the poll, then $P(E_2|E'_1) = 0.3$. It is required to determine the proportion of voters who supported $A$, given that they responded to the exit poll, that is, to determine $P(E_1|E_2)$.

In order to compute $P(E_1|E_2)$ using (3.41), first determine the numerator by means of (3.43), so that

$$P(E_1 \cap E_2) = 0.45 \times 0.4 = 0.18 \tag{3.44}$$

Evaluating the denominator in (3.41) is more complicated, and requires the observation that any voter who participates in the exit poll has voted either for $A$ or for $B$, so that

$$\underset{\substack{\text{probability of} \\ \text{responding to} \\ \text{exit poll}}}{P(E_2)} \quad = \quad \underset{\substack{\text{probability of} \\ \text{voting for } A \\ \text{and responding} \\ \text{to exit poll}}}{P(E_1 \cap E_2)} \quad + \quad \underset{\substack{\text{probability of} \\ \text{voting for } B \\ \text{and responding} \\ \text{to exit poll}}}{P(E'_1 \cap E_2)} \tag{3.45}$$

The second term on the right in (3.45) is obtained from (3.43) with $E_1$ replaced by $E'_1$, so that

$$P(E'_1 \cap E_2) = P(E'_1) \, P(E_2|E'_1)$$
$$= 0.55 \times 0.3 = 0.165 \tag{3.46}$$

Hence (3.44), (3.45) and (3.46) give $P(E_2) = 0.18 + 0.165 = 0.345$. Finally, substituting this value and (3.44) into (3.41) produces

$$P(E_1|E_2) = \frac{0.18}{0.345} = 0.522$$

This means that 52.2% of those in the exit poll sample supported candidate $A$. On this basis the TV interviewer predicts that $A$ will win. However, this prediction is false, since candidate $B$ actually receives a majority (55%) of the total vote and so wins the ballot. The reason that the exit poll produces an incorrect prediction is that unequal proportions of supporters of the two candidates responded to the exit poll (see Exercises 3.134 and 3.135). In symbolic form, the required probability was computed from (3.41) and (3.45) as

$$P(E_1|E_2) = \frac{P(E_1 \cap E_2)}{P(E_2)}$$

$$= \frac{P(E_1 \cap E_2)}{P(E_1 \cap E_2) + P(E_1' \cap E_2)}$$

Substituting for $P(E_1 \cap E_2)$ using (3.43) and for $P(E_1' \cap E_2)$ using (3.46) gives

$$P(E_1|E_2) = \frac{P(E_1)\,P(E_2|E_1)}{P(E_1)\,P(E_2|E_1) + P(E_1')\,P(E_2|E_1')} \tag{3.47}$$

and (3.47) is known as the **Bayes Formula**.

Example 3.34 **Genetic profiling**

Legal cases which aim to identify the father of a child often use genetic testing. Suppose a man accused in a paternity suit possesses a certain genetic marker which occurs in only 2% of the male population. The male child in question also has this genetic marker, which can only be transmitted to the child through his father. Moreover, the child is 100% certain to acquire the marker if his father has it. The problem is to determine the probability that the accused man is the father, given that the child has the marker.

Let $E_1$ be the event that the man is the father, and $E_2$ the event that the child has the marker. It is required to compute $P(E_1|E_2)$ using the Bayes formula (3.47). In this expression, $P(E_2|E_1) = 1$ since a father always transmits his marker to his child. Also, if the accused man is *not* the father (the event $E_1'$) then the probability that the child has the marker is the same as that for the male population, so that $P(E_2|E_1') = 0.02$. Finally, suppose that other evidence suggests that it is reasonable to assume that the man is equally likely either to be or not to be the father, that is, $P(E_1) = 0.5$, $P(E_1') = 0.5$. Substituting these values into (3.47) gives

$$P(E_1|E_2) = \frac{0.5 \times 1}{0.5 \times 1 + 0.5 \times 0.02}$$

$$= 0.98$$

showing that there is a 98% probability that the man is indeed the father. However, a weakness in the argument lies in the initial assumed value for $P(E_1)$. If there is no supporting evidence and a smaller value is used for $P(E_1)$, then the probability of the man's guilt reduces considerably (see Exercise 3.136).

**Exercise**  **3.128** In part (a) of Example 3.31, what is the probability that a student (selected at random) who fails mathematics also fails physics? What is the probability that a student fails at least one of mathematics or physics?

**Exercise**  **3.129** If $P(E_1) = 0.3$, $P(E_2) = 0.2$ and $P(E_1 \cap E_2) = 0.15$, find:
(a) $P(E_1|E_2)$ (b) $P(E_2|E_1)$
(c) $P(E_1 \cup E_2)$ (d) $P(E_1'|E_2')$ (use the result in Problem 3.40).

**Exercise**  **3.130** If $P(E_1) = 0.4$, $P(E_2) = 0.5$ and $P(E_1 \cup E_2) = 0.7$, find $P(E_1|E_2)$.

**Exercise**  **3.131** A batch of 100 electronic components is known to contain six defective items. Two components are selected without replacement. What is the probability that both are perfect?

**Exercise**  **3.132** A single card is taken out of a well-shuffled pack of 52 playing cards. Let $E_1$ be the event that a club is chosen, and $E_2$ the event that a queen is chosen. Are $E_1$ and $E_2$ independent?

**Exercise**  **3.133** A box contains seven white balls, eight black balls and nine red balls. Three balls are taken out without replacement, and are seen to be all the same colour. What is the probability that all three balls are white?

**Exercise**  **3.134** Suppose that in Example 3.33 exactly half the supporters of each candidate responded to the exit poll. What prediction is now produced?

**Exercise**  **3.135** In Example 3.33, let $P(E_1) = a$, $P(E_2|E_1) = b$, $P(E_2|E_1') = b$. Use (3.47) to show that $P(E_1|E_2) = a$.

**Exercise**  **3.136** Suppose that in Example 3.34 the supporting evidence for the man's paternity is weak. Show that if $P(E_1) = 0.001$ then $P(E_1|E_2) = 4.8\%$ What is the result in this case if the incidence of the genetic marker in the male population is only 1%?

**Exercise**  **1.137** A medical test for detecting a certain disease gives a positive reaction in 98% of cases where a person actually has the disease. Unfortunately, it also gives a positive reaction in 5% of cases where a person does not have the disease. It is known that 1% of the population has the disease. Your doctor tells you that your test was positive. What is the probability that you have the disease? Should you be worried? (Hint: let $E_1$ = event that a person has the disease; $E_2$ = event that a person reacts positively.)

## Miscellaneous problems

**3.1**  In the Morse code, each symbol is represented by at most five dots or dashes. For example, the first few letters of the alphabet are coded as follows:

A     B     C     D     E

·—   —···   —·—·   —··   ·

How many symbols can be represented by the code?

**3.2**  Motor vehicle registration plates in Great Britain consisted for many years of seven symbols, for example J750 WAR. The first letter relates to the date of initial registration, for example J was issued in the period August 1991 to July 1992. This first letter can be any letter of the alphabet except I, O, Q, U or Z. The digital part can be any number from 1 to 999. The remaining three letters relate to the district of initial registration, and again to avoid confusion the first of these three letters cannot be I, Q or Z. In practice some combinations of these three letters are not used, but ignoring this restriction, how many different registration plates are there?

  The issuing authority withholds all registrations having a digital part from 1 to 20, and sells them at a premium price. How many are there of these special plates?

**3.3**  How many 5-digit numbers are there whose digits are distinct (that is, all different) integers from 1 to 9, and contain both 3 and 4 as non-consecutive digits (in either order)?

**3.4**  Prove that

$$C(n,r) + 2C(n, r-1) + C(n, r-2) = C(n+2, r)$$

**3.5**  Six-letter words are to be constructed using the 26 letters of the alphabet. Repetition of letters is allowed, and the 'words' need not make linguistic sense, for example *aabbbb* is a valid word.

(a)  How many words are there?

(b)  How many words contain at least one vowel?

(c)  How many words contain two, three or four vowels?

**3.6**  Suppose that there are $n$ objects of which three are identical and the others are all different. Show that the number of combinations of three objects is

$$\tfrac{1}{6}(n^3 - 9n^2 + 32n - 36)$$

**3.7**  A group of 12 people is to be split up into two teams in order to play a game. How many different teams are there (a) when the teams are equal in size? (b) when the teams are unequal in size?

**3.8**  Evaluate the sum

$$1 + 3C(n, 1) + 3^2 C(n, 2) + \ldots + 3^r C(n, r) + \ldots + 3^n$$

where $n$ is a positive integer.

**3.9** Prove that

$$C(n + 1, 3) - C(n - 1, 3) = (n - 1)^2$$

**3.10** In how many ways can one 50p coin, one 20p coin, one 10p coin and thirty 2p coins be distributed among six children (a) with no restrictions? (b) so that the youngest child gets either 20p or 50p?

**3.11** In how many ways can ten complimentary football tickets be distributed amongst five club members so that no member gets more than six tickets?

**3.12** Show that the number of ways in which $r$ identical objects can be distributed amongst $n$ different boxes, with empty boxes not allowed, is $C(r - 1, n - 1)$.

**3.13** How many solutions are there of the equation

$$x_1 + x_2 + x_3 + x_4 + x_5 = 12$$

where each $x_i$ is a non-negative integer?

To find the number of non-negative integer solutions of the inequality

$$x_1 + x_2 + x_3 + x_4 + x_5 < 12$$

show that this is the same as the number of non-negative integer solutions of the equation

$$x_1 + x_2 + x_3 + x_4 + x_5 + x_6 = 11$$

and hence determine the number.

**3.14** A king is placed on the top right-hand corner of a chessboard (which consists of $8 \times 8$ squares). Suppose that the king moves only one square at a time, either downwards or to the left. How many possible paths are there to the bottom left-hand square?

**3.15** Consider the coefficient of $x^n$ on both sides of the identity

$$(1 + x)^{2n} \equiv (1 + x)^n (1 + x)^n$$

Hence prove that

$$C(2n, n) = 2 + [C(n, 1)]^2 + [C(n, 2)]^2 + \ldots + [C(n, n - 1)]^2$$

**3.16** Use the binomial theorem to evaluate

$$C(n, 0) + 2C(n, 1) + 2^2 C(n, 2) + \ldots + 2^n C(n, n)$$

**3.17** Prove the following, where $n$ and $r$ are positive integers:

(a) $C(n, r + 1) = \dfrac{n - r}{r + 1} C(n, r)$

(b) $rC(n, r) = nC(n - 1, r - 1)$

(c) $C(n, r) = C(n - 2, r) + 2C(n - 2, r - 1) + C(n - 2, r - 2), \quad 2 \leqslant r \leqslant n - 2.$

**3.18** Use the results in Exercises 3.58 and 3.54 to evaluate

$$C(n, 1) + 2C(n, 2) + 3C(n, 3) + \ldots + nC(n, n)$$

**3.19** Show that for a given positive integer

(a) $C(r, r) + C(r + 1, r) = C(r + 2, r + 1)$

(b) $C(r, r) + C(r + 1, r) + C(r + 2, r) = C(r + 3, r + 1)$.

Hence prove the result (3.21) by the method of induction.

**3.20** Consider the harmonic triangle defined in Exercise 3.67. Show that the $r$th element in the second row is $1/r(r + 1)$. Prove that the sum of the terms in the second row approaches 1 as the number of terms increases. Similarly, determine a general expression for elements in the third row, and the corresponding value which the row sum approaches.

**3.21** Using the results in the previous problem, show that the numbers along the $k$th diagonal of the harmonic triangle are equal to the reciprocals of the numbers along the $k$th diagonal of the third form of Pascal's triangle (given in Exercise 3.67) each divided by $k$. For example, when $k = 4$

$$\left(\tfrac{1}{4}, \tfrac{1}{12}, \tfrac{1}{12}, \tfrac{1}{4}\right) \quad = \quad \tfrac{1}{4}\left(\tfrac{1}{1}, \tfrac{1}{3}, \tfrac{1}{3}, \tfrac{1}{1}\right)$$

harmonic triangle       reciprocals of Pascal's
diagonal          triangle diagonal

**3.22** Generalize the results of Example 3.13(d) and Exercise 3.73 as follows. Show that if $n + 2$ integers are selected from the set $\{1, 2, 3, \ldots, 2n + 1\}$ then there will always be one pair whose sum is $2n + 2$.

**3.23** Generalize the results of Example 3.13(f) and Exercise 3.76 as follows. Show that if at most $n$ events are scheduled over $m$ days, with at least one event taking place each day, then there will be a period of consecutive days in which $2m - n - 1$ events occur.

**3.24** Let $n$ be any positive odd integer. Show that amongst the $n + 1$ positive integers $2 - 1, 2^2 - 1, 2^3 - 1, \ldots, 2^{n+1} - 1$, there are two (at least) which have the same remainder after division by $n$. Hence show that there is a positive integer $m$ such that $n$ divides $2^m - 1$.

**3.25** Show that if any seven integers are selected from the set $\{1, 2, 3, 4, \ldots, 64\}$ then there will be at least one pair ($m$ and $n$, say) such that $|\sqrt{m} - \sqrt{n}| < 1$.

**3.26** How many positive integers must be selected between 1 and 35 (inclusive) so as to ensure that at least two of them have a greatest common divisor greater than 1?

**3.27** Let $B$ be any subset of 30 integers chosen from the set $A = \{1, 2, 3, 4, \ldots, 57, 58\}$. Prove that $B$ must contain two numbers such that one divides the other. What is the most general result of this type?

**3.28** How many 4-digit decimal numbers (with non-zero first digit)

(a) contain at least one of the digits 0 and 1?

(b) do not contain at least one of the digits 0 and 1?

**3.29** Consider permutations of the nine letters *AABBBCCCC*. Use the inclusion–exclusion principle to count how many permutations do *not* have identical letters occurring all together in a single block. (*Hint*: let $X$ be the set of permutations containing two consecutive $A$'s, $Y$ the set of permutations containing three consecutive $B$'s, and $Z$ the set of permutations containing four consecutive $C$'s, and use the inclusion–exclusion principle.)

**3.30** How many permutations are there of the 26 letters of the alphabet which do not include the words *CAT, DOG, FISH*? (*Hint*: consider permutations which do include one of these words, and use the inclusion–exclusion principle.)

**3.31** Let $d_n$ be the number of derangements of $n$ objects, given by the formula in Exercise 3.99. Prove that

$$d_n = (n-1)(d_{n-1} + d_{n-2}), \qquad n = 3, 4, 5, 6, \dots$$

(This is an example of a recurrence relation, to which Chapter 5 is devoted.) Since $d_1 = 0$ and $d_2 = 1$, use the formula to compute $d_3$, $d_4$, $d_5$ and $d_6$.

Verify also a curious result involving suffices:

$$d_1 + d_4 + d_8 + d_3 + d_4 + d_9 = \mathbf{148349}$$

**3.32** There are six different pairs of gloves in a drawer. Six people choose a left-hand glove at random, and a right-hand glove at random. How many selections are there in which

(a) no person has a matching pair of gloves?

(b) exactly one person has a matching pair?

(c) at least two people have a matching pair?

**3.33** Use the inclusion–exclusion principle to count how many positive integers less than 1000 have digits whose sum is 22. List all these integers.

**3.34** Generalize the result in part (c) of Exercise 3.101 to ternary sequences of length $n$.

**3.35** If the probability that a certain event happens is $p = a/b < 1/2$, what are the odds that the event (a) happens? (b) does not happen?

**3.36** Consider the situation described in Exercise 3.115 of transmitting a message consisting of 7-bit bytes. Suppose it is possible to correct a single error in any received byte.

(a) What percentage of bytes will be incorrectly decoded?

(b) What is the percentage if *two* errors can be corrected?

**3.37** A drugs smuggler transports boxes, each containing 100 cartons labelled 'cigarettes'. One carton in each box contains illegal drugs. A customs officer selects for inspection one carton at random from each of 100 boxes. Show that the probability that the smuggler successfully escapes detection is 0.366.

Suppose that there are $n$ cartons per box, and $n$ boxes are inspected by selecting one carton at random from each. Show that the probability of the smuggler's success is $\left(1 - \dfrac{1}{n}\right)^n$.

Use your calculator to investigate what happens as $n$ gets larger and larger (use the result in Exercise 3.102).

**3.38** In the survey of cola-drinking preferences in Exercise 3.90, what is the probability that two people selected at random from the group both like both brands?

**3.39** The Perfect Homes Company on its latest development offers the options of a conservatory ($E_1$); a landscaped garden ($E_2$) or a swimming pool ($E_3$). On the basis of past experience the marketing manager estimates the probabilities that a buyer will take up various options as follows:

| Option | $E_1$ | $E_2$ | $E_3$ | $E_1 \cap E_2$ | $E_1 \cap E_3$ | $E_2 \cap E_3$ | $E_1 \cap E_2 \cap E_3$ |
|---|---|---|---|---|---|---|---|
| Probability | 0.70 | 0.60 | 0.25 | 0.45 | 0.15 | 0.15 | 0.05 |

What is the probability that a purchaser of one of the houses will not take up any of the three options?

**3.40** Use appropriate Venn diagrams to show that for two sets $A$ and $B$,

$$(A \cup B)' = A' \cap B', \quad (A \cap B)' = A' \cup B'$$

where $'$ denotes complement. Hence deduce that for two events $E_1$, $E_2$ then

$$P(E_1' \cap E_2') = 1 - P(E_1 \cup E_2)$$

**3.41** Let $A$ and $B$ be events such that

$$P(A) = \tfrac{5}{8}, \quad P(B) = \tfrac{1}{4}, \quad P(A \cap B) = \tfrac{3}{8}$$

Use the results in the preceding problem to find

(a) $P(A \cup B)$  (b) $P(A')$  (c) $P(A' \cap B')$  (d) $P(A' \cup B')$.

**3.42** Use the result (3.42) to prove that for three events $E_1$, $E_2$ and $E_3$ then

$$P(E_1 \cap E_2 \cap E_3) = P(E_1) \, P(E_2|E_1) \, P[E_3|(E_1 \cap E_2)]$$

A box contains 5 white and 95 black balls. What is the probability that three balls taken out one after the other without replacement will all be black?

**3.43** Two unbiased coins are tossed. Let $E_1 = $ tails on first coin, $E_2 = $ tails on second coin, $E_3 = $ tails on exactly one coin. Show that each pair of events is independent, but $P(E_1 \cap E_2 \cap E_3) \neq P(E_1) \, P(E_2) \, P(E_3)$.

**3.44** Suppose $F_1, F_2, \ldots, F_m$ are pairwise mutually exclusive events (that is, $F_i \cap F_j = \varnothing$, all $i \neq j$) and let $F_1 \cup F_2 \cup F_3 \cup \ldots \cup F_m = S$. Then a generalization of the Bayes formula (3.47) is that for any event $E$ in $S$ then

$$P(F_i|E) = \frac{P(F_i)\, P(E|F_i)}{\displaystyle\sum_{j=1}^{m} P(F_j)\, P(E|F_j)}$$

(in (3.47), $m = 2$, $E_2 = E$, $E_1 = F_1$, $E_1' = F_2$).

At the Interplanetary University a module in English language contains 60% Venusian students, 25% Martians and 15% Jovians. On the basis of past records it is expected that 7% of Venusians will fail the module, compared with 4% of Martians and 3% of Jovians.

(a) If a student fails, what is the probability that they come from Mars?

(b) If a student passes, what is the probability that they are Jovian?

## Student project

**3.45** It can be shown that the number of ways $S(m,n)$ of distributing $m$ different objects into $n$ identical containers with no container left empty is

$$S(m, n) = \frac{1}{n!} \sum_{k=0}^{n-1} (-1)^k\, C(n, k)(n - k)^m$$

The numbers $S(m,n)$ are called **Stirling numbers of the second** kind, named after the mathematician whose approximation for $n!$ is given in (3.3).

(a) Deduce that $S(m, 1) = 1$ and $S(m, m) = 1$. Evaluate the Stirling numbers for $m = 3, 4, 5$ and $n = 2, 3, 4$.

(b) List all the different ways of distributing five different objects between three identical boxes, with no box left empty.

(c) Prove that for $n > 1$

$$S(m + 1, n) = S(m, n - 1) + nS(m, n)$$

Hence compute the values of $S(m,n)$ for $m = 6, 7, 8$.

(d) What is the number of ways of distributing the objects if empty containers are allowed?

## Student project

**3.46** Prove the general form (3.29) of the inclusion–exclusion principle by taking an element $x \in A_1 \cup A_2 \cup \ldots \cup A_m$ and counting how many times $x$ is included and excluded by the formula on the right-hand side of (3.29).

**3.47**   The most general form of the pigeonhole principle is as follows. Let $m_1, m_2, \ldots, m_n$ be positive integers. If $m_1 + m_2 + \ldots + m_n - n + 1$ (or more) objects are put into $n$ boxes, then either the first box contains at least $m_1$ objects, or the second box contains at least $m_2$ objects, ..., or the $n$th box contains at least $m_n$ objects.

(a) Prove this by investigating what happens if the $i$th box contains fewer than $m_i$ objects for every value of $i = 1, 2, \ldots, n$.

(b) Deduce the **extended** pigeonhole principle as stated in Section 3.3.

(c) Consider any sequence of $k$ positive integers. Any selection of fewer than $k$ integers from these (keeping the same order) is called a **subsequence**. If the values of the integers in a subsequence increase successively then the subsequence is called **increasing**; if the values decrease it is called **decreasing**. For example, take a sequence of ten integers

   6, 3, 5, 4, 12, 19, 11, 9, 10, 13

The subsequences 3, 4, 12, 19 and 4, 9, 10, 13 are increasing, and the subsequences 6, 5, 4 and 19, 11, 9 are decreasing.

   A theorem states that given *any* sequence of $n^2 + 1$ distinct integers, there is either an increasing subsequence of $n + 1$ terms or a decreasing subsequence of $n + 1$ terms. In the example above, $n = 3$.

   By constructing examples, show that it is possible to have an increasing or decreasing subsequence of more than $n + 1$ terms; or to have both an increasing and a decreasing sequence of $n + 1$ terms. Show also that if only $n^2$ terms are taken then there may not be either an increasing or a decreasing subsequence of $n + 1$ terms.

   The theorem is proved using the extended form of the pigeonhole principle given in Section 3.3. Write out a detailed proof with the aid of appropriate textbooks.

**3.48**   *The Birthday Problem*: For a group of people chosen at random, how many people do you need so that the probability that at least two of them have the same birthday (day and month) is greater than 50%? (Ignore leap years, and assume birthdays are uniformly distributed throughout the year.)

   Since there are 365 possible dates, then if you take 366 people it is certain that two will have the same birthday (see Exercise 3.74). You may well feel that to get a probability of greater than 1/2 then you'll need about 183 people (that is, half as many) but in fact only the much smaller number of 23 is required.

   To investigate this problem, proceed as follows: for two people, the probability that they have the same birthday is $1/365$, so the probability that they have different birthdays is $364/365$. The probability that a third person also has a different birthday is $363/365$, so the probability that *all three* have *different* birthdays is

$$\tfrac{364}{365} \times \tfrac{363}{365} = 0.992$$

Hence the probability that at least two of these three people have the *same* birthday is $1 - 0.992 = 0.008$. Continue this process for $n = 4, 5, 6, \ldots, 60$ people and draw a graph of the probability $P$ of coinciding birthdays against $n$. In particular, show that $P = 0.507$ when $n = 23$.

An interesting alternative interpretation of the birthday problem is the following (see Garza, 1987). Suppose that there are at least 23 accidents per year along a certain stretch of road. Then the probability that at least two of them occur on the same day is greater than 50%. (This assumes the accidents are uniformly spread across the year – in practice, there are usually more accidents in winter.) How many accidents do there have to be in a year before you can conclude that it's virtually certain that two of them will happen on the same day?

Incidentally, this interpretation is an illustration of **Murphy's Law**, in the version which states: 'Things go wrong in batches' (the original law is: 'If anything can go wrong, it will').

## Further reading

Sections
3.1–3.2
Brualdi R.A. (1977). *Introductory Combinatorics*. New York: Elsevier North-Holland, Chapters 3 and 4

Finkbeiner II D.T. and Lindstrom W.D. (1987). *A Primer of Discrete Mathematics*. New York: W.H. Freeman, Chapter 3

Grimaldi R.P. (1994). *Discrete and Combinatorial Mathematics, An Applied Introduction*, 3rd edn. Reading MA: Addison-Wesley, Chapter 1

Jackson B.W. and Thoro D. (1990). *Applied Combinatorics with Problem Solving*. Reading MA: Addison-Wesley, Chapter 2

Molluzzo J.C. and Buckley F. (1986). *A First Course in Discrete Mathematics*. Belmont CA: Wadsworth, Chapter 3

Roberts F.S. (1984). *Applied Combinatorics*. Englewood Cliffs NJ: Prentice-Hall, Chapters 1 and 2

Section
3.2.2
Colledge T. (1992). *Pascal's Triangle*. Diss, Norfolk: Tarquin Publications

Edwards A.W.F. (1987). *Pascal's Arithmetical Triangle*. London: Griffin

Gardner M. (1976). Pascal's Triangle. In *Mathematical Carnival*. London: George Allen and Unwin, p. 194

Honsberger R. (1976). *Mathematical Gems II*. Washington DC: Mathematical Association of America, p.3

Section 3.3    Brualdi R.A. (1977). *op. cit.*, Chapter 2
               Grimaldi R.P. (1994). *op cit.*, p.275
               Jackson B.W. and Thoro D. (1990). *op. cit.*, p.34
               Roberts F.S. (1984). *op. cit.*, p.319

Section 3.4.1  Finkbeiner II D.T. and Lindstrom W.D. (1987). *op. cit.*, Chapter 1
               McNeill D. and Freiberger P. (1994). *Fuzzy Logic*. New York: Simon and Schuster
               Molluzzo J.C. and Buckley F. (1986). *op. cit.*, Chapter 2

Section 3.4.2  Grimaldi R.P. (1994). *op. cit.*, Chapter 8
               Jackson B.W. and Thoro D. (1990). *op. cit.*, Chapter 3
               Roberts F.S. (1984). *op. cit.*, Chapter 6
               Slomson A. (1991). *An Introduction to Combinatorics*. London: Chapman and Hall, Chapter 2

Section 3.5    Dierker P.F. and Voxman W.L. (1986). *Discrete Mathematics*. San Diego: Harcourt Brace Jovanovich, Chapter 9
               Garza G.G. (1987). Murphy's Law and Probability, or How to Compute Your Misfortune. *Mathematics Magazine*, **60**, 159–65
               Molluzzo J.C. and Buckley F. (1986). *op. cit.*, Chapter 4

Section 3.5.2  Isaac R. (1996). *The Pleasures of Probability*. New York: Springer, Chapters 3, 4

# 4 Codes

Most products you buy in a supermarket carry a barcode, like that shown in Figure 4.1. The bars are read at the checkout desk by a laser scanning system which converts the black and white bars of different widths to the numbers printed underneath. Each item is then identified by this string of digits, called a **codeword**. On the back cover of most books you will find another barcode, as illustrated in Figure 4.2. The upper number is called the **International Standard Book Number (ISBN)**, and every published book can be identified in this way.

5 000127 163195 >

Kellogg's Special K

*Figure 4.1*   A barcode

ISBN 0-13-834094-3

9 780138 340940

*Figure 4.2*   ISBN

195

When processing or transmitting information in the form of codewords, errors can occur because of electrical faults, external interference such as lightning or radiation, human mistakes, or other technical problems. For this reason, some of the digits in a codeword are there to ensure that such errors are corrected, or at least detected. You can then be certain that you are correctly charged for an item you buy, or that the bookshop gets the right book you have ordered. These so-called **check digits** (for example, the right-most digits 5 and 3 in Figures 4.1 and 4.2 respectively) are chosen so as to utilize some underlying properties of numbers. You are already familiar with this idea from verbal or written language, which contains so much structure that you can often guess correctly what is meant even if a message contains spelling or other mistakes. For example, if you receive a fax from a friend stating 'Meat me at fore p.n. tomorrow', you will be able to make sense of this (see below), despite the errors.

| Example 4.1 | A simple code |
| --- | --- |

To illustrate how numbers can be used to incorporate structure like that which is inherent in language, suppose you and a friend wish to communicate by a rather unreliable channel. You agree on nine different messages, such as 'Meet me at four p.m. tomorrow' (the correct version of the fax above), or 'See you at dinner tonight', which you label 1, 2, 3, ..., 9 respectively. Unfortunately, it is useless to transmit one of these integers, because technical problems can cause errors of up to ±2 during transmission. Thus for example, if you receive the digit 3 then there is no way of knowing what your friend sent – two possibilities are 1 (with an error of 2), or 4 (with an error of −1), for example. However, you have a brainwave: multiply the number of the message by 5 and transmit this product. For example, if the message you wish to send to your friend is labelled 4, then you transmit $4 \times 5 = 20$. If the maximum transmission error remains ±2, then the message can always be correctly interpreted, or **decoded**, as follows. Suppose your friend receives 22, then he or she correctly deduces that you must have sent 20, with a transmission error of +2, so the actual message must be $20 \div 5 = 4$. Similarly, if you receive 38 then you decide that your friend must have sent 40 with an error of −2, so the actual message must be $40 \div 5 = 8$. You can see that any received message can be uniquely and correctly decoded according to the rules: round the received number up or down to the nearest multiple of 5, and then divide by 5.

The use of codes to transmit and process information as accurately as possible is a vital ingredient of contemporary life. For example, in addition to product barcodes, PINs (Personal Identification Numbers) are used with automatic cashcard machines; European Union passports carry identity numbers which protect against forgery; and the transmission of data from interplanetary probes uses error-correcting codes so as to compensate for the vast distances and the limited power of the transmitters. Last but not least, every compact disc (CD) carries the inscription 'DIGITAL AUDIO'. Compact discs were introduced in 1982 and are used for audio reproduction, and to store information in digital form. The sounds are first decomposed into very many tiny components which are converted into binary numbers. To play back your favourite music, the bits are read off a CD by a laser, and 1,460,000 bits of audio information are processed every second. A lengthy recording contains around 20 billion bits, and even with the most careful manufacturing procedures, faults do occur on CDs. The reason why these flaws do not affect the music, which sounds very authentic and free from 'clicks', 'hiss' and other unwanted background noise, is that about two-thirds of the information content of a CD is devoted to non-audio information. This extra information is used to process the music before it reaches your ears, so that it ends up sounding virtually perfect. In fact data groups consisting of 588 bits are used, of which 192 bits contain the audio information, and no less than 64 bits are check bits which correct errors.

In this chapter some basic concepts of error-correcting codes are developed, using the ideas of modular arithmetic and finite fields from Chapter 2. A very brief introduction is also given to another aspect of coding, called **cryptology**, where the objective is to transmit messages in secret.

**Exercise** **4.1** Rail tickets in France have to be stamped by a machine before boarding a train, and are subsequently checked by an inspector. On a trip in April 1990 my ticket was stamped 54700099 in Paris on the 9th, and 57100105 in Tours on the return journey on the 15th. What simple code was being used to detect fraudulent ticket use?

**Exercise** **4.2** Suppose I have selected one number $i$ from the set of eight numbers $\{0, 1, 2, 3, \ldots, 6, 7\}$. Your objective is to determine $i$ by asking me questions, each of which I will answer with either 'yes' or 'no'. By representing the numbers in 3-bit binary form from $(000)_2$ to $(111)_2$, show that you can determine $i$ with three questions.

This problem is analyzed in Example 6.15.

## 4.1 Examples of codes

### 4.1.1 Repetition code

**Example 4.2** **Four messages**

Suppose you want to send four different commands to your video cassette recorder (VCR) using a remote controller. These instructions could be represented by **binary codewords** as follows:

| Command | Stop | Play | Fast forward (FF) | Rewind (REW) |
|---|---|---|---|---|
| Codeword | 00 | 01 | 10 | 11 |

However, if even a single error occurs in transmitting the instruction to the device (for example, 0 is replaced by 1, or 1 by 0), then the wrong command will be executed because the VCR has no way of knowing that an error has occurred. For example, if you send 10 but this is corrupted to 00 then the VCR stops instead of fast forwarding.

In everyday speech if you don't understand what someone says, you ask them to say it again. A natural way of trying to correct errors is therefore to simply repeat each transmitted word, so the list of codewords becomes:

| Command | Stop | Play | FF | REW |
| --- | --- | --- | --- | --- |
| Codeword | 0000 | 0101 | 1010 | 1111 |

If you now transmit 1010 and a single error occurs in the first bit, say, then the VCR receives 0010. This is not a codeword, and is simply called a **word**. Since the first two digits 00 are different from the second pair 10, the VCR detects that an error has occurred during transmission. However, the VCR cannot correct the error, because there is no way of deciding whether 0010 came from 1010 with an error in the first bit, or from 0000 with an error in the third bit. There is an improvement in performance, since the VCR no longer performs an incorrect operation, but instead registers an 'error' message. You will probably have seen a form of repetition code in action at the supermarket checkout. If for some reason the scanner fails to read the barcode correctly, an 'error' message is shown on the checkout display, and the operator then repeats the process, usually entering the numerical codeword by hand.

The next obvious step is to repeat the message *twice*, so the codewords are:

| Command | Stop | Play | FF | REW |
| --- | --- | --- | --- | --- |
| Codeword | 000000 | 010101 | 101010 | 111111 |

If you now transmit 101010 and again a single error occurs, say in the second bit so 111010 is received, not only is this error detected as before, but it can also be corrected. This is done by using a 'majority count' for each of the repetitions of the first pair of bits, as follows:

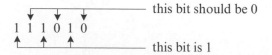

The second bit is deduced to be 0 (correctly) because it is 0 two out of three times. The received word is therefore decoded as 101010. The process of obtaining the transmitted codeword from the received word is called **decoding**. In this example the VCR can now decide that the **information message** was 10, and fast forwards, as required.

This repetition code corrects any single transmission error in the same way as in the specific example. However, it suffers from the obvious disadvantage that three times the original information has to be transmitted. This is not only expensive, but may well be impractical because of, for example, limited capacity of the communication channel.

You must realize that since the extra bits (that is, the repetitions) are themselves also subject to errors, correct decoding cannot be guaranteed. For example, the above received message 111010 might have come from the transmitted message 111111 with two errors, occurring in the fourth and sixth bits. It is assumed throughout this chapter that the probability of a single error is very small, thanks to the reliability of electronic equipment, so that a single transmission error is much more likely than two or more such errors. The objective is to make the probability of correctly decoding the received message as high as possible.

This idea can be extended to what is called **nearest-neighbour decoding**. The receiver holds a complete list of codewords. If the received word is not a codeword then one or more errors are immediately detected. To attempt error-correction, the selection of the codeword most likely to have been transmitted is done by comparing the received word with the list, and choosing the codeword which differs from the received word with the *smallest* number of errors. For example, with the 6-bit repetition code above, if 101111 is received then comparison with the four codewords gives:

| Codeword | 000000 | 010101 | 101010 | 111111 |
|---|---|---|---|---|
| Errors | 101111 | 101111 | 101111 | 101111 |
| Number of errors | 5 | 4 | 2 | 1 |

The message is therefore decoded as 111111, since this is the nearest neighbour to the received word, differing from it in only one bit.

**Exercise**

**4.3** A message to be transmitted consists of a single bit, 0 ($=$ YES) or 1 ($=$ NO). The following repetition code is used:

| Message | 0 | 1 |
|---|---|---|
| Codeword | 000 | 111 |

By considering all the eight possible received words, show that with nearest-neighbour decoding, this code corrects all single errors.

**Exercise**    **4.4** For the 6-bit code in Example 4.2, list the received messages which can be correctly decoded, assuming at most a single error occurs in transmission. If more than one error can occur, how many different received words are possible?

**Exercise**    **4.5** A code consists of four codewords 00110, 01011, 10001, 11100. Use the nearest-neighbour scheme to decode each of the following received words:

(a) 10000    (b) 01100    (c) 00111    (d) 11010.

**Exercise**    **4.6** A binary code consists of the four codewords 010101, 101010, 011011, 001010. Show that a received word 110110 cannot be decoded using the nearest-neighbour principle. What can be concluded?

**Exercise**    **4.7** A coin is tossed twice, and the results (where $H$ = head, $T$ = tail) are transmitted using the following binary codewords:

| HH | HT | TH | TT |
|----|----|----|----|
| 00 | 01 | 10 | 11 |

Suppose that the probability of an error in a single bit is 0.001. Show that the probability that a correct message is received is 99.8%.

**Exercise**    **4.8** Suppose that in the preceding exercise the codewords are now:

| HH | HT | TH | TT |
|------|------|------|------|
| 0000 | 0111 | 1001 | 1110 |

Suppose that during transmission no error can occur in the fourth bit, but a single error may occur in any one of the first three bits. List all the 16 possible received binary words, and show how each one can be uniquely decoded.

If again the probability of an error in a single bit is 0.001, show that the probability of a correct message, after decoding, is increased to 99.8997%.

## 4.1.2   Parity codes

**Example 4.3**    **Binary codes**

A simple but useful code is obtained by appending a single extra bit to each message containing the information which is to be transmitted. This bit is chosen so that the resulting codeword contains an *even* number of ones. This produces the so-called **even parity code** (the parity of an integer is its property of being even or odd). Similarly, if the extra bit, called the **parity-check bit**, or simply the **check bit**, is chosen so that the total number of ones is odd, then this results in the **odd parity code**.

For example, consider the first code in Example 4.2, and put either 0 or 1 onto the end of each codeword so that each new codeword contains an even number of ones. Thus 01 becomes 011, and the complete even parity code is:

| Command | Stop | Play | FF | REW |
|---|---|---|---|---|
| Codeword | 000 | 011 | 101 | 110 |

If, say, 101 is sent and a single error occurs then the received word will be one of 001, 111, 100. Each of these words contains an odd number of ones, so an error is detected. However, it is not possible to determine which particular bit is incorrect.

In general, suppose the original message contains $n - 1$ bits, $x_1, x_2, \ldots, x_{n-1}$, called the **information bits**. An extra bit $x_n$, called the check bit, is chosen so that the parity of the codeword is even. That is, codewords contain an even number of ones, so in the notation of Section 2.4.1

$$x_1 + x_2 + x_3 + \ldots + x_n \equiv 0 \,(\mathrm{mod}\ 2) \tag{4.1}$$

The transmitted codewords are $x_1 x_2 \ldots x_n$, where each $x_i$ is 0 or 1, and the **length** of codewords, or of the code itself, is defined to be $n$. Suppose a single error occurs during transmission in the $i$th bit. This means that if $x_i = 0$ then the $i$th bit $y_i$ of the received word is $y_i = 1$, and similarly $y_i = 0$ if $x_i = 1$. This can be expressed as

$$y_i \equiv (x_i + 1)(\mathrm{mod}\ 2) \tag{4.2}$$

and the received word is $r = x_1 x_2 \ldots x_{i-1} y_i x_{i+1} \ldots x_n$. The parity of $r$ is obtained by summing its digits modulo 2, which gives

$$x_1 + x_2 + \ldots + x_{i-1} + y_i + x_{i+1} + \ldots + x_n$$
$$\equiv x_1 + \ldots + x_{i-1} + (x_i + 1) + x_{i+1} + \ldots + x_n \quad (\mathrm{mod}\ 2)$$
$$\equiv 1 \,(\mathrm{mod}\ 2)$$

showing that $r$ has odd parity. You can verify (see Exercise 4.10) that in general, if an odd number of transmission errors occurs, then parity $(r) = 1$. In other words, if the received word has odd parity then the code detects that an odd number of errors has occurred. You can show in a similar fashion (see Exercise 4.11) that if the received word has even parity then there must be an even number of errors (including no errors).

Assuming that the probability of any single error is very small, then if the received word has odd parity it is most likely that only one error has occurred; and if the received word has even parity then it is most likely that there has been no error. Hence, under the above assumption, the even parity check code detects all single errors. It cannot, however, correct a single error since there is no way of determining which bit in the received word is incorrect.

As a further example, suppose an even parity check code of length 5 is used, and the information message to be sent is 0111. In order to satisfy (4.1) the check bit must be $x_5 = 1$, so the codeword to be transmitted is 01111, which has even parity. Assuming that at most one transmission error occurs, then if 01111 is received, this word can be taken as correct because it has even parity, so on dropping the check bit the decoded information message is correctly obtained as 0111. However, if the received word was, say, 10101 then since this has odd parity an error is detected. This received word is decoded by reporting 'error'.

As another example, consider the 7-bit ASCII code which was introduced in Section 2.1.3. This is often extended to an 8-bit code by appending an even parity check bit. Referring to Table 2.1, the 7-bit codeword 1000001 for A becomes 10000010, and a few more 8-bit codewords are listed below:

| Character | B | C | a | b | 1 |
|-----------|---|---|---|---|---|
| Codeword | 10000100 | 10000111 | 11000011 | 11000101 | 01100011 |

**Example 4.4**  **Ternary code**

In this case codewords are $x_1 x_2 \ldots x_{n-1} x_n$ where each $x_i$ is 0, 1 or 2 and arithmetic is performed modulo 3. The check digit $x_n$ is chosen so that (4.1) is replaced by

$$x_1 + x_2 + \ldots + x_n \equiv 0 \,(\mathrm{mod}\ 3) \tag{4.3}$$

An error $e_i$ in any digit $x_i$ can now be 1 or 2. In particular, if there is only a single error in the $i$th digit, then the received word is $r = x_1 x_2 \ldots x_{i-1} y_i x_{i+1} \ldots x_n$, where

$$y_i \equiv (x_i + e_i) \,(\mathrm{mod}\ 3)$$

which is the expression replacing (4.2). In this case, the sum of the digits of $r$ is

$$x_1 + x_2 + \ldots + x_{i-1} + y_i + x_{i+1} + \ldots + x_n$$
$$\equiv x_1 + \ldots + x_{i-1} + (x_i + e_i) + x_{i+1} + \ldots + x_n \,(\mathrm{mod}\ 3)$$
$$\equiv e_i \,(\mathrm{mod}\ 3)$$

Hence, assuming at most a single error occurs in transmission, then the code detects all such errors because the sum of the digits of the received word is not divisible by 3.

For example, suppose that the information message is 10210. The check digit $x_6$ is chosen so that (4.3) is satisfied, that is,

$$1 + 0 + 2 + 1 + 0 + x_6 \equiv 0 \,(\mathrm{mod}\ 3)$$

which has the solution $x_6 = 2$. Hence the transmitted codeword is 102102. If the received word is, say, 112102 then the sum of its digits is

$$1 + 1 + 2 + 1 + 0 + 2 \equiv 1 \,(\mathrm{mod}\ 3)$$

showing that an error has occurred.

Strictly speaking, the term 'parity' only applies to 'even-ness' or 'odd-ness', but it is a natural extension to apply it to codes with any base $p$, with the sum of the digits of codewords being $0(\bmod\ p)$, as in (4.3).

Exercise **4.9** List the 8-bit codewords including an even parity check bit for the remaining characters in Table 2.1.

Exercise **4.10** Show that if an odd number of transmission errors occurs for the even parity code in Example 4.3 then the received word has odd parity.

Exercise **4.11** Show that if a received word for the even parity code in Example 4.3 has even parity then either no errors or an even number of errors have occurred.

Exercise **4.12** For the odd parity binary code, the check bit is chosen so that (4.1) is replaced by $x_1 + x_2 + \ldots + x_n \equiv 1 \pmod 2$. Show that if the parity of a received word $r$ is even then an odd number of errors has occurred, and if parity $(r)$ is odd then either no errors or an even number of errors have occurred.

Exercise **4.13** Using the even parity binary code, the following words are received, where the last bit is the check bit:

(a) 0101011 (b) 1100001 (c) 1111000.

Assuming at most one error has occurred, determine the information message, where possible.

Exercise **4.14** Using a ternary parity code satisfying (4.3), received words are as follows:

(a) 1221012 (b) 0102021 (c) 2002001.

Determine if possible the information messages, assuming at most one transmission error has occurred.

Exercise **4.15** For a ternary parity code satisfying (4.3), determine all possible transmitted codewords when the received word is 2120, assuming

(a) a single transmission error (b) two transmission errors.

Exercise **4.16** Consider a ternary parity code satisfying (4.3), and assume more than one transmission error may occur. Show that if the sum of the digits of a received word is not divisible by 3, then all that can be deduced is that at least one error has occurred. What is the conclusion if the sum *is* divisible by 3?

Exercise **4.17** For a decimal parity code, each codeword $x_1 x_2 \ldots x_n$ satisfies the condition

$$\sum_{i=1}^{n} x_i \equiv 0 \pmod{10}$$

where $x_i \in \{0, 1, 2, \ldots, 8, 9\}$ and $x_n$ is the check digit.

(a) Determine the information message for the received words 175142, 275143.

(b) Decode the received word 920438 if it is known that only the fourth bit was subject to error during transmission.

### 4.1.3 Barcodes

Example 4.5 **European Article Number**

The barcodes shown in Figure 4.1 are examples of a widely-used product identification code called the **European Article Number (EAN)**, which was adopted as a standard in 1976. It was developed from the Universal Product Code (UPC) which was introduced in the United States in 1973 (see Exercise 4.25). Some examples of EANs are shown in Table 4.1, the first one referring to Figure 4.1.

Each number consists of 13 decimal digits. The first two digits are allocated to countries, with for example 30 belonging to France, 49 to Japan, 50 to the United Kingdom, 80 to Italy and 93 to Australia. The next five digits are the manufacturer's number, and you can see from Table 4.1 that, for example, 00127 identifies Kellogg's and 00119 belongs to Tesco. The next five digits give the unique number identifying the product, so for example 16319 and 01699 refer respectively to Special K and Corn Flakes made by Kellogg's. The last digit is the check digit, which enables the store's computer to check that it has the correct identification number for the product being scanned. Notice that the price of the product is not part of the barcode. This information is held in the memory of the store's computer which informs the electronic cash register of the price of each item as its barcode is scanned. The EAN for books is based on the ISBN and begins with the digits 978, the next nine digits being the first nine digits of the ISBN – see Figure 4.2 and Exercise 4.93.

If an EAN is denoted by $x_1x_2x_3 \ldots x_{11}x_{12}x_{13}$ where each $x_i$ is a decimal digit, that is, $x_i \in \{0, 1, 2, 3, \ldots, 7, 8, 9\}$, then the check digit $x_{13}$ is calculated so that the **check sum**

$$S = x_1 + x_3 + x_5 + x_7 + x_9 + x_{11} + 3(x_2 + x_4 + x_6 + x_8 + x_{10} + x_{12}) + x_{13}$$

(4.4)

is a multiple of 10, that is, $S$ satisfies the **check equation**

$$S \equiv 0 \,(\mathrm{mod}\ 10)$$

(4.5)

*Table 4.1* Some EANs

| Product | EAN |
|---|---|
| Kellogg's Special K | 5000127163195 |
| Tea bags | 5000183069516 |
| Kellogg's Corn Flakes | 5000127016996 |
| Tesco plain flour | 5000119101594 |
| Abbot Ale | 5010549000213 |
| Maille Provençale mustard | 3036817800295 |
| Sacla pesto sauce | 8001060375109 |
| Book in Figure 4.2 | 9780138340940 |

For example, the first item in Table 4.1 when substituted into (4.5) gives

$$5 + 0 + 1 + 7 + 6 + 1 + 3(0 + 0 + 2 + 1 + 3 + 9) + 5 = 70 \equiv 0 \,(\text{mod } 10)$$

You can verify that the other entries in Table 4.1 are all correct (see Exercise 4.18).

If an error of magnitude $e$ occurs in any one of the digits $x_i$ then $S$ in (4.4) changes by $e$ or $3e$ according to whether $i$ is odd or even. In either case (since $0 \leqslant e \leqslant 9$) the change in $S$ will not be 0(mod 10), so (4.5) will be violated. Hence the EAN code detects all single errors. However, unless it is known which particular digit of an EAN is incorrect, the code cannot correct single errors.

A common type of error which can occur when entering digits onto a keyboard, or when reading them aloud, is inadvertently to **transpose** (that is, interchange) two adjacent digits. If, for example, the fifth and sixth digits $x_5$ and $x_6$ in an EAN are transposed, then in the check sum (4.4) $x_5$ is replaced by $x_6$ and $3x_6$ by $3x_5$. The net change in the check sum is therefore

$$-x_5 + x_6 - 3x_6 + 3x_5 = 2(x_5 - x_6)$$

Hence if $x_5 - x_6 = \pm 5$, then the change in the check sum will be $\pm 10$, so the check sum will remain congruent to 0(mod 10) and the transposition error will go undetected. Clearly, the same applies for the transposition of any two adjacent digits which differ by 5. However, all other errors of transposition of adjacent digits will be detected. For example, if the code for the tea bags in Table 4.1 was incorrectly read as 5000183065916 (the digits $x_{10} = 9$ and $x_{11} = 5$ having been transposed) then the check sum (4.4) becomes

$$S = 5 + 0 + 1 + 3 + 6 + 9 + 3(0 + 0 + 8 + 0 + 5 + 1) + 6 = 72$$

and since $S \equiv 2(\text{mod } 10)$ the error is detected. However if, for example, the code for Kellogg's Special K was incorrectly read as 5000172163195, the check sum (4.4) becomes

$$S = 5 + 0 + 1 + 2 + 6 + 1 + 3(0 + 0 + 7 + 1 + 3 + 9) + 5 = 80 \equiv 0 \,(\text{mod } 10)$$

so the transposition of the sixth and seventh digits ($x_6 = 2$, $x_7 = 7$, $x_6 - x_7 = -5$) is not detected.

Shortened 12-digit and 8-digit forms of EAN are also in use, and some retail chainstores have their own in-house barcode systems.

| Example 4.6 | **United States money order** |

Money orders issued by the United States Postal Service carry an identification number $x_1 x_2 \ldots x_9 x_{10} x_{11}$ where again each $x_i$ is a decimal digit, but the check digit $x_{11}$ is now defined as the remainder modulo 9 of the 10-digit number, that is,

$$x_1 x_2 x_3 \ldots x_9 x_{10} \equiv x_{11} \,(\text{mod } 9) \tag{4.6}$$

so that $0 \leqslant x_{11} \leqslant 8$. For example, if the ten digits are 3844809642 then division by 9 gives

$$3844809642 = 9 \times 427201071 + 3$$

so $x_{11} = 3$ and the codeword is 38448096423. An easier way of computing the check digit is to replace (4.6) by the equivalent congruence

$$\sum_{i=1}^{10} x_i \equiv x_{11}(\text{mod } 9) \tag{4.7}$$

(see Exercise 4.26). Using (4.7) with the 10-digit number above gives

$$3 + 8 + 4 + 4 + 8 + 0 + 9 + 6 + 4 + 2 = 48 \equiv 3(\text{mod } 9)$$

showing that $x_{11} = 3$, as before.

If an error occurs in any one of the first ten digits then it is easy to see that the congruence (4.7) will no longer be valid, *except* when 0 is replaced by 9, or 9 by 0. Hence the code detects all single errors in the digits $x_1$ to $x_{10}$ except for these two cases. For example, if $x_6 = 0$ in the 10-digit example above is replaced by 9 then for this new number the sum on the left in (4.7) is $48 + 9 = 57 \equiv 3(\text{mod } 9)$, showing that the check digit is unaltered, so the error in the sixth digit is undetected. However, if an error affects only the check digit $x_{11}$ then this will be detected since (4.7) will be violated. In fact, around 2% of single errors go undetected (see Problem 4.5).

However, the ability of the code to detect errors involving the transposition of two adjacent digits $x_i$ and $x_{i+1}$ is very poor. Indeed, if $i \leqslant 9$ and two such digits are interchanged then the sum on the left side in (4.7) is unaltered, so the check digit does not change and the error is undetected.

The remaining possibility is when $x_{10}$ and $x_{11}$ are interchanged. In this case the new sum of digits on the left in (4.7) is

$$x_1 + x_2 + \ldots + x_9 + x_{11} = x_1 + x_2 + \ldots + x_9 + x_{10} + (x_{11} - x_{10})$$
$$\equiv x_{11}(\text{mod } 9) + (x_{11} - x_{10})(\text{mod } 9)$$
$$\equiv x_{10}(\text{mod } 9) + (2x_{11} - 2x_{10})(\text{mod } 9)$$

This last expression cannot be congruent to $x_{10}(\text{mod } 9)$ unless $x_{11} = x_{10}$, because $x_{11} \leqslant 8$ so $2(x_{11} - x_{10}) \not\equiv 0(\text{mod } 9)$. This shows that the word $x_1x_2x_3 \ldots x_9x_{11}x_{10}$ is not a codeword unless $x_{10} = x_{11}$, in which case transposition of $x_{10}$ and $x_{11}$ makes no difference. Hence the code only detects transposition errors when they involve the check digit.

| Example 4.7 | **United States post code** |

The United States Postal Service introduced the 5-digit **ZIP code** (abbreviated from Zone Improvement Plan) in 1963, and extended it by four digits ('ZIP + 4') in 1983 to facilitate computer sorting. An example of a business return envelope is shown in Figure 4.3, the ZIP code being sandwiched between

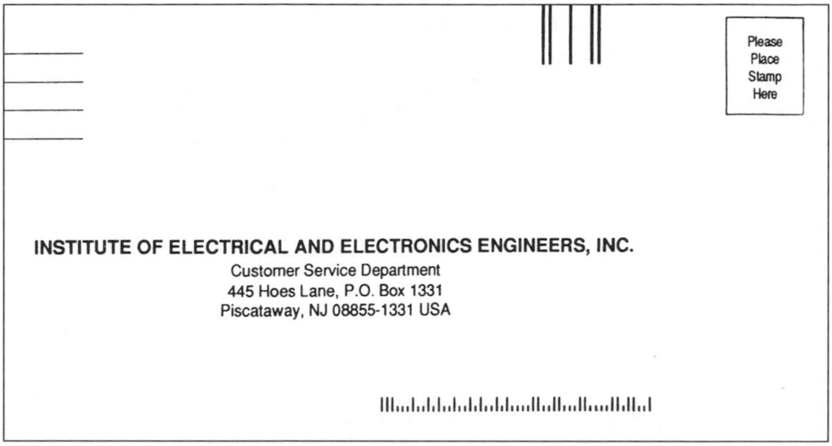

*Figure 4.3* Barcoded envelope

'NJ' (the abbreviation for New Jersey) and 'USA'. The first digit represents one of ten geographical areas, usually a group of states, from 0 in the north-east (including New Jersey in Figure 4.3) to 9 in the far west. The next two digits (88 in Figure 4.3) identify a mail distribution centre, followed by two digits indicating the town or local post office (55 in Figure 4.3).

In the extra four digits (1331 in Figure 4.3), the first two represent a delivery sector – for example, a group of streets or several office buildings; and the final two digits further narrow down the area – for example, one floor in a large office building or a department in a large company.

The barcode below the address actually represents a 10-digit codeword, the extra digit being the check digit. For a ZIP + 4 code there are 52 bars, as shown in Figure 4.3. The single long bars at either end do not form part of the codeword, and simply act as a 'frame'. Each digit is represented by a block of five bars, two long and three short. If 0 represents a short bar and 1 a long bar, then this gives the 'two-out-of-five' 5-bit code introduced in Problem 2.3. Each block *abcde* (where two of the bits are 1 and the other three are zero) is converted to a decimal digit by the expression $7a + 4b + 2c + d$, except for 11000 which represents 0. For example, **▮▮▮▮▮** becomes 10010, which gives

$$7 \times 1 + 4 \times 0 + 2 \times 0 + 1 = 8$$

The complete table of conversions is:

| *Barcode* | 00011 | 00101 | 00110 | 01001 | 01010 |
|---|---|---|---|---|---|
| *Decimal digit* | 1 | 2 | 3 | 4 | 5 |
| *Barcode* | 01100 | 10001 | 10010 | 10100 | 11000 |
| *Decimal digit* | 6 | 7 | 8 | 9 | 0 |

There is no other possible arrangement of three zeros and two ones. The barcode in Figure 4.3, ignoring the frame bars, can now be interpreted as follows:

| 11000 | 10010 | 10010 | 01010 | 01010 | 00011 | 00110 | 00110 | 00011 | 01100 |
|-------|-------|-------|-------|-------|-------|-------|-------|-------|-------|
| 0 | 8 | 8 | 5 | 5 | 1 | 3 | 3 | 1 | 6 |

The code is a decimal parity code, defined in Exercise 4.17, so a codeword $x_1 x_2 \ldots x_{10}$ satisfies the condition that the sum of the digits is a multiple of 10, that is,

$$\sum_{i=1}^{10} x_i \equiv 0 (\text{mod } 10) \tag{4.8}$$

where $x_{10}$ is the check digit. For the example in Figure 4.3 the check digit is 6, and (4.8) becomes

$$0 + 8 + 8 + 5 + 5 + 1 + 3 + 3 + 1 + 6 = 40 \equiv 0 (\text{mod } 10)$$

If the barcode scanner makes a single error when reading a barcode (that is, a long bar is read as a short bar, or a short bar as a long bar) then a 5-bit block which breaks the two-out-of-five pattern will be produced, so such an error is always detected. If there is a single error which occurs in the $i$th block, then the corresponding decimal digit $x_i$ will be undefined. It can, however, be determined uniquely from (4.8), so all such errors can be corrected. For example, the barcode

| 00101 | 11000 | 10001 | 10010 | 00101 | 10000 | 10100 | 01100 | 11000 | 10001 |
|-------|-------|-------|-------|-------|-------|-------|-------|-------|-------|
| 2 | 0 | 7 | 8 | 2 | $x_6$ | 9 | 6 | 0 | 7 |

contains an error in the sixth block, which has only one long bar. In (4.8) the sum of digits is

$$2 + 0 + 7 + 8 + 2 + x_6 + 9 + 6 + 0 + 7 = 41 + x_6$$
$$\equiv 0 (\text{mod} 10)$$

showing that $x_6 = 9$.

Furthermore, the code can detect two errors provided they occur in the same block of five bits. Such pairs of errors can occur in two ways. Firstly, if a short bar is incorrectly read as 1, and a long bar as 0, then the block of five bits still contains two 1's and so is decoded as a decimal digit which is different from what it should be. This error is detected because (4.8) will not hold. However, the error cannot be corrected since there is no way of knowing which block of five bits is incorrect. The second way in which a pair of errors can occur in a block is when two short bars are read as 1's, or two long bars are read as 0's. In either case the block no longer contains two 1's, so the errors are detected. The correct decimal digit for this block can then be determined from (4.8) in exactly the same way as for a single error.

| | |
|---|---|
| Exercise | **4.18** Verify that the entries in Table 4.1 satisfy the condition (4.5). |
| Exercise | **4.19** Determine the check digit for Tesco's canned peaches, for which the EAN product identity number is 10540. |
| Exercise | **4.20** Determine whether the following are EANs:<br>(a) 5011346500121   (b) 5018374314156   (c) 8002210500204. |
| Exercise | **4.21** For the *Guardian* newspaper on Wednesdays the EAN defined by (4.4) and (4.5) is 9770261307330, where the first three digits denote the product type (that is, newspaper or magazine), $x_4 = 0$ retrieves the price from the memory of the retailer's computer, the next seven digits are the identifier for the *Guardian* and $x_{12}$ is the day of the week (Monday $= 1$, Tuesday $= 2$, Wednesday $= 3$, and so on). |

(a) Check that this EAN is correct.

(b) The identifier for the *Yorkshire Post* is 9631494. Determine the EAN for this newspaper on Thursdays.

| | |
|---|---|
| Exercise | **4.22** Suppose that you receive a fax containing an EAN, but one of the digits is illegible. Show that this error can be corrected. Determine the correct EAN if $30576x0100178$ is received (actually Volvic bottled water). |
| Exercise | **4.23** Suppose that in (4.4) the factor 3 multiplying the sum of the even-suffix digits is replaced by 4 or 5. Explain why in each of these cases some single-digit errors are not detected. What happens in each case for transposition errors involving adjacent digits? |
| Exercise | **4.24** Two kinds of Safeway canned fruits carry the barcodes 05022632 and 05034673, and a pack of Safeway ground coffee has 05038473. The last digit is the check digit. Find a check sum like that in (4.4) which is 0(mod 10) in each case. |
| Exercise | **4.25** The **Universal Product Code** (UPC) consists of codewords $x_1 x_2 \ldots x_{11} x_{12}$ with 12 decimal digits. The first digit identifies the type of product according to the scheme shown in the table: |

| First digit | Product type |
|---|---|
| 2 | random weight item, e.g. cheese |
| 3 | pharmaceutical |
| 4 | reduced-price |
| 5 | cents-off coupon |
| 9 | book, newspaper, magazine |
| 0, 6, 7 | all other US brands |

The digits $x_2 x_3 x_4 x_5 x_6$ represent the manufacturer, the next five digits identify the product, and $x_{12}$ is the check digit. This is determined by

$$3(x_1 + x_3 + x_5 + x_7 + x_9 + x_{11}) + x_2 + x_4 + x_6 + x_8 + x_{10} + x_{12} \equiv 0(\text{mod } 10)$$

which replaces the condition (4.5).

For example, a manufacturer's coupon which gives 50 cents off a box of Kellogg's Rice Crispies has UPC $53800051150x_{12}$. Determine the check digit. Show that the UPC has the same error detection properties as the EAN code.

**Exercise**    **4.26**   Prove by writing

$$x_1x_2\ldots x_{10} = x_1 10^9 + x_2 10^8 + \ldots + x_{10}$$

that (4.7) is equivalent to (4.6). (Compare with the results in Sections 1.3.7 and 2.2.6 on divisibility by 9.)

**Exercise**    **4.27**   Codewords $x_1x_2x_3x_4x_5$ with decimal digits are defined by $x_1x_2x_3x_4 \equiv x_5(\mathrm{mod}\ 9)$, where $x_5$ is the check digit. Show that:

(a) $x_1 + x_2 + x_3 + x_4 \equiv x_5(\mathrm{mod}\ 9)$

(b) 1.8% of single-digit errors are undetected

(c) for transposition errors, only those involving the check digit are detected.

**Exercise**    **4.28**   Show that if two non-adjacent digits in the 11-digit money order code in Example 4.6 are interchanged, then only those transpositions involving the check digit are detected.

**Exercise**    **4.29**   Several companies (including Federal Express and United Parcel Services) use a code with ten decimal digits. The check digit $x_{10}$ is the remainder of $x_1x_2\ldots x_9(\mathrm{mod}\ 7)$, so $0 \leqslant x_{10} \leqslant 6$.

(a) Determine the condition for single errors to be undetected.

(b) Show that 6.0% of single errors are undetected.

(c) Show that (unlike the code in Example 4.6) most errors involving transposition of adjacent digits are detected.

**Exercise**    **4.30**   A code consists of codewords $x_1x_2x_3x_4x_5$ with five decimal digits, and the check digit $x_5$ is defined by

$$x_1 + 3x_2 + 7x_3 + 9x_4 + x_5 \equiv 0(\mathrm{mod}\ 10)$$

(a) Determine the codeword when the first four digits are 4193.

(b) Show that all errors in a single digit are detected.

(c) Show that transposition errors involving adjacent digits are detected unless $|x_i - x_{i+1}| = 5$, $i \leqslant 4$.

(d) Show that the *weights* 1, 3, 7, 9 multiplying $x_1$, $x_2$, $x_3$, $x_4$ respectively are the only possible integers less than 10 which can be used if the code is to detect all single errors.

**Exercise**    **4.31**   The **International Standard Serial Number** (ISSN) applies to publications which are issued in successive parts and are intended to be continued indefinitely. Examples are newspapers, magazines, annual yearbooks and academic journals. The ISSN is an 8-digit codeword with decimal digits, but the check digit $x_8$ can take in addition the value 10, which is denoted by the Roman numeral X. This check digit is defined by $S \equiv 0(\text{mod } 11)$, where

$$S = 8x_1 + 7x_2 + 6x_3 + 5x_4 + 4x_5 + 3x_6 + 2x_7 + x_8$$

(a) Determine the check digit for the *Journal of Classical Music*, for which the first seven digits of the ISSN are 0317847.

(b) Show that the code detects all single errors. (*Hint*: find the change in $S$ if $x_i$ changes by an amount $e$, and use the fact that 11 is a prime number.)

**Exercise** **4.32** Determine the check digit for a ZIP code 80321-1432.

**Exercise** **4.33** Determine the ZIP code and check digit for the barcode

‖₁₁₁‖‖₁ₗₗₗₗₗₗ‖‖‖₁ₗₗₗₗₗₗ‖‖ₗₗₗₗₗₗₗ‖ₗ‖ₗₗₗₗₗₗₗ‖ₗ‖

**Exercise** **4.34** The following barcode contains exactly one error. Determine the correct ZIP code.

‖ₗₗₗₗₗₗ‖ₗ‖ₗ‖‖‖ₗₗₗₗₗ‖‖ₗₗₗₗ‖ₗ‖ₗₗₗ‖ₗₗₗₗ‖‖ₗₗₗₗ‖ₗₗₗ‖

**Exercise** **4.35** The following barcode contains exactly two errors. Determine the correct ZIP code.

‖ₗ‖ₗₗₗₗₗₗ‖‖ₗ‖ₗ‖ₗ‖ₗₗ‖‖‖ₗₗ‖ₗ‖‖ₗₗₗ‖‖ₗₗ‖ₗₗₗ‖ₗₗₗ‖‖

**Exercise** **4.36** Alternatives to the 'two-out-of-five' code (used in Example 4.7 to represent the decimal digits 0 to 9), use 4-bit codewords *abcd* as follows:

(a) $8a + 4b + 2c + d$ ('8-4-2-1 code')

(b) $6a + 3b + c + d$ ('6-3-1-1 code')

(c) $(0011)_2$ added to codewords in (a).

For example, 0111 corresponds to

$$8 \times 0 + 4 \times 1 + 2 \times 1 + 1 \times 1 = 7 \quad \text{in scheme (a)}$$
$$6 \times 0 + 3 \times 1 + 1 \times 1 + 1 \times 1 = 5 \quad \text{in scheme (b)}$$

and 7 in scheme (c) has representation

$$(0111)_2 + (0011)_2 = (1010)_2$$

Construct a table of codewords for each of these three cases. What can you say about (a) and (c)?

## 4.2 Hamming distance

In Example 4.2 where four commands are represented by 00, 01, 10 and 11, if a single transmission error occurs then the wrong command is executed. This is because an error in just one bit of a codeword turns it into a different codeword

– the codewords are 'too close together'. This idea was quantified by the American mathematician and computer scientist R.W. Hamming in 1950.

Recall from Section 2.4.4, that $\mathbb{Z}_p$ is the set of integers $0, 1, 2, \ldots, p-1$ subject to modulo $p$ arithmetic. A string of digits $x = x_1 x_2 \ldots x_{n-1} x_n$, where each belongs to $\mathbb{Z}_p$, is called a **word** of **length** $n$. Since each $x_i$ can take $p$ values there are $p^n$ such words in total. A $p$-**ary code** $C$ of length $n$ is a set of such words, and if $x \in C$ then $x$ is called a **codeword**. In particular, if $p = 2$ then $x_i = 0$ or 1 and $C$ is a **binary code**; if $p = 3$ then $x_i = 0$, 1 or 2 and $C$ is a **ternary code**. In this chapter, only codes where all the codewords have the same length will be considered. Some codes where this is not the case are in Problems 3.1, 6.20 and 6.21.

Let $x$ and $y$ be two words of length $n$. Then the **Hamming distance** $d(x, y)$ between them is the number of places in which $x$ and $y$ differ. For example, for a ternary code of length 5 then $d(01221, 10211) = 3$ since the two words differ in the first, second and fourth places.

Another way of looking at $d(x, y)$ is to regard it as the smallest number of digits of $x$ which have to be changed in order to produce $y$. The use of the term 'distance' is justified, because $d(x, y)$ has three of the most important mathematical properties of the geometrical distance between two points in space. The first two of these properties are obvious consequences of the definition:

(1) $d(x, y) = 0$ if and only if $x = y$, that is, $x_i = y_i$ for $i = 1, 2, \ldots, n$. Otherwise $d(x, y) > 0$.

(2) $d(x, y) = d(y, x)$.

The third property is named after the following geometrical fact concerning a plane triangle: if $X$, $Y$, $Z$ are three points in a plane, with $D(X, Y)$ denoting the distance between $X$ and $Y$ then

$$D(X, Y) \leqslant D(X, Z) + D(Z, Y) \quad \text{(see Figure 4.4)}$$

The corresponding result for the Hamming distance has the same form:

(3) **Triangle inequality** – for any third word $z = z_1 z_2 \ldots z_n$, then

$$d(x, y) \leqslant d(x, z) + d(z, y) \tag{4.9}$$

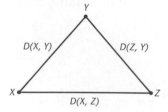

*Figure 4.4* Triangle inequality

To prove (4.9), suppose $x$ is changed into $y$ by going via $z$. The number of digit changes involved is $d(x, z)$ from $x$ to $z$, and $d(z, y)$ from $z$ to $y$, giving a total of $t = d(x, z) + d(z, y)$ digit changes. However, $d(x, y)$ is the *smallest* possible number of digit changes needed to go from $x$ to $y$, and so cannot exceed $t$. In other words, the inequality (4.9) must be satisfied.

The concept of nearest-neighbour decoding introduced in Example 4.2 can now be re-interpreted: select as the transmitted codeword the one which is *nearest* to the received word, in the sense that the Hamming distance between the two words is least.

A crucial parameter which affects the error detection and correction properties of a code $C$ is the 'overall closeness' of words, as measured by the **minimum distance** $\delta(C)$. This is defined as the smallest of the values $d(x, y)$ for all $x$, $y$ in $C$ with $x \neq y$.

| Example 4.8 | **Minimum distances for Example 4.2** |
|---|---|

The first code used in Example 4.2 is $C_1 = \{00, 01, 10, 11\}$, and the distances between all possible pairs of codewords are

$$d(00, 01) = 1, \quad d(00, 10) = 1, \quad d(00, 11) = 2$$
$$d(01, 10) = 2, \quad d(01, 11) = 1, \quad d(10, 11) = 1$$

The minimum distance is therefore $\delta(C_1) = 1$, and it can now be explained why the code $C_1$ is useless, being unable even to detect single errors. Indeed, for *any* code $C$ with $\delta(C) = 1$ there will be at least two codewords $a$ and $b$ for which $d(a, b) = 1$. If $a$ is transmitted, and an error occurs in that one particular digit $a_i$ in which $a$ and $b$ differ, changing it to $b_i$, then the codeword $b$ will be received. This will be assumed correct, so there is no way of detecting that an error has occurred.

Now turn to the second code in Example 4.2, where each message was transmitted twice, so that $C_2 = \{a, b, c, e\}$ where $a = 0000$, $b = 0101$, $c = 1010$, $e = 1111$. The Hamming distances between codewords are now $d(a, b) = 2$, $d(a, c) = 2$, $d(a, e) = 4$, $d(b, c) = 4$, $d(b, e) = 2$, $d(c, e) = 2$. The minimum distance is therefore $\delta(C_2) = 2$, and the code was seen to detect all single errors.

The third repetition code in Example 4.2 is $C_3 = \{a_1, a_2, a_3, a_4\}$ where $a_1 = 000000$, $a_2 = 010101$, $a_3 = 101010$, $a_4 = 111111$. The distances between codewords are now $d(a_1, a_2) = 3$, $d(a_1, a_3) = 3$, $d(a_1, a_4) = 6$, $d(a_2, a_3) = 6$, $d(a_2, a_4) = 3$, $d(a_3, a_4) = 3$, and the minimum distance is $\delta(C_3) = 3$. This code was seen to correct all single errors.

This example illustrates the following important result, which shows how the minimum distance determines the error detecting and correcting properties of a code.

## Theorem: Error detection and correction

Let $C$ be a code whose minimum distance is $\delta$, and assume received words are decoded using the nearest-neighbour principle. Then

(1) $C$ will detect $e$ errors provided

$$\delta \geqslant e + 1 \tag{4.10}$$

(2) $C$ will correct $e$ errors provided

$$\delta \geqslant 2e + 1 \tag{4.11}$$

**Proof**

(1) By the definition of minimum distance, any pair of codewords differs in at least $\delta$ places. In view of (4.10), this means that any pair of codewords differs in at least $e + 1$ places. Suppose a codeword $x$ is transmitted and at most $e$ errors occur during transmission. Then the received word will differ from $x$ in at most $e$ places, and so will not be a codeword. Hence up to $e$ transmission errors are detected.

(2) Suppose again that a codeword $x$ is transmitted and at most $e$ errors occur, so that the received word $z$ differs from $x$ in at most $e$ places, that is,

$$d(x, z) \leqslant e \tag{4.12}$$

If $y$ is any codeword different from $x$, then by the definition of minimum distance

$$\begin{aligned} d(x, y) &\geqslant \delta \\ &\geqslant 2e + 1, \quad \text{using (4.11)} \end{aligned} \tag{4.13}$$

Substituting (4.12) into the triangle inequality (4.9) gives

$$d(x, y) \leqslant e + d(z, y)$$

Combining this with (4.13) produces

$$\begin{aligned} d(z, y) &\geqslant d(x, y) - e \\ &\geqslant 2e + 1 - e = e + 1 \end{aligned}$$

This shows that the distance between the received word $z$ and any codeword $y (\neq x)$ is at least $e + 1$. In view of (4.12), the nearest-codeword to $z$ is therefore $x$, so the nearest-neighbour principle correctly decodes the received word as $x$. Hence up to $e$ transmission errors are corrected.

Rearranging (4.10) and (4.11) gives respectively $e \leqslant \delta - 1$, and $e \leqslant (1/2)(\delta - 1)$. The theorem can therefore be reformulated as follows:

If a code $C$ has minimum distance $\delta$, then $C$ can be used *either* to detect at most $\delta - 1$ errors; *or*, to correct at most $(1/2)(\delta - 1)$ errors if $\delta$ is odd, or $(1/2)(\delta - 2)$ errors if $\delta$ is even.

Remember that when errors are detected but cannot be corrected, then the receiver has to request retransmission of the message.

**Example 4.9** **Applications of the theorem**

(a) Applying the theorem to the three codes in Example 4.8 gives the following:

| Code | $\delta$ | Number of errors detected | Number of errors corrected |
|------|----------|---------------------------|----------------------------|
| $C_1$ | 1 | 0 | 0 |
| $C_2$ | 2 | 1 | 0 |
| $C_3$ | 3 | 2 | 1 |

These results agree with what was found when the repetition codes were discussed in Example 4.2. Notice that it is now revealed in addition that code $C_3$ (when the message is repeated twice) can either correct one error or detect *two* errors, because codewords are a distance at least 3 apart.

(b) Consider a binary code of length 5

$$C = \{11001, 01110, 10100, 00011\}$$

It is easy to show that the minimum distance is $\delta = 3$ (see Exercise 4.37). Hence the code can either detect up to two errors, or correct one error. To illustrate these two cases, suppose firstly that a word $r = 10010$ is received. Its distances from the four codewords are $d(r, 11001) = 3$, $d(r, 01110) = 3$, $d(r, 10100) = 2$, $d(r, 00011) = 2$. Since the third and fourth codewords are equally close to the received word, the nearest-neighbour principle cannot determine which of these two codewords was transmitted, that is, the errors cannot be corrected. However, two errors have been detected, so for example $r$ could have come from 10100 with errors in the second and third bits, or from 00011 with errors in the first and last bits.

Alternatively, suppose 11100 is received. The distances from the codewords are respectively 2, 2, 1, 5 (see Exercise 4.38), showing that the third codeword is nearest to the received word. By the nearest-neighbour principle it is concluded that the third codeword 10100 was transmitted, thereby correcting the single error (in the second bit).

| Exercise | **4.37** Determine the minimum distance for each of the following binary codes: |
|---|---|

(a) $C = \{0111, 0100, 1011, 0001\}$

(b) the code in Example 4.9(b)

(c) $C = \{111111, 010101, 101010, 000110, 100001\}$.

| Exercise | **4.38** In Example 4.9(b), determine the distances of the received word 11100 from each of the four codewords. |
|---|---|

| Exercise | **4.39** For the code in part (c) of Exercise 4.37, determine the distances of each of the following received words from each of the codewords. Hence decode the received words using the nearest-neighbour principle. |
|---|---|

(a) 111010   (b) 111001   (c) 011101.

| Exercise | **4.40** For the code $C_3$ in Example 4.8, decode the following received words using the nearest-neighbour principle: |
|---|---|

(a) 110101   (b) 100100.

| Exercise | **4.41** What is the smallest minimum distance that a code must have in order to correct up to three errors? How many errors can it detect? |
|---|---|

| Exercise | **4.42** What is the minimum distance for the two-out-of-five code used in Example 4.7? |
|---|---|

| Exercise | **4.43** An important early application of error-correcting binary codes was to the transmission of data from the Mariner spacecraft missions to Mars from 1969 to 1972. A code of length 32 with 26 check bits and minimum distance 16 was used. How many errors could it correct? |
|---|---|

| Exercise | **4.44** The 'three-out-of-seven' code consists of all possible binary words of length 7 containing exactly three 1's. |
|---|---|

(a) How many words does the code contain?

(b) What is the minimum distance?

(c) How many errors can the code detect or correct?

| Exercise | **4.45** Show that a code having minimum distance 4 can be used to correct all single errors and simultaneously to detect all double errors. |
|---|---|

| Exercise | **4.46** A code has minimum distance 3. Show that it is not possible to correct all single errors and simultaneously to detect all double errors. That is, show that there are codewords $x$ and $y$ and a received word $r$ such that $r$ comes from $x$ via one transmission error, and from $y$ via two errors. |
|---|---|

## 4.3   Linear codes

### 4.3.1   Definitions for binary codes

Define the *sum* of two binary codewords $x = x_1 x_2 \ldots x_n$ and $y = y_1 y_2 \ldots y_n$ as $z = x + y$, where $z_i \equiv (x_i + y_i)(\bmod 2)$, $i = 1, 2, \ldots, n$. That is, the bits are added term-by-term according to the rules set out in equation (2.41) for modulo 2 addition, namely

$$0 + 0 = 0, \quad 1 + 1 = 0, \quad 1 + 0 = 1 = 0 + 1 \tag{4.14}$$

It follows from (4.14) that $z_i = 0$ if $x_i = y_i$, and $z_i = 1$ if $x_i \neq y_i$.

A **linear binary code** $C$ has the property that the sum of *any* two codewords is also a codeword. That is, if $x$ and $y$ are in $C$ then so is $z = x + y$. In particular, if $y = x$ then the $i$th bit of $x + x$ is either

$$x_i + x_i = 1 + 1 = 0, \quad \text{or} \quad x_i + x_i = 0 + 0 = 0$$

showing that any linear code always contains the **zero word 0**, all of whose bits are zero.

## Example 4.10  Linear and non-linear codes

(a) Consider the binary code $C = \{000, 100, 101, 001\}$. The bits of the sum $100 + 101$ are respectively $1 + 1 = 0, 0 + 0 = 0, 0 + 1 = 1$, so $100 + 101 = 001$ which is also in $C$. Similarly, the other sums of pairs of codewords are

$$000 + 100 = 100, \quad 000 + 101 = 101, \quad 000 + 001 = 001,$$
$$100 + 001 = 101, \quad 101 + 001 = 100$$

all of which are also codewords, so $C$ is linear. It is worth emphasizing that the codewords are *not* binary numbers, so their sums are *not* obtained using binary addition (Section 2.2.2.1, page 54). Simply add the bits in corresponding positions and apply (4.14).

(b) The even parity binary code in Example 4.3 is a linear code. To see why this is so, recall that any two codewords $x$ and $y$ satisfy the condition (4.1), namely $\Sigma x_i \equiv 0, \Sigma y_i \equiv 0 \pmod 2$. Hence for $z = x + y$, the sum of bits of $z$ is

$$\Sigma z_i = \Sigma(x_i + y_i) = \Sigma x_i + \Sigma y_i \equiv 0 \pmod 2$$

This shows that $z$ also satisfies (4.1), and so is a codeword.

(c) The code $C = \{0000, 1010, 0111\}$ is *not* linear, because $1010 + 0111 = 1101$, which is not a codeword.

In view of the theorem in Section 4.2, it is necessary to compute the minimum distance $\delta$ for a code in order to determine its error-handling properties. One reason linear codes are important is because the value of $\delta$ can be found much more easily than by calculating the distances between all possible pairs of codewords (see Example 4.8). Another definition is required: the **weight** $w(x)$ of a binary codeword $x$ is the number of 1's in $x$. Consider now the sum $z$ of two codewords $x$ and $y$. It was seen that $z_i = 1$ when $x_i \neq y_i$, so the number of 1's in $z$ is equal to the number of places where $x$ and $y$ differ. From the definition of $d(x, y)$ it therefore follows that $w(z) = w(x + y)$ satisfies

$$w(x + y) = d(x, y) \tag{4.15}$$

In particular, setting $y = \mathbf{0}$ in (4.15) shows that for any codeword $x$, its weight is given by

$$w(x) = d(x, \mathbf{0}) \tag{4.16}$$

The key relationship between minimum distance and weight can now be established.

## Theorem

The minimum distance $\delta$ for a linear binary code $C$ is equal to the smallest non-zero weight of codewords, that is,

$$\delta = \min_{x \neq 0} w(x) \tag{4.17}$$

**Proof** Denote by $w^*$ the minimum value of $w(x)$ in (4.17), and let $u$ be a codeword for which $w(u) = w^*$. From (4.16) it follows that $w^* = d(u, \mathbf{0})$. From the definition of minimum distance it also follows that $\delta \leqslant d(u, \mathbf{0})$ since both $u$ and $\mathbf{0}$ are codewords. Combining these results produces $\delta \leqslant w^*$.

However, it is easy to show (see Exercise 4.50) that $\delta \geqslant w^*$. The only way that $\delta$ can be both greater than or equal to $w^*$, and less than or equal to $w^*$ is for $\delta$ to actually equal $w^*$, which is the desired result (4.17).

## Example 4.11 Applications of the theorem

(a) The weights of the non-zero codewords in the code in Example 4.10(a) are $w(100) = 1$, $w(101) = 2$, $w(001) = 1$. Hence by (4.17) the minimum distance is $\delta = 1$.

(b) It is easy to verify (Exercise 4.47) that the code $C = \{000000, 010101, 101010, 111111\}$ is linear. The weights are $w(010101) = 3$, $w(101010) = 3$, $w(111111) = 6$, so by (4.17) the code has minimum distance $\delta = 3$.

---

**Exercise** **4.47** Verify that the code $C$ in Example 4.11(b) is linear.

**Exercise** **4.48** Determine which of the following sets of codewords constitutes a linear code:

(a) 000, 011, 100, 111

(b) 000, 110, 100, 111

(c) 00000, 10111, 01110, 11001

(d) 00000, 11100, 10011, 01111.

In each case find the minimum distance.

**Exercise**   **4.49**   Why can you deduce immediately that neither the two-out-of-five code in Example 4.7, nor the three-out-of-seven code in Exercise 4.44 is linear?

**Exercise**   **4.50**   For a linear binary code $C$ with minimum distance $\delta$, let $a$ and $b$ be two codewords such that $d(a, b) = \delta$. Use (4.15) to prove that $\delta \geqslant w^*$, where $w^*$ is the minimum of the weights of all non-zero codewords in $C$.

**Exercise**   **4.51**   For a code $C$ containing $N$ codewords, how many comparisons do you have to make in order to determine the minimum distance

(a) when $C$ is linear?

(b) when $C$ is not linear?

**Exercise**   **4.52**   Prove that if $x = x_1 x_2 \ldots x_n$ is a codeword for a linear binary code, then $x_1^2 + x_2^2 + \ldots + x_n^2 \equiv 0 \pmod 2$ if and only if $w(x)$ is even.

**Exercise**   **4.53**   For two binary codewords $x = x_1 \ldots x_n$, $y = y_1 \ldots y_n$, define $x * y$ to be the codeword whose bits are $x_1 y_1, x_2 y_2, \ldots, x_n y_n$. Prove that

$$w(x + y) = w(x) + w(y) - 2w(x * y).$$

(*Hint*: let $s$ = number of bits for which $x$ has 1, $y$ has 0
$\phantom{(Hint: let} t$ = number of bits for which $x$ has 0, $y$ has 1
$\phantom{(Hint: let} p$ = number of bits for which $x$ has 1, $y$ has 1

and express each side of the identity in terms of these numbers.)

**Exercise**   **4.54**   Let $C$ be the code consisting of all possible binary words of length $n$. Use the result in Exercise 3.54 to show that exactly half the codewords have even weight.

## 4.3.2   Matrix representation of binary codes

**Example 4.12**   **Check equations**

Consider a binary code $C_1$ of length 3 whose codewords $x_1 x_2 x_3$ satisfy the condition $x_1 + x_2 = 0$. There are eight binary words of length 3, and selecting those satisfying this condition gives $C_1 = \{000, 001, 110, 111\}$. Since $\delta(C_1) = 1$, the code does not even detect single errors. Now suppose that an extra condition $x_1 + x_3 = 0$ is applied. Only the first and last codewords in $C_1$ satisfy this, so the code which satisfies *both* conditions is $C_2 = \{000, 111\}$. Clearly $\delta(C_2) = 3$, so this code can correct all single errors. The two conditions are called **parity-check** (or simply **check**) **equations**.

This example illustrates a general idea in the construction of linear codes. The objective is to add conditions in the form of linear check equations so as to increase the minimum distance between codewords, and hence improve the error handling properties. However, this will reduce the number of codewords, and therefore reduce the information-carrying capacity of the code. For example, $C_1$ contains four messages, but only two messages can be sent using $C_2$ (for example, 'yes' and 'no').

In general, because the conditions to be satisfied by the bits of codewords take the form of linear equations, these are conveniently handled using the language of matrices. Strictly speaking, the 'equations' are actually congruences modulo 2, but it is often convenient to express them as ordinary equalities, except where ambiguities may arise. It is assumed that you have encountered previously the notation of matrices and vectors, and the concept of their product. A check equation

$$a_1 x_1 + a_2 x_2 + \ldots + a_n x_n = 0$$

can be written in the compact form $\mathbf{ax} = \mathbf{0}$, where $\mathbf{a} = [a_1, a_2, \ldots, a_n]$ is a **row vector** with **components** $a_1, a_2, \ldots, a_n$, $\mathbf{x}$ is the **code vector**

$$\mathbf{x} = \begin{bmatrix} x_1 \\ x_2 \\ \vdots \\ x_n \end{bmatrix}$$

(4.18)

and

$$\mathbf{ax} = a_1 x_1 + a_2 x_2 + \ldots + a_n x_n$$

(4.19)

is the **scalar product** of $\mathbf{a}$ and $\mathbf{x}$. Notice that the code vector in (4.18) is written as a single column, with the digits of the codeword $x = x_1 x_2 \ldots x_n$ read from left to right, and stacked from top to bottom. It is sometimes convenient to refer to the scalar product of two vectors in row form, this still being defined as the element-by-element product displayed in (4.19).

If there are several check equations, then they can be written in the combined form

$$\underbrace{\begin{bmatrix} a_1 & a_2 & \ldots & a_n \\ b_1 & b_2 & \ldots & b_n \\ \cdot & \cdot & \ldots & \cdot \\ \cdot & \cdot & \ldots & \cdot \end{bmatrix}}_{H} \mathbf{x} = \begin{bmatrix} 0 \\ 0 \\ \vdots \\ 0 \end{bmatrix} = \mathbf{0}$$

Here $H$ is the **check matrix**, whose rows consist of the coefficients of the check equations, and $\mathbf{0}$ is the **zero vector** all of whose entries are zero. It is always clear from the context whether $\mathbf{0}$ represents the zero vector or zero word. Notice that the **product** $H\mathbf{x}$ is defined as a column vector whose $i$th component is the scalar product of the $i$th row of $H$ with the code vector $\mathbf{x}$. For binary codes all the entries in $H$ are 0 or 1, and $H$ is called a **binary matrix**.

**Example 4.13**  **Check matrix**

Using (4.19) the two check equations in Example 4.12, namely

$$1x_1 + 1x_2 + 0x_3 = 0, \quad 1x_1 + 0x_2 + 1x_3 = 0$$

can be written as the scalar products

$$[1, 1, 0]\mathbf{x} = \mathbf{0}, \quad [1, 0, 1]\mathbf{x} = \mathbf{0} \tag{4.20}$$

where

$$\mathbf{x} = \begin{bmatrix} x_1 \\ x_2 \\ x_3 \end{bmatrix}$$

Combining the two expressions in (4.20) gives

$$\begin{bmatrix} 1 & 1 & 0 \\ 1 & 0 & 1 \end{bmatrix} \mathbf{x} = \begin{bmatrix} 0 \\ 0 \end{bmatrix} = \mathbf{0} \tag{4.21}$$

It should be stressed that (4.21) is simply an equivalent way of writing (4.20). The expression (4.21) can be expressed compactly as

$$H\mathbf{x} \equiv \mathbf{0}(\text{mod } 2) \tag{4.22}$$

where the binary matrix

$$H = \begin{bmatrix} 1 & 1 & 0 \\ 1 & 0 & 1 \end{bmatrix} \tag{4.23}$$

is the check matrix. Codewords $x_1 x_2 x_3$ are defined as solutions of the check equations (4.22) – these were found in Example 4.12 to be 000 and 111.

There would be little point in introducing check matrices merely as a tidy way of expressing check equations. The important reason is that the linear nature of a code is a built-in property of the check matrix approach. To see this, suppose $x$ and $y$ are any two codewords in a code $C$ satisfying the same set of check equations, so that in matrix form

$$H\mathbf{x} = \mathbf{0}, \quad H\mathbf{y} = \mathbf{0}$$

If $z = x + y$, then

$$H\mathbf{z} = H(\mathbf{x} + \mathbf{y}) = H\mathbf{x} + H\mathbf{y} = \mathbf{0} + \mathbf{0} = \mathbf{0}$$

showing that $z$ also satisfies the check equations, and so belongs to $C$. Therefore, by definition, $C$ is a linear code. In other words, a linear binary code $C$ is precisely the set of codewords which satisfy the condition

$$H\mathbf{x} \equiv \mathbf{0}(\text{mod } 2) \tag{4.24}$$

where $\mathbf{x}$ is the code vector in (4.18).

If there are $m$ check equations, then $H$ in (4.24) has $m$ rows and $n$ columns, and is said to have **dimensions** $m \times n$, or to be an $m \times n$ matrix.

---

**Example 4.14**   **Generating a linear code**

In Examples 4.12 and 4.13, the codewords were obtained by simply finding from amongst the total set of words of length 3, those words which satisfied the check equations (4.22). This is a lengthy procedure if $n$ is large, and a more systematic way of finding the codewords is now demonstrated. Suppose a check matrix is

$$H = \begin{bmatrix} 1 & 0 & 1 & 0 \\ 0 & 1 & 1 & 1 \end{bmatrix}$$

(4.25)

The corresponding check equations (4.24) are obtained by writing the rows of $H$ as the coefficients of the equations. For example, the first row of $H$ gives $1x_1 + 0x_2 + 1x_3 + 0x_4 = 0$, or

$$x_1 + x_3 = 0$$

(4.26)

and similarly the second row of $H$ gives

$$x_2 + x_3 + x_4 = 0$$

(4.27)

The codewords are the solutions of (4.26) and (4.27). Recall from Section 2.4.4 that additive inverses for $\mathbb{Z}_2$ are $-0 = 0$, $-1 = 1$, so since $x_i \in \mathbb{Z}_2$ it follows that $-x_i = x_i$. Equation (4.26) can therefore be rearranged as

$$x_1 = -x_3 = x_3$$

(4.28)

and (4.27) as

$$x_2 = -x_3 - x_4 = x_3 + x_4$$

(4.29)

The bits $x_1$ and $x_2$ have now been expressed in terms of $x_3$ and $x_4$, which can be regarded as two *independent* bits, each taking the values 0 or 1. There are four possibilities in all, as set out in the table.

| $x_1$ | $x_2$ | $x_3$ | $x_4$ |
|-------|-------|-------|-------|
| 0 | 0 | 0 | 0 |
| 0 | 1 | 0 | 1 |
| 1 | 1 | 1 | 0 |
| 1 | 0 | 1 | 1 |

For each pair of values of $x_3$ and $x_4$, the corresponding values of $x_1$ and $x_2$ in the first two columns are obtained from (4.28) and (4.29). For example, when $x_3 = 1$, $x_4 = 0$ then $x_1 = x_3 = 1$, $x_2 = x_3 + x_4 = 1$. From the table you can see that there are four codewords 0000, 0101, 1110, 1011. Using (4.17), the minimum distance is $\delta = \min[w(0101), w(1110), w(1011)] = 2$.

Since $x_3$ and $x_4$ can be chosen arbitrarily they are called the **information bits**. The other two bits are called the **check bits**, and these are determined uniquely by the information bits, as shown in the table above. The information message $x_3x_4$ is said to be **encoded** into the codeword $x_1x_2x_3x_4$.

---

The preceding example illustrates that the problem of constructing a linear code relies on choosing a suitable check matrix. Notice that the equations (4.26) and (4.27) were easy to solve because of the two check bits $x_1$ and $x_2$, only $x_1$ appeared in the first equation and only $x_2$ in the second equation. To preserve this pattern, if there are, say, three check bits and equations for a code of length 5, write the check equations in the form

$$
\begin{aligned}
x_1 \qquad\quad &+ a_1x_4 + a_2x_5 = 0 \\
x_2 \quad &+ b_1x_4 + b_2x_5 = 0 \\
x_3 &+ c_1x_4 + c_2x_5 = 0
\end{aligned}
\tag{4.30}
$$

where $x_4$ and $x_5$ are the information bits. As in Example 4.14, these can be rewritten using modulo 2 arithmetic to express the check bits as

$$
\left.
\begin{aligned}
x_1 &= a_1x_4 + a_2x_5 \\
x_2 &= b_1x_4 + b_2x_5 \\
x_3 &= c_1x_4 + c_2x_5
\end{aligned}
\right\}
\tag{4.31}
$$

In matrix notation (4.30) becomes $H\mathbf{x} = \mathbf{0}$, where

$$
H = \begin{bmatrix} 1 & 0 & 0 & a_1 & a_2 \\ 0 & 1 & 0 & b_1 & b_2 \\ 0 & 0 & 1 & c_1 & c_2 \end{bmatrix}, \quad
\mathbf{x} = \begin{bmatrix} x_1 \\ x_2 \\ x_3 \\ x_4 \\ x_5 \end{bmatrix}
\tag{4.32}
$$

$$\underleftarrow{\phantom{aa}} I_3 \underrightarrow{\phantom{aa}} \underleftarrow{\phantom{aa}} A \underrightarrow{\phantom{aa}}$$

The first three columns of $H$ in (4.32) have been denoted by $I_3$, which is the $3 \times 3$ **unit matrix** having ones along the **principal diagonal** (top left corner to bottom right corner), and zeros elsewhere. The remaining columns of (4.32) form a $3 \times 2$ matrix $A$.

In a similar way, going back to the check matrix in (4.25), this can be written as $H = [I_2\ A]$ where

$$
I_2 = \begin{bmatrix} 1 & 0 \\ 0 & 1 \end{bmatrix}, \quad A = \begin{bmatrix} 1 & 0 \\ 1 & 1 \end{bmatrix}
$$

In general, to construct a linear binary code of length $n$ with $m$ check bits, the check matrix is

$$
H = [I_m\ A]
$$

where $A$ is an $m \times (n - m)$ matrix. Codewords $x = x_1 x_2 \ldots x_n$ satisfy the check equations $H\mathbf{x} = 0$, where $\mathbf{x}$ is defined in (4.18). The check bits $x_1, x_2, \ldots, x_m$ are expressed as

$$
\begin{bmatrix} x_1 \\ x_2 \\ \vdots \\ x_m \end{bmatrix} = A \begin{bmatrix} x_{m+1} \\ x_{m+2} \\ \vdots \\ x_n \end{bmatrix}
\tag{4.33}
$$

For example, the expressions (4.31) can be written as

$$
\begin{bmatrix} x_1 \\ x_2 \\ x_3 \end{bmatrix} = A \begin{bmatrix} x_4 \\ x_5 \end{bmatrix}
$$

where from (4.32)

$$
A = \begin{bmatrix} a_1 & a_2 \\ b_1 & b_2 \\ c_1 & c_2 \end{bmatrix}
$$

There are $k = n - m$ information bits $x_{m+1}, \ldots, x_n$. Since these can each independently take the value 0 or 1, there are a total of $2^k$ codewords. The number $k$ is called the **dimension** of the code, which is itself referred to as an $[n, k]$ code. To encode an information message consisting of $x_{m+1} x_{m+2} \ldots x_n$, simply prefix it with the check bits $x_1 x_2 \ldots x_m$ computed using (4.33). Some authors put the check bits at the end of the codeword, instead of at the beginning, but this is simply a trivial reordering of the bits.

## Example 4.15  Construction of a linear code

To construct a code of length 7 with three check bits, take as a check matrix

$$
H = \begin{bmatrix} 1 & 0 & 0 & 1 & 1 & 0 & 1 \\ 0 & 1 & 0 & 1 & 1 & 1 & 0 \\ 0 & 0 & 1 & 0 & 1 & 1 & 1 \end{bmatrix}
\tag{4.34}
$$

$$
\longleftarrow I_3 \longrightarrow \longleftarrow A \longrightarrow
$$

The rows of the matrix $A$ in (4.34) are simply the coefficients in the check equations represented by the right-hand side of (4.33). Since $m = 3$, the check bits are $x_1$, $x_2$, $x_3$, so for example the first row of $A$ in (4.34) gives $x_1 = 1x_4 + 1x_5 + 0x_6 + 1x_7$, that is,

$$
x_1 = x_4 + x_5 + x_7
$$

Similarly, the second and third rows of $A$ in (4.34) give

$$
x_2 = x_4 + x_5 + x_6, \quad x_3 = x_5 + x_6 + x_7
$$

As in Example 4.14, selecting all possible values of the information bits $x_4$, $x_5$, $x_6$, $x_7$ gives the total set of codewords. For example, if $x_4 = 1$, $x_5 = 0$, $x_6 = 1$, $x_7 = 0$ then

$$x_1 = 1 + 0 + 0 = 1, \quad x_2 = 1 + 0 + 1 = 0, \quad x_3 = 0 + 1 + 0 = 1$$

so the corresponding codeword is 1011010. The dimension of the code is $k = 7 - 3 = 4$, and there are $2^4 = 16$ codewords. Four more codewords are listed in the table.

| Check bits | | | Information bits | | | |
|---|---|---|---|---|---|---|
| $x_1$ | $x_2$ | $x_3$ | $x_4$ | $x_5$ | $x_6$ | $x_7$ |
| 1 | 1 | 0 | 1 | 0 | 0 | 0 |
| 1 | 1 | 1 | 0 | 1 | 0 | 0 |
| 0 | 1 | 1 | 0 | 0 | 1 | 0 |
| 0 | 0 | 1 | 1 | 1 | 0 | 0 |

No reason was given in the preceding example for the particular choice of the check matrix $H$ in (4.34). It's now time to investigate how $H$ can be constructed so as to produce a code with desired error-handling properties. A major advantage of this approach is that the minimum distance can be determined *without* having to find codewords.

### Single-error detection

For a binary code, a single error means that a bit transmitted as 0 is received as 1, or a transmitted 1 is received as 0. Recall from Section 4.2 that for a code to detect all single errors it must have a minimum distance $\delta$ of at least 2. However it was seen in (4.17) that for a linear binary code, $\delta$ is equal to the smallest of all the weights of non-zero codewords. Hence in this case the requirement $\delta \geqslant 2$ implies that there must be no codewords of weight 1. If $e$ is a word of weight 1, it has all bits zero except for the $i$th (say). For such a word $e$ *not* to be a codeword it must satisfy $He \neq \mathbf{0}$. However, the product $He$ is just the $i$th column of $H$ (see Exercise 4.59), so it follows that no column of $H$ can be zero. In other words, $H$ is the check matrix for a single-error-detecting linear binary code provided it does not contain a column consisting entirely of zeros.

Notice that this result does *not* mean that if $H$ has no zero column then the code only detects errors – it may well have $\delta > 2$ and so also be capable of correcting errors.

**Example 4.16**  **Two check matrices**

(a) The check matrix $H$ in (4.25) does not have a zero column, and therefore generates a code which detects all single errors. It was in fact found that $\delta = 2$ for this code.

(b)  The check matrix

$$H = \begin{bmatrix} 1 & 0 & 0 & 1 \\ 0 & 0 & 1 & 1 \\ 1 & 0 & 1 & 0 \end{bmatrix}$$

has its *second* column zero. Hence 0100, with only the *second* bit non-zero, is a codeword of weight 1. The minimum distance is therefore $\delta = 1$, and the code whose check matrix is $H$ does not detect single errors.

*Single-error correction*
Again referring to Section 4.2, it is now necessary to have $\delta \geqslant 3$, which requires that there must be no codewords with weight 1 or 2. For the latter case, this means that if $e$ is a binary word of weight 2 with just two non-zero bits in positions $i$ and $j$ (say) then $He \neq 0$. However, this implies that the $i$th and $j$th columns $\mathbf{h}_i, \mathbf{h}_j$ of $H$ satisfy $\mathbf{h}_i + \mathbf{h}_j \neq 0$ (see Exercise 4.60). This is equivalent to $\mathbf{h}_i \neq \mathbf{h}_j$, because $-\mathbf{h}_j = \mathbf{h}_j \pmod 2$. In other words, $H$ is the check matrix for a single-error-correcting (s.e.c.) linear binary code provided no column of $H$ is zero, and no two columns of $H$ are equal.

| Example 4.17 | s.e.c. codes |

(a)  The check matrix $H$ in (4.25) has the second and fourth columns equal, and so is not the check matrix for a s.e.c. code. This agrees with the earlier finding that the minimum distance for the code is 2.

(b)  If a code has only two check bits, then the only possible check matrix satisfying the conditions for single-error correction is

$$H = \begin{bmatrix} 1 & 0 & 1 \\ 0 & 1 & 1 \end{bmatrix}$$

since the three columns shown are the only different non-zero columns with two elements. As remarked previously, permutation of the columns of $H$ (for example, to produce the matrix $H$ in (4.23)) does not generate a different code.

(c)  When there are three check bits, the only choices for columns of $A$ which are different from the columns of $I_3$ are those shown in (4.34). The matrix $H$ in (4.34) is therefore the check matrix for a s.e.c. code of length 7. Similarly, selecting any one, two or three of the columns of $A$ in (4.34), and appending them to the columns of $I_3$, results in check matrices for s.e.c. codes of lengths 4, 5 or 6 respectively. For example,

$$H_1 = \begin{bmatrix} 1 & 0 & 0 & 1 & 1 \\ 0 & 1 & 0 & 1 & 1 \\ 0 & 0 & 1 & 0 & 1 \end{bmatrix}, \quad H_2 = \begin{bmatrix} 1 & 0 & 0 & 0 & 1 \\ 0 & 1 & 0 & 1 & 1 \\ 0 & 0 & 1 & 1 & 1 \end{bmatrix}$$

$$(4.35)$$

each produce a s.e.c. code of length 5 with three check bits and two information bits. Notice that in each case the pattern of the first three columns (that is, $I_3$) is retained.

(d) The check matrix

$$H = \begin{bmatrix} 1 & 0 & 0 & 1 & 1 & 0 \\ 0 & 1 & 0 & 1 & 1 & 1 \\ 0 & 0 & 1 & 0 & 1 & 0 \end{bmatrix}$$

has the second and sixth columns equal. You can check that $x = 010001$ with 1's in the second and sixth positions satisfies $Hx = \mathbf{0}$, so $x$ is a codeword of weight 2. The code therefore has minimum distance 2, and so does not correct single errors. Since $H$ does not have a zero column, the code does detect single errors.

In part (c) of the preceding example, it was found that for a s.e.c. code with three check bits, the maximum possible length of the code is 7. In general, for a code of length $n$ with $m$ check bits, the check matrix will be

$$H = [\underset{\substack{m \\ \text{columns}}}{I_m} \quad \underset{\substack{n-m \\ \text{columns}}}{A}] \, {}_{m \, \text{rows}}$$

Since each of the $m$ elements in a column of $A$ can be 0 or 1, there are at most $2^m$ different possible columns which can be chosen for $A$. However, in order for the code to have the s.e.c. property, the zero column and the $m$ columns of $I_m$ must be excluded from $A$, reducing the total number of possible column choices to $2^n - m - 1$. That is, for a s.e.c. binary code, the number $(n - m)$ of columns of $A$ must satisfy

$$n - m \leqslant 2^m - m - 1$$

Hence the length of such codes satisfies the condition $n \leqslant 2^m - 1$. When $n = 2^m - 1$ the code is called **perfect**. For example, when there are four check bits, a perfect s.e.c. code has length $2^4 - 1 = 15$, and the number of information bits is $15 - 4 = 11$. Non-perfect s.e.c. codes with four check bits have fewer than 11 information bits.

**Exercise**    **4.55**   Compute the following products using modulo 2 arithmetic:

(a) $\begin{bmatrix} 0 & 1 & 1 & 0 \\ 1 & 0 & 1 & 1 \end{bmatrix} \begin{bmatrix} 0 \\ 1 \\ 0 \\ 1 \end{bmatrix}$

(b) $\begin{bmatrix} 1 & 1 & 0 & 0 & 1 \\ 1 & 0 & 1 & 0 & 1 \\ 1 & 1 & 1 & 1 & 1 \end{bmatrix} \begin{bmatrix} 1 \\ 0 \\ 1 \\ 1 \\ 0 \end{bmatrix}$

**Exercise** **4.56** Determine the remaining 11 codewords in Example 4.15. Verify that each codeword $x$ satisfies $Hx = 0$. What is the minimum distance for this code?

**Exercise** **4.57** Linear binary codes are determined by the following check matrices:

(a) the two matrices in (4.35)    (b) $\begin{bmatrix} 1 & 0 & 0 & 0 & 1 & 1 & 0 \\ 0 & 1 & 0 & 0 & 1 & 1 & 1 \\ 0 & 0 & 1 & 0 & 1 & 1 & 0 \\ 0 & 0 & 0 & 1 & 1 & 0 & 1 \end{bmatrix}$

In each case find the set of codewords, the dimension and minimum distance.

**Exercise** **4.58** Can any of the following be used as check matrices for single-error-detecting or single-error-correcting binary codes?

(a) $\begin{bmatrix} 0 & 1 & 0 & 0 \\ 0 & 0 & 0 & 1 \\ 1 & 1 & 0 & 0 \end{bmatrix}$    (b) $\begin{bmatrix} 0 & 1 & 0 & 0 \\ 0 & 1 & 0 & 0 \\ 1 & 0 & 0 & 1 \\ 1 & 0 & 1 & 0 \end{bmatrix}$    (c) $\begin{bmatrix} 1 & 1 & 0 & 0 & 1 & 0 & 1 \\ 1 & 1 & 1 & 0 & 0 & 1 & 0 \\ 1 & 0 & 0 & 0 & 1 & 0 & 0 \\ 1 & 0 & 1 & 0 & 0 & 1 & 0 \\ 1 & 0 & 0 & 0 & 1 & 0 & 0 \\ 0 & 0 & 1 & 1 & 1 & 1 & 1 \end{bmatrix}$

**Exercise** **4.59** If $H$ is a check matrix having columns $\mathbf{h}_1, \mathbf{h}_2, \ldots, \mathbf{h}_n$, and $e$ is a binary word of length $n$ all of whose bits are zero except $e_i = 1$, show that $He = \mathbf{h}_i$.

**Exercise** **4.60** If $H$ is the check matrix in the previous exercise, but $e$ is now a binary word of length $n$ with exactly two non-zero bits $e_i = 1$, $e_j = 1$, show that $He = \mathbf{h}_i + \mathbf{h}_j$.

**Exercise** **4.61** Confirm that

$$H = \begin{bmatrix} 1 & 0 & 0 & 1 & 0 \\ 0 & 1 & 0 & 1 & 0 \\ 0 & 0 & 1 & 1 & 1 \end{bmatrix}$$

cannot be used as the check matrix for a single-error-correcting binary code by

(a) giving a codeword of weight 2

(b) finding two non-zero codewords whose Hamming distance is 2.

**Exercise** **4.62** (a) What is the maximum possible number of information bits for a linear s.e.c. binary code with five check bits?

(b) How many check bits are needed for a linear s.e.c. binary code with 30 information bits?

**Exercise** **4.63** Using a suitable check matrix, list the codewords for a s.e.c. binary code with two information bits and the smallest possible number of check bits.

**Exercise**   **4.64**   The Faculty of Difficult Studies contains 1920 students, each of whom is assigned a unique identity number in the form of a binary codeword.

(a) What is the least possible number of information bits of an appropriate linear code? How many codewords will be unassigned?

(b) If the code is to be capable of correcting all single errors, what is the smallest possible length of codewords?

**Exercise**   **4.65**   How many different messages can be sent using a perfect linear binary code with five check bits?

## 4.3.3   Syndrome decoding for binary codes

If $r$ is a received word, then up to now this has been decoded by finding the Hamming distance between $r$ and each of the codewords. The nearest-neighbour principle then selects the codeword nearest to $r$ as the most likely to have been transmitted. This decoding scheme is impractical if there are a large number of codewords. Instead, a direct approach using the check matrix $H$ for the code is greatly preferable. However, the 'maximum likelihood' rule remains in force – it is assumed that a received word with no errors is more likely than anything else, and a received word with one error is more likely than a word with two or more errors, and so on.

If $r$ is a codeword, then by definition it satisfies $H\mathbf{r} = \mathbf{0}$. If there is a single transmission error in the $i$th position (say) then $r = c + e$, where $e$ is a word having $e_i = 1$, all other $e_i = 0$. In this case again $H\mathbf{c} = \mathbf{0}$, but by Exercise 4.59 $H\mathbf{e} = \mathbf{h}_i$ which is the $i$th column of $H$. Hence

$$H\mathbf{r} = H(\mathbf{c} + \mathbf{e}) = H\mathbf{c} + H\mathbf{e}$$
$$= \mathbf{0} + \mathbf{h}_i = \mathbf{h}_i$$

showing that if a single error occurs in transmission then $H\mathbf{r}$ is equal to a column of $H$. Because of its role in determining the transmitted codeword, the column vector $H\mathbf{r}$ (obtained by multiplying the check matrix and the column vector $\mathbf{r}$ formed from the received word $r$) is called the **syndrome** of $r$. This name comes from the medical usage of the term, where it means 'symptom'. To decode a received binary word $r$, use the following algorithm.

---

**Syndrome decoding algorithm for binary codes**

**Step 1:**   Compute the syndrome $\mathbf{s} = H\mathbf{r}(\bmod 2)$.

**Step 2:**   If $\mathbf{s} = \mathbf{0}$, then assume $r$ is a codeword and no transmission error occurred.

**Step 3:**   If $\mathbf{s} = i$th column of $H$, assume a single transmission error occurred in the $i$th bit, and the transmitted codeword was $r + e$, obtained by correcting $r_i$ to $r_i + 1$.

**Step 4:**   If $\mathbf{s} \neq \mathbf{0}$, $\mathbf{s} \neq i$th column of $H$, then assume more than one error occurred, and retransmission is requested.

| Example 4.18 | **Application of the algorithm** |
| --- | --- |

Consider the s.e.c. code of length 5 whose check matrix is the first matrix $H_1$ in (4.35).

(a) If a received word is $r = 11001$ then the syndrome $s = H_1 r$ is

$$s = \begin{bmatrix} 1 & 0 & 0 & 1 & 1 \\ 0 & 1 & 0 & 1 & 1 \\ 0 & 0 & 1 & 0 & 1 \end{bmatrix} \begin{bmatrix} 1 \\ 1 \\ 0 \\ 0 \\ 1 \end{bmatrix}$$

$$= \begin{bmatrix} 1+0+0+0+1 \\ 0+1+0+0+1 \\ 0+0+0+0+1 \end{bmatrix} = \begin{bmatrix} 0 \\ 0 \\ 1 \end{bmatrix}$$

Comparing $s$ with $H_1$ in (4.35) shows that it is equal to the third column of $H_1$. Hence by Step 3, a transmission error occurred in the third bit, so $e = 00100$. The third bit of $r$ is corrected to 1, so the transmitted codeword was $c = 11101$. You can check that $H_1 c = 0$, confirming that $c$ is indeed a codeword.

(b) If a received word is $r = 10100$ then the syndrome is

$$s = H_1 \begin{bmatrix} 1 \\ 0 \\ 1 \\ 0 \\ 0 \end{bmatrix} = \begin{bmatrix} 1 \\ 0 \\ 1 \end{bmatrix}$$

Clearly $s$ is not a column of $H_1$, so by Step 4 more than one transmission error has been detected. In fact, using the codewords found in part (a) of Exercise 4.57, you can check that in this case possible transmitted codewords are 00000 with errors in the first and third bits; or 11101 with errors in the second and last bits; or 00111 with errors in the first, fourth and fifth bits. The code cannot correct these multiple errors.

Part (b) of the preceding example shows that it's possible to obtain a syndrome which is not a column of the check matrix. This means that there is no codeword whose Hamming distance from the received word is 1. However, if the code is *perfect* then this cannot happen – every possible syndrome is equal to a column of $H$. In other words, every received word can be regarded as coming from a unique codeword with at most one error. To see why this is so, recall that a perfect code with $m$ check bits has length $n = 2^m - 1$, and contains $2^k$ codewords, where $k = n - m$. There are exactly $n$ words which are Hamming distance 1 from the zero codeword 00...0, differing from it in just one bit. The same applies for each of the codewords. Therefore the total number of words of length $n$ which are distance 1 from a codeword is

$$2^k \times n = 2^k(2^m - 1) = 2^n - 2^k$$

so that

$$\underset{\substack{\text{total number} \\ \text{of words of} \\ \text{length } n}}{2^n} = \underset{\substack{\text{number of} \\ \text{codewords}}}{2^k} + \underset{\substack{\text{number of words} \\ \text{distance 1 from} \\ \text{a codeword}}}{2^k \times n}$$

Stated in a different way, with a perfect code every word of length $n$ is either itself a codeword, or is distance 1 from a codeword. In this case Step 4 of the algorithm does not apply, which explains why such codes are called 'perfect'. All this assumes that at most one transmission error occurs. It's possible, for example, for a codeword to be transmitted, and to be received as a different codeword due to several errors occurring. The syndrome decoding algorithm can only conclude 'no error' in such circumstances.

**Exercise**  **4.66**  Consider the code in Example 4.15.

(a) If a single error occurs in the fourth bit of a transmitted codeword, what is the syndrome?

(b) Decode the received word 1011000.

**Exercise**  **4.67**  For the code with check matrix $H_2$ in (4.35), decode the received words:

(a) 11001  (b) 11000  (c) 11111.

**Exercise**  **4.68**  For the code with check matrix

$$H = \begin{bmatrix} 1 & 0 & 0 & 1 & 0 & 1 \\ 0 & 1 & 0 & 1 & 1 & 1 \\ 0 & 0 & 1 & 1 & 1 & 0 \end{bmatrix}$$

show that if 110010 is received then more than one error has occurred in transmission. Find two codewords which could have been sent with two errors occurring, and a codeword which could have been sent with three errors occurring.

**Exercise**  **4.69**  If $H$ is not the check matrix for a s.e.c. code, then the syndrome decoding algorithm breaks down. Verify this by considering the check matrix

$$H = \begin{bmatrix} 1 & 0 & 0 & 1 & 0 \\ 0 & 1 & 0 & 1 & 1 \\ 0 & 0 & 1 & 0 & 0 \end{bmatrix}$$

and received words 10010, 01000.

### 4.3.4 Binary Hamming codes

A disadvantage of the algorithm of the preceding section is that a syndrome has to be compared with the columns of the check matrix. Hamming devised a scheme in 1950 whereby the position of single errors can be obtained directly from the syndrome. Recall that the check matrix for a s.e.c. binary code must not contain a zero column, or two identical columns. Hamming pointed out that a natural way of satisfying this condition is to select as the columns of $H$ binary representations of the integers 1, 2, 3, 4, ... .

**Example 4.19** **Hamming code of length five**

Using Table 2.2, the binary representations of the integers 1 to 5 using *three* bits are:

| Decimal | 1 | 2 | 3 | 4 | 5 |
|---------|-----|-----|-----|-----|-----|
| Binary | 001 | 010 | 011 | 100 | 101 |

The notation using a suffix 2 has been omitted here. These binary numbers exclude 000 and are all different, so writing them as columns of a check matrix will indeed produce a s.e.c. code, having *three* check bits.

As before, the bits are read from left to right, and stacked from top to bottom, so for example 001 becomes $\begin{bmatrix} 0 \\ 0 \\ 1 \end{bmatrix}$. Doing this for each binary number produces

$$H = \begin{bmatrix} 0 & 0 & 0 & 1 & 1 \\ 0 & 1 & 1 & 0 & 0 \\ 1 & 0 & 1 & 0 & 1 \end{bmatrix} \tag{4.36}$$

This is the check matrix for a binary Hamming [5, 2] code. Notice that $H$ in (4.36) no longer has the form where the first three columns comprise the $3 \times 3$ unit matrix $I_3$, as in (4.32). Instead, it is the first, second and fourth columns in (4.36) which contain a single 1, corresponding to the binary representations of 1, 2 and 4. The check bits are therefore $x_1$, $x_2$ and $x_4$. Writing down the check equations $H\mathbf{x} = \mathbf{0}$, where $x = x_1 x_2 x_3 x_4 x_5$, gives

$$x_4 + x_5 = 0, \quad x_2 + x_3 = 0, \quad x_1 + x_3 + x_5 = 0$$

These can be rearranged as

$$x_1 = x_3 + x_5, \quad x_2 = x_3, \quad x_4 = x_5$$

Selecting all possible values for the information bits $x_3$ and $x_5$ gives the table shown.

| Check bits | | | Information bits | |
|---|---|---|---|---|
| $x_1$ | $x_2$ | $x_4$ | $x_3$ | $x_5$ |
| 0 | 0 | 0 | 0 | 0 |
| 1 | 1 | 0 | 1 | 0 |
| 1 | 0 | 1 | 0 | 1 |
| 0 | 1 | 1 | 1 | 1 |

Writing the bits in the correct order gives the four codewords 00000, 11100, 10011, 01111.

Now consider decoding using the syndrome algorithm in Section 4.3.3. If a single error occurs then the syndrome is equal to a column of $H$. However, there is no need to examine $H$ in (4.36) to see which column this is. For example, suppose a received word is $r = 01101$. The syndrome is

$$\mathbf{s} = H\mathbf{r} = \begin{bmatrix} 1 \\ 0 \\ 0 \end{bmatrix}$$

Converting this column vector $\mathbf{s}$ back to its binary number form $s = (100)_2$ shows that the decimal equivalent is 4, so the error is in the *fourth* bit, and the transmitted codeword was $r + 00010 = 01111$. Deleting the check bits $x_1$, $x_2$ and $x_4$ gives the decoded information message 11. This illustrates the key property of binary Hamming codes: the binary number equivalent of the syndrome immediately reveals which bit is in error.

If more than one error occurs then Step 4 of the syndrome decoding algorithm states that the syndrome is not a column of the check matrix. For example, if $r = 11010$ then

$$\mathbf{s} = H\mathbf{r} = \begin{bmatrix} 1 \\ 1 \\ 1 \end{bmatrix}$$

which converts to the binary number $(111)_2 = 7$. The code has length only 5, so there cannot be a seventh bit in error: the conclusion is that more than one error has occurred.

Incidentally, the code in this example can be used to send the four information messages 00, 10, 01, 11 and is an improvement over the repetition code in Example 4.2 which required six bits to correct single errors, compared with only five here.

### 4.3.4.1 Properties of binary Hamming [n, k] code

(1) The code has length $n$, dimension $k$ and $m = n - k$ check bits. If $n = 2^m - 1$ the code is **perfect**, if $n < 2^m - 1$ the code is **shortened**.

(2) For $i = 1, 2, \ldots, n$, the $i$th column of the $m \times n$ check matrix $H$ is the column vector form of the binary number representation of $i$ using $m$ bits.

(3) The check bits are in the positions where the columns of $H$ contain a single 1, namely

$$x_1, x_2, x_4, x_8, \ldots, x_p \text{ where } p = 2^{m-1}.$$

(4) By construction the code corrects all single errors, and so has minimum distance at least 3. In fact $\delta$ is actually equal to 3 (see Exercise 4.75).

(5) To *decode* a received word $r$, compute the syndrome $\mathbf{s} = H\mathbf{r}$. If $\mathbf{s} = \mathbf{0}$, assume the codeword $r$ was transmitted. Otherwise, if $(s)_2 = (j)_{10} \leqslant n$, assume a single error occurred in the $j$th bit; if $j > n$ then more than one error occurred.

(6) For perfect Hamming codes, every syndrome (except $\mathbf{0}$) occurs as a column of the check matrix, and so corresponds to a single correctable error. For shortened Hamming codes, some syndromes indicate multiple errors.

**Example 4.20**  **Perfect Hamming [7, 4] code**

Here the number of check bits is $m = 7 - 4 = 3$, and $n = 2^3 - 1$. The binary representations $6 = (110)_2$ and $7 = (111)_2$ give two extra columns to be appended to those in (4.36). The $3 \times 7$ check matrix is therefore

$$H = \begin{bmatrix} 0 & 0 & 0 & 1 & 1 & 1 & 1 \\ 0 & 1 & 1 & 0 & 0 & 1 & 1 \\ 1 & 0 & 1 & 0 & 1 & 0 & 1 \end{bmatrix} \tag{4.37}$$

For example, suppose a received word is $r = 0110111$, and at most one error occurred in transmission. The syndrome is

$$\mathbf{s} = H\mathbf{r} = H \begin{bmatrix} 0 \\ 1 \\ 1 \\ 0 \\ 1 \\ 1 \\ 1 \end{bmatrix} = \begin{bmatrix} 1 \\ 0 \\ 1 \end{bmatrix}$$

This produces $(101)_2 = 5$, showing that an error occurred in the fifth bit, and the transmitted codeword is therefore 0110011. Discarding the check bits $x_1$, $x_2$, $x_4$ leaves the decoded information message 1011.

For this particular code a simple representation involves **Venn diagrams**, introduced in Section 3.4.1 (page 158). Using (4.37) the check equations $H\mathbf{x} = \mathbf{0}$ are

$$x_4 + x_5 + x_6 + x_7 = 0, \quad x_2 + x_3 + x_6 + x_7 = 0, \quad x_1 + x_3 + x_5 + x_7 = 0$$

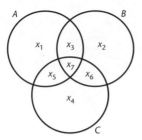

*Figure 4.5*   Venn diagram for Hamming [7, 4] code

The sum of the bits inside circle $A$ in Figure 4.5 is precisely $x_1 + x_3 + x_5 + x_7$, which is zero by the third check equation. Similarly, you can see that the sums of the bits inside each of the circles $B$ and $C$ in Figure 4.5 are zero, using the other two check equations. Hence for codewords, each circle contains an even number of 1's.

To apply the Venn diagram to encoding, suppose the information message is $x_3 x_5 x_6 x_7 = 0101$. The information bits occupy appropriate intersections of the circles, as shown in Figure 4.6(a). It is easy to see that the *only* way that each circle can contain an even number of 1's is that shown in Figure 4.6(b). Hence $x_1 = 0$, $x_2 = 1$, $x_4 = 0$ and the codeword is 0100101.

To illustrate the Venn diagram method for decoding, use the same received word as above, $r = 0110111$. Entering the bits onto a Venn diagram in Figure 4.5 produces the result shown in Figure 4.7. You can see that circle $B$ contains an even number of 1's but circles $A$ and $C$ contain an odd number of 1's. The error must therefore lie at the intersection of $A$ and $C$ only, so from Figure 4.5 there must be an error in $x_5$. The decoded message is therefore 0110011, as before.

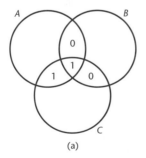

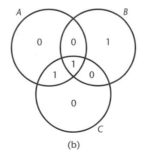

(a)                                    (b)

*Figure 4.6*   Encoding

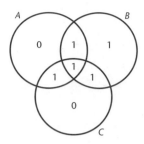

*Figure 4.7* Decoding

If there is a single error in one of the check bits then this is immediately recognized, since you can see from Figure 4.5 that just one of the circles has odd parity in this case.

The Venn diagram approach is only applicable to the [7, 4] code.

**Exercise**  **4.70**  For the Hamming code in Example 4.19, decode the received words
(a) 11011  (b) 01010.

**Exercise**  **4.71**  For the [7, 4] Hamming code in Example 4.20, construct the set of codewords using the check matrix $H$ in (4.37).

**Exercise**  **4.72**  For the [7, 4] Hamming code in Example 4.20, determine the information message when 1011101 is received, using both the syndrome and Venn diagram methods (assuming at most one transmission error).

**Exercise**  **4.73**  Give the check matrix for the binary Hamming code with six information bits and four check bits.

(a) Encode the information message 110101.

(b) Decode the received words (i) 1001010101, (ii) 1000101100, giving the information message where possible.

**Exercise**  **4.74**  Starting with the check matrix in the preceding exercise, give the check matrix for the perfect Hamming code with four check bits. If a received word is 101111111100010, determine the information bits of the transmitted message.

**Exercise**  **4.75**  Prove that for any binary Hamming code of length $n \geqslant 4$, the word whose first three bits are 1 and remaining $n - 3$ bits are zero, is a codeword. Hence deduce that Hamming codes have minimum distance exactly 3.

**Exercise**  **4.76**  Use the Venn diagram method to construct the set of codewords for the perfect [7, 4] Hamming code. Suppose that the codeword 1001101 is transmitted and errors occur in the third and fifth bits. Investigate what happens when you try to apply the Venn diagram to decode the received word.

#### 4.3.5    *p*-ary Hamming codes

A $p$-ary code $C$ of length $n$ was defined in Section 4.2 as a set of codewords $x = x_1 x_2 \ldots x_n$ with $x_i \in \mathbb{Z}_p$. Only codes with $p$ a prime number are considered, so that $\mathbb{Z}_p$ is a finite field (the reasons for this will become apparent). The error detection and correction theorem in Section 4.2 applies for any $p$-ary code, but the definition of linearity in Section 4.3.1 needs modifying in the following way. For two $p$-ary codewords $x$ and $y = y_1 \ldots y_n$, the **sum** $z = x + y$ has digits $z_i \equiv (x_i + y_i)(\mathrm{mod}\ p)$, $i = 1, \ldots, n$. Then $C$ is **linear** if and only if (i) $x + y \in C$ for all $x$, $y \in C$ *and* (ii) $ax \in C$ for all $a \in \mathbb{Z}_p$. For binary codes, this condition (ii) becomes irrelevant since $\mathbb{Z}_2 = \{0, 1\}$. The **weight** of a $p$-ary codeword is now the number of non-zero digits it contains. With these extended definitions, the important result (4.17), namely that the minimum distance for a linear code is equal to the smallest of the weights of non-zero codewords, still holds (see Exercise 4.79). For a linear code to be capable of correcting all single errors it must have minimum distance 3 (at least). This in turn requires firstly that the check matrix $H$ must not have a zero column, as for binary codes; secondly, however, the previous condition, that no two columns of the check matrix can be equal, is now replaced by the requirement that no column can be a non-zero scalar *multiple* of another column; such a multiple is defined by multiplying each element of the vector by $c \in \mathbb{Z}_p$, that is,

$$c \begin{bmatrix} x_1 \\ x_2 \\ \vdots \end{bmatrix} = \begin{bmatrix} cx_1 \\ cx_2 \\ \vdots \end{bmatrix} (\mathrm{mod}\ p)$$

The check equations (4.24) now become $H\mathbf{x} \equiv \mathbf{0}(\mathrm{mod}\ p)$.

| Example 4.21 | **Ternary linear code** |
|---|---|

The check matrix

$$\begin{bmatrix} 0 & 1 & 1 & 2 \\ 1 & 0 & 2 & 1 \end{bmatrix}$$

is *not* the matrix for a ternary s.e.c. code, because the fourth column is equal to 2 times the third column, that is,

$$2 \begin{bmatrix} 1 \\ 2 \end{bmatrix} = \begin{bmatrix} 2 \\ 4 \end{bmatrix} = \begin{bmatrix} 2 \\ 1 \end{bmatrix} \tag{4.38}$$

using modulo 3 arithmetic. However, the check matrix

$$H = \begin{bmatrix} 0 & 1 & 1 & 1 \\ 1 & 0 & 1 & 2 \end{bmatrix} \tag{4.39}$$

does define a s.e.c. ternary linear code, since no column is a multiple of any other. To verify this, simply write down all possible non-zero multiples of columns and confirm that none is repeated. For this code, codewords $x = x_1 x_2 x_3 x_4$ are obtained by solving the check equations $H\mathbf{x} \equiv \mathbf{0}(\text{mod } 3)$ with $H$ in (4.39). This gives

$$x_2 + x_3 + x_4 = 0, \quad x_1 + x_3 + 2x_4 = 0$$

Taking $x_3$ and $x_4$ as the information digits, the check digits can be expressed as

$$\left. \begin{array}{l} x_1 = -x_3 - 2x_4 = 2x_3 + x_4 \\ x_2 = -x_3 - x_4 \; = 2x_3 + 2x_4 \end{array} \right\} \tag{4.40}$$

since $-1 = 2$, $-2 = 1$ using modulo 3 arithmetic. The digits $x_3$ and $x_4$ can independently take the values 0, 1, 2 so there are $3^2 = 9$ codewords. As for binary linear codes, the values of the check digits are obtained from (4.40), but remember to use modulo 3 arithmetic. For example, when $x_3 = 1$, $x_4 = 2$ then $x_1 = 2 + 2 = 1$, $x_2 = 2 + 4 = 0$, and the corresponding codeword is 1012. Some more codewords are given in the following table:

| Check digits | | Information digits | |
|:---:|:---:|:---:|:---:|
| $x_1$ | $x_2$ | $x_3$ | $x_4$ |
| 1 | 2 | 0 | 1 |
| 2 | 1 | 0 | 2 |
| 2 | 2 | 1 | 0 |
| 1 | 0 | 1 | 2 |

The code defined by the check matrix $H$ in (4.39) is in fact a ternary Hamming code of length 4 with two check bits and two information bits. Writing the columns of $H$ as ternary numbers:

$$(01)_3 = 1, \quad (10)_3 = 3, \quad (11)_3 = 4, \quad (12)_3 = 5 \tag{4.41}$$

confirms that the columns of $H$ satisfy the stated conditions for a s.e.c. code.

### 4.3.5.1 Properties of p-ary Hamming [n, k] code

(1) The code has length $n = (p^m - 1)/(p - 1)$, $m$ check digits, dimension $k = n - m$, and contains $p^k$ codewords (see Exercise 4.83).

(2) The columns of the $m \times n$ check matrix $H$ are the column vector forms of the base-$p$ numbers with $m$ digits whose first non-zero digit is 1. These numbers are taken in the order of increasing magnitude, that is,

$$00 \ldots 01, 00 \ldots 010, 00 \ldots 011, 00 \ldots 012, \ldots, 00 \ldots 01(p-1),$$
$$00 \ldots 0100, \ldots \tag{4.42}$$

(3) The check digits are in the positions where the columns of $H$ have a single non-zero entry equal to 1.

(4) The code has a minimum distance 3 (see Problem 4.15), corrects all single errors, and is perfect.

(5) To *decode* a received $p$-ary word $r = r_1 r_2 \ldots r_n$, compute the syndrome $\mathbf{s} = H\mathbf{r} \pmod{p}$. If $\mathbf{s} = \mathbf{0}$, assume the codeword $r$ was transmitted. Otherwise, $\mathbf{s} = e\mathbf{h}_i$, where $e \neq 0$ and $e \in \mathbb{Z}_p$, and $\mathbf{h}_i$ is the $i$th column of $H$. A single error of magnitude $e$ is assumed in the $i$th digit, and the corrected digit is $r_i - e$ (see Exercise 4.81).

The assertion in (4) that the code is perfect needs to be justified. It is necessary to show that every non-zero syndrome is equal to $e$ times some column of $H$, where $e \neq 0$ and $e \in \mathbb{Z}_p$. For convenience on the printed page, write the column vectors as rows (the technical term for this is 'transposing'), so for example

$$\begin{bmatrix} 1 \\ 1 \\ 0 \end{bmatrix} \quad \text{becomes} \quad [1, \quad 1, \quad 0]$$

Any non-zero syndrome $\mathbf{s}$ has $m$ elements, and can be written in the transposed form

$$[0, 0, \ldots, 0, s_1, s_2, \ldots, s_{m-q}] \tag{4.43}$$
$$\underleftarrow{\quad q \text{ zeros} \quad}$$

where $0 \leqslant q < m$ and $s_1$ is the first non-zero entry of $\mathbf{s}$, with all $s_i \in \mathbb{Z}_p$. For $\mathbf{s}$ to be equal to $e$ times a column of $H$ means that there must be a column of $H$ which has the transposed form

$$[0, 0, \ldots, 0, 1, c_2, c_3, \ldots, c_{m-q}] \tag{4.44}$$
$$\underleftarrow{\quad q \text{ zeros} \quad}$$

Equating each element in (4.43) to $e$ times the corresponding element in (4.44) gives

$$s_1 = e, \quad s_2 = ec_2, \quad s_3 = ec_3, \ldots, s_{m-q} = ec_{m-q} \tag{4.45}$$

Since $p$ is a prime number, every member of $\mathbb{Z}_p$ has a multiplicative inverse (see Section 2.4.4). In particular, $e^{-1} = s_1^{-1} \in \mathbb{Z}_p$, so (4.45) can be solved as

$$c_2 = s_1^{-1} s_2, \quad c_3 = s_1^{-1} s_3, \ldots, c_{m-q} = s_1^{-1} s_{m-q}$$

showing that (4.44) does indeed represent a column of $H$, as required.

---

**Example 4.22** *p*-ary Hamming codes

(a) When $p = 5$ and $m = 3$, the length of the code is given by property (1) as $n = (5^3 - 1)/(5 - 1) = 31$, and $k = 31 - 3 = 28$. The base-5 numbers with three digits are obtained from (4.42) as:

001, 010, 011, 012, 013, 014, 100, 101, 102, 103, 104, 110, 111, 112, 113, 114,
120, 121, 122, 123, 124, 130, 131, 132, 133, 134, 140, 141, 142, 143, 144

You can verify that the decimal equivalents of these 31 numbers are 1, 5, 6, 7, 8, 9, 25, 26, ..., 48, 49, confirming that the numbers have been set out in increasing order. They are then used as the columns of the $3 \times 31$ check matrix for a ternary Hamming [31, 28] code. The check digits are $x_1, x_2, x_7$, corresponding to 001, 010 and 100, and there are $5^{28} \approx 3.73 \times 10^{19}$ codewords.

(b) When $p = 3$, $m = 2$ the code has length $n = (3^2 - 1)/(3 - 1) = 4$. The numbers (4.42) are listed in (4.41), and the check matrix $H$ is given in (4.39). Suppose a received word is $r = 1200$. The syndrome is

$$s = H \begin{bmatrix} 1 \\ 2 \\ 0 \\ 0 \end{bmatrix} = \begin{bmatrix} 0 + 2 + 0 + 0 \\ 1 + 0 + 0 + 0 \end{bmatrix} \text{(mod 3)}$$

$$= \begin{bmatrix} 2 \\ 1 \end{bmatrix}$$

$$= 2 \begin{bmatrix} 1 \\ 2 \end{bmatrix}, \qquad \text{by (4.38)}$$

This syndrome is 2 times the *fourth* column of $H$, so by the decoding scheme (5) there is an error $e = 2$ in the *fourth* digit. The corrected digit of the received word is $r_4 - 2 = 0 - 2 = 1 \pmod 3$, so the transmitted codeword was 1201.

(c) When $p = 5$ and $m = 2$ the length of the code is $n = (5^2 - 1)/(5 - 1) = 6$, and $k = 6 - 2 = 4$. The base-5 numbers in (4.42) with two digits are, in increasing order, 01, 10, 11, 12, 13, 14. Converting these into columns in the usual way produces the check matrix

$$H = \begin{bmatrix} 0 & 1 & 1 & 1 & 1 & 1 \\ 1 & 0 & 1 & 2 & 3 & 4 \end{bmatrix} \tag{4.46}$$

Suppose a received word is $r = 202123$. Using $H$ in (4.46), the syndrome is

$$s = H \begin{bmatrix} 2 \\ 0 \\ 2 \\ 1 \\ 2 \\ 3 \end{bmatrix} = \begin{bmatrix} 0 + 0 + 2 + 1 + 2 + 3 \\ 2 + 0 + 2 + 2 + 6 + 12 \end{bmatrix} \text{(mod 5)}$$

$$= \begin{bmatrix} 3 \\ 4 \end{bmatrix}$$

It is easy to see that

$$s = 3 \begin{bmatrix} 1 \\ 3 \end{bmatrix} = 3h_5$$

so there is an error $e = 3$ in the fifth digit of $r$, which is corrected to $2 - 3 = -1 = 4$. The transmitted codeword was 202143.

(d) When $p = 3$, $m = 3$ the ternary Hamming code has length $(3^3 - 1)/(3 - 1) = 13$, and dimension $k = 13 - 3 = 10$. Converting the numbers in (4.42) to columns produces the check matrix

$$H = \begin{bmatrix} 0 & 0 & 0 & 0 & 1 & 1 & 1 & 1 & 1 & 1 & 1 & 1 & 1 \\ 0 & 1 & 1 & 1 & 0 & 0 & 0 & 1 & 1 & 1 & 2 & 2 & 2 \\ 1 & 0 & 1 & 2 & 0 & 1 & 2 & 0 & 1 & 2 & 0 & 1 & 2 \end{bmatrix}$$

The check digits are $x_1$, $x_2$ and $x_5$, and there are $3^{10} = 59,049$ codewords. If a word $r_1 = 1102112100112$ is received then you can verify that $s = Hr_1 = 0$, so the codeword $r_1$ was transmitted. If another word $r_2 = 1000101220120$ is received you can compute that

$$s = Hr_2 = \begin{bmatrix} 0 \\ 1 \\ 1 \end{bmatrix} = h_3$$

The third digit of $r_2$ is corrected to $0 - 1 = 2$, and the transmitted codeword was $1020101220120$.

(e) When $p = 11$, $m = 2$ the check matrix is

$$\begin{bmatrix} 0 & 1 & 1 & 1 & 1 & 1 & 1 & 1 & 1 & 1 & 1 & 1 \\ 1 & 0 & 1 & 2 & 3 & 4 & 5 & 6 & 7 & 8 & 9 & 10 \end{bmatrix} \qquad (4.47)$$

---

**Exercise**    **4.77**   Determine the remaining five codewords for the ternary code with check matrix $H$ in (4.39). Verify that all nine codewords $x$ satisfy $Hx = 0$, and that the minimum distance is 3.

**Exercise**    **4.78**   For the ternary code in the preceding exercise, use syndrome decoding to decode the received words 0120, 2110.

**Exercise**    **4.79**   Show that the proof of (4.17) still holds for a $p$-ary linear code.

**Exercise**    **4.80**   Prove that any $p$-ary linear code contains a codeword consisting entirely of zeros.

**Exercise**    **4.81**   Suppose that for a $p$-ary Hamming code a codeword $x$ is transmitted, and a single error occurs in the $i$th digit which changes $x_i$ into $x_i + e$. Show that the syndrome is $eh_i$, where $h_i$ is the $i$th column of the check matrix $H$.

**Exercise**    **4.82**   For each of the following codes, what is the length of the $p$-ary Hamming code, and how many codewords does it contain?

(a) $p = 7$, $m = 3$    (b) $p = 11$, $m = 2$    (c) $p = 13$, $m = 4$.

**Exercise**    **4.83**   Show that the number of columns of the check matrix $H$ defined by (4.42) is

$$1 + p + p^2 + \ldots + p^{m-1} = \frac{p^m - 1}{p - 1}$$

**Exercise**　**4.84**　Consider a non-perfect ternary code with check matrix

$$\begin{bmatrix} 1 & 0 & 0 & 2 & 0 & 1 \\ 0 & 1 & 0 & 1 & 2 & 0 \\ 0 & 0 & 1 & 0 & 2 & 2 \end{bmatrix}$$

(a) Encode the information messages 100, 010, 001, 200, 201, 221.

(b) Decode the received words 112020, 201102.

**Exercise**　**4.85**　For the 5-ary code with check matrix (4.46):

(a) How many codewords does this code contain?

(b) Encode the information messages 1224, 4321.

(c) Decode the received word 323440.

**Exercise**　**4.86**　In what positions are the check digits for a 5-ary Hamming code with four check digits?

## 4.4　Decimal codes

Some decimal codes already encountered include the European Article Number (Example 4.5), the United States money order (Example 4.6), the United States ZIP postcode (Example 4.7), the Universal Product Code (Exercise 4.25), and the International Standard Serial Number (ISSN) in Exercise 4.31. The feature they have in common is that codeword digits belong to $\{0, 1, 2, \ldots, 9\}$ (with the possible exception of the check digit) although the arithmetic varies amongst modulo 9, 10 or 11 according to the particular code.

### 4.4.1　International Standard Book Number (ISBN)

As mentioned at the beginning of this chapter, a rare example of international collaboration is provided by the ISBN. Every published book is identified by its ISBN, an example of which was displayed in Figure 4.2. The ISBN is a 10-digit codeword $x_1 x_2 \ldots x_9 x_{10}$ where the digits $x_1$ to $x_9$ are decimal digits, but like the ISSN in Exercise 4.31 the check digit $x_{10}$ can in addition take the value 10, which is denoted by the Roman numeral X. This is because both the ISSN and ISBN are defined with modulo 11 arithmetic, so that once again the code is defined over a finite field, $\mathbb{Z}_{11}$. The lower number in Figure 4.2 is the European Article Number (EAN) for the book – see Exercise 4.93. The initial digits of an ISBN, called the 'Group Identifier', denote the country, or group of countries. For example, $x_1 = 0$ is used for all books (whether in English or not) published in the United States, United Kingdom, Canada, Australia and a few other countries, $x_1 = 3$ indicates a book published in the German-speaking world, Denmark has $x_1 x_2 = 87$, and for Sweden $x_1 x_2 = 90$. Generally speaking, the smaller a country's annual output of published books the more digits its identifier contains.

This is because the second part of the ISBN, called the 'Publisher Prefix' may consist of two, three, four, five, six or seven digits. This identifies the publisher, for example in Figure 4.2 the digits $x_2x_3$ are 13, which is Prentice Hall's identifier. The third part of the ISBN can be from one to six digits in length and is the 'Title Number', which is the number assigned to the particular book by the publisher – for example, 834094 in Figure 4.2. The length of the title number depends on the length of the previous parts of the ISBN, but the Group Identifier, Publisher Prefix and Title Number always total nine digits. The last digit $x_{10}$ is the check digit, and (as seen in Exercise 2.104) is chosen so that the check sum

$$S = \sum_{i=1}^{10} ix_i = x_1 + 2x_2 + 3x_3 + \ldots + 9x_9 + 10x_{10}$$

(4.48)

is a multiple of 11, that is,

$$S \equiv 0 \pmod{11}$$

(4.49)

Hyphens are often inserted between the various parts of the ISBN, but have no mathematical significance.

For the ISSN in Exercise 4.31 the check sum is defined differently from (4.48), but satisfies the same condition (4.49). Incidentally, the ISSN does not incorporate any information about the publisher.

Setting the sum (4.48) equal to zero, and using the fact that $-10 \equiv 1 \pmod{11}$ produces the expression

$$x_{10} \equiv \sum_{i=1}^{9} ix_i \pmod{11}$$

(4.50)

This enables the check digit to be computed from the first nine digits of the ISBN. In fact it is easy to show (see Problem 4.19) that if the sum on the right in (4.50) is a 3-digit decimal number $abc$ then (4.50) simplifies to

$$x_{10} \equiv (a - b + c) \pmod{11}$$

(4.51)

Similarly, if in (4.48) $S = (s_1 s_2 s_3)_{10}$, then (4.49) is equivalent to $s_1 - s_2 + s_3 \equiv 0 \pmod{11}$. Of course, it is easy to compute the sums in (4.48) and (4.50) using a calculator. However, the ISBN was introduced around 1968, before the ready availability of cheap calculators, so a simple tabular array for use by librarians was introduced. This involves only successive additions, with no multiplications required. The condition (4.49) is tested using the following rules:

(1) Construct an array with three rows and ten columns. The entries in the first row are $x_1, x_2, x_3, \ldots, x_{10}$, and in the first column are $x_1, x_1, x_1$.

(2) At any stage, if the entries in a column are $a_1, a_2, a_3$, then the entries $b_2, b_3$ in the next column to the right are given by following the arrows:

$$\text{first row} \longrightarrow \quad a_1 \quad \nearrow \quad b_1$$
$$a_2 \longrightarrow b_2 = a_2 + b_1$$
$$a_3 \longrightarrow b_3 = a_3 + b_2$$

(3) The condition (4.49) is equivalent to the last entry in the bottom row of the array being 0(mod 11).

To verify the statement in (3), construct the array as follows:

$$
\begin{array}{llllll}
x_1 & x_2 & x_3 & x_4 & \cdots & x_{10} \\
x_1 \rightarrow (x_1+x_2) \rightarrow (x_1+x_2+x_3) & (x_1+x_2+x_3+x_4) & \cdots & T_1 \\
x_1 \rightarrow (2x_1+x_2) \rightarrow (3x_1+2x_2+x_3) & (4x_1+3x_2+2x_3+x_4) & \cdots & T_2
\end{array}
$$

You should check that $T_1 = x_1 + x_2 + x_3 + \ldots + x_{10}$, and that the last entry in the bottom row is

$$T_2 = 10x_1 + 9x_2 + 8x_3 + \ldots + 2x_9 + x_{10}$$

Moreover, it is easy to see that $T_2 + S = 11T_1$, where $S$ is the sum in (4.48), so $T_2 \equiv 0(\text{mod } 11)$ if and only if $S \equiv 0(\text{mod } 11)$, as required.

The application of rule (2) above can be made even simpler by performing each individual addition modulo 11 during the construction of the array.

**Example 4.23**  **Checking an ISBN**

Consider the ISBN 0138340943 in Figure 4.2. The check sum $S$ in (4.48) is

$$
\begin{aligned}
S &= 1 \times 0 + 2 \times 1 + 3 \times 3 + 4 \times 8 + 5 \times 3 + 6 \times 4 + 7 \times 0 + 8 \times 9 \\
&\quad + 9 \times 4 + 10 \times 3 \\
&= 0 + 2 + 9 + 10 + 4 + 2 + 0 + 6 + 3 + 8 \quad (\text{mod } 11) \\
&= 44 \equiv 0(\text{mod } 11)
\end{aligned}
$$

verifying that the ISBN is indeed correct. Alternatively, using rules (1) and (2), the array is constructed as:

$$
\begin{array}{lllllllllll}
0 & 1 & 3 & 8 & 3 & 4 & 0 & 9 & 4 & 3 & \leftarrow\text{ISBN} \\
0 \rightarrow 1 \rightarrow 4 \rightarrow 12 & 15 & 19 & 19 & 28 & 32 & 35 \\
0 \rightarrow 1 \rightarrow 5 \rightarrow 17 & 32 & 51 & 70 & 98 & 130 & 165
\end{array}
$$

The last entry in the bottom row is 165 which is also 0(mod 11), as expected. The simpler version of this table uses mod(11) addition throughout: for example, the entries in the fourth column reduce to $8 + 4 = 1$, $1 + 5 = 6$. The complete simplified array is:

$$
\begin{array}{llllllllll}
0 & 1 & 3 & 8 & 3 & 4 & 0 & 9 & 4 & 3 \\
0 & 1 & 4 & 1 & 4 & 8 & 8 & 6 & 10 & 2 \\
0 & 1 & 5 & 6 & 10 & 7 & 4 & 10 & 9 & 0
\end{array} \tag{4.52}
$$

and the last entry is 0, again as expected. This procedure also gives an easy way of finding the check digit if the values of $x_1$ to $x_9$ are known. In this example, suppose $x_{10}$ is unknown. The last two columns of the simplified array would then be:

$$
\begin{array}{ll}
4 & x_{10} \\
10 & x_{10} + 10 \\
9 & x_{10} + 8
\end{array}
$$

and requiring $x_{10} + 8 \equiv 0 \pmod{11}$ gives the correct value $x_{10} = 3$.

The ISBN code detects all single errors. To see this, suppose a correct ISBN is $x_1 x_2 \ldots x_{10}$, but the $i$th digit is incorrectly recorded as $x_i + e$. Then the check sum $S$ in (4.48) contains an extra term $ie$ and this product cannot be zero over the finite field $\mathbb{Z}_{11}$ since $i \neq 0$, $e \neq 0$ (see Section 2.4.4). This shows why it is essential to work over $\mathbb{Z}_{11}$. If, for example, ordinary decimal arithmetic was used, so (4.49) was replaced by $S \equiv 0 \pmod{10}$, then because of the term $2x_2$ in the sum $S$ in (4.48), an error of 5 in $x_2$ would change $S$ by 10, and so would go undetected; similarly errors of $\pm 2$ in $x_5$ would also be undetected.

If it is known which digit is in error, then the value of this digit can be found by solving the congruence (4.49), or an equivalent form, as was done for $x_{10}$ in Example 4.23.

**Example 4.24** **Correcting an ISBN**

Suppose that the sixth digit (equal to 4) in the ISBN in Example 4.23 has been accidentally lost, so the ISBN is recorded as $01383x0943$. Either compute $S$ from (4.48), or use the simplified array. The first five columns in (4.52) are the same as before, but the remainder of the array is now

$$
\begin{array}{ccccc}
x & 0 & 9 & 4 & 3 \\
x+4 & x+4 & x+2 & x+6 & x+9 \\
x+3 & 2x+7 & 3x+9 & 4x+4 & 5x+2
\end{array}
$$

where all additions have been performed modulo 11. The usual requirement on the last entry in the bottom row gives

$$5x + 2 \equiv 0 \pmod{11} \tag{4.53}$$

This has solution (see Section 2.4.4, page 110)

$$5x = -2 = 9$$

so that

$$x = 5^{-1} \times 9 = 9 \times 9 = 81 = 4$$

which is the value of the missing digit. Notice that again the existence of multiplicative inverses over $\mathbb{Z}_{11}$ has been invoked. Alternatively, you can solve (4.53) simply by trying $x = 1, 2, 3, \ldots$ until you find the correct value. Working over $\mathbb{Z}_{11}$ ensures that the solution of (4.53) is unique.

Suppose a received 10-digit decimal number is found *not* to be an ISBN. If it is not known which digits are incorrect there will be many possible ISBNs which may have been transmitted. Even the usual assumption that an error in a single digit is most likely, is not sufficient to enable such an error to be corrected (see Exercise 4.91). However, the ISBN code does detect all transposition errors in which two digits (usually, but not necessarily, adjacent) are accidentally interchanged. To see this, suppose a correct ISBN $x_1x_2\ldots x_{10}$ is transmitted, and the received word has $x_j$ and $x_k$ interchanged, with $j < k$ and $x_j \neq x_k$ (if $x_j = x_k$ then there is no error). The check sum $S$ in (4.48) for the received word contains terms $jx_k + kx_j$ instead of $jx_j + kx_k$. It is easy to verify (Exercise 4.92) that the change in the check sum is $(x_j - x_k)(k - j)$ and again this product cannot be 0(mod 11) because each term is non-zero. Hence the check sum for the received word is not 0(mod 11), so the transposition error is detected. If it is known that two *adjacent* digits have been transposed then this error can be corrected (see Exercise 4.90(b.ii)).

**Exercise**  **4.87**  Determine which of the following ISBNs is correct. Use the check sum (4.48) and the simplified version of the array.

(a) 0854297383  (b) 2-85120-460-2  (c) 0852457106.

**Exercise**  **4.88**  Verify that 0883855119 is a valid ISBN. Confirm by using the simplified array that if any two adjacent unequal digits are accidentally interchanged, then in each case the error is detected.

**Exercise**  **4.89**

(a) An ISBN is received with one digit illegible, namely $08247x6114$. Determine $x$.

(b) For the ISSN described in Exercise 4.31, show that the array defined for the ISBN still works (with eight columns instead of ten). Hence determine the missing digit for an ISSN $02650x54$.

**Exercise**  **4.90**  An ISBN is recorded as 0-19-588327-3.

(a) Show that this is incorrect.

(b) Correct this number if it is assumed that either (i) there is a single error in the sixth digit from the left, or (ii) there has been an interchange of two adjacent digits.

**Exercise**  **4.91**  An ISBN is received as 0671875356.

(a) Show that this is incorrect.

(b) Determine the check digit, assuming the other digits are correct.

(c) Suppose the original check digit is correct, and the book was published in the USA. If there is a single error in one of the other digits, determine three possible ISBNs.

**Exercise**   **4.92** Show that interchanging $x_j$ in position $j$ with $x_k$ in position $k$ in an ISBN $(k > j)$ produces an increase in the check sum (4.48) of $(x_j - x_k)(k - j)$.

**Exercise**   **4.93** To obtain the European Article Number for a book, take the ISBN, delete the check digit, add the prefix 978, and compute the new check digit $x_{13}$ according to (4.4) and (4.5). Determine the EAN for a book whose ISBN is 0201600447.

## 4.4.2   Single-error-correcting code

An improvement over the ISBN is now introduced which *corrects* all single errors. This code consists of 10-digit codewords $x_1 x_2 \ldots x_{10}$ which satisfy the same check equation (4.49) as the ISBN, namely

$$S_1 \equiv \sum_{i=1}^{10} i x_i \equiv 0 (\text{mod } 11)$$

(4.54)

together with an additional overall parity check equation (see Section 4.1.2)

$$S_2 \equiv \sum_{i=1}^{10} x_i \equiv 0 (\text{mod } 11)$$

(4.55)

Since there are two check equations there are now *two* check digits $x_9$ and $x_{10}$. These check equations define an 11-ary code, but now *omit* all codewords which contain the digit X ($=10$), the resulting code containing only true decimal digits.

Suppose a single error $e$ occurs in the $i$th digit of a transmitted codeword, so that the received word $r$ has $r_i = x_i + e$. As for the ISBN, the check sum in (4.54) for the received word $r$ contains an extra term $ie$, so that

$$S_1 \equiv ie (\text{mod } 11)$$

(4.56)

The second check sum in (4.55) becomes $S_2 \equiv e (\text{mod } 11)$. Hence the magnitude of the error is $e \equiv S_2 (\text{mod } 11)$, and from (4.56) the position of the error is

$$i \equiv S_1 e^{-1} (\text{mod } 11) \equiv S_1 S_2^{-1} (\text{mod } 11)$$

The **decoding algorithm** for this code is as follows, where all arithmetic is modulo 11:

(1) For a received word $r$, compute the syndrome

$$H\mathbf{r} = \begin{bmatrix} S_2 \\ S_1 \end{bmatrix}$$

where

$$H = \begin{bmatrix} 1 & 1 & 1 & 1 & 1 & 1 & 1 & 1 & 1 & 1 \\ 1 & 2 & 3 & 4 & 5 & 6 & 7 & 8 & 9 & 10 \end{bmatrix}$$

(4.57)

(2) If $S_1 = 0$, $S_2 = 0$ then assume there is no error, and the codeword $r$ was transmitted.

(3) If $S_1 \neq 0$, $S_2 \neq 0$ then assume a single error occurred in digit $i = S_1 S_2^{-1}$, and the corrected digit is $r_i - S_2$.

(4) If $S_1 = 0$ or $S_2 = 0$ (not both) then at least two errors have been detected.

In fact (4) always occurs when two digits have been transposed (see Exercise 4.96), but unlike the ISBN this code can detect *all* errors occurring in two digits. This is because the code has minimum distance 3. The easiest way to demonstrate this is to notice that the check matrix in (4.57) is just the check matrix in (4.47) for an 11-ary Hamming code, but with the first two columns in (4.47) deleted. This does not alter the minimum distance, which for $p$-ary Hamming codes is 3 (see Problems 4.15 and 4.27).

| Example 4.25 | **Applications of the decoding scheme** |

(a) For a received word 0206211909 it is easy to compute using (4.54) and (4.55) that

$$S_1 = 213 \equiv 4(\text{mod } 11), \quad S_2 = 30 \equiv 8(\text{mod } 11)$$

Hence by Step (3) above, a single error has occurred in digit $i$, where

$$i = 4 \times 8^{-1} = 4 \times 7 = 28 \equiv 6(\text{mod } 11)$$

Therefore the corrected sixth digit of the received word is $r_6 - 8 = 1 - 8 = -7 \equiv 4(\text{mod } 11)$, so the transmitted codeword was 0206241909.

(b) For a received word 5764013052 you can check that $S_1 = 1$ and $S_2 = 0$, modulo 11. Hence by Step (4) there are at least two errors in the received word, and retransmission is requested.

| Exercise | **4.94** Using the code of this section, decode the received words: |

(a) 9104536138   (b) 2104136538   (c) 0612960587

(d) 3104893144   (e) 3717425859.

| Exercise | **4.95** Using the result in Problem 4.20, determine the check digits when the information digits are 40731255. |

| Exercise | **4.96** Show that for the code of this section, when two different digits of a codeword are interchanged during transmission, then $S_2 = 0$ and all such errors are detected. |

**Exercise** **4.97** Show that a 10-digit decimal linear code with $x_i \in \{0,1,2,\ldots,8,9\}$ and check equations

$$\sum_{i=1}^{10} x_i \equiv 0(\text{mod } 10), \quad \sum_{i=1}^{10} ix_i \equiv 0(\text{mod } 10)$$

is *not* a single-error-correcting code, by finding two codewords with Hamming distance 2.

**Exercise** **4.98** Since 1966 every Norwegian citizen has been allocated an identity number consisting of 11 decimal digits. The first six digits represent the date of birth ($x_1x_2$ = day, $x_3x_4$ = month, $x_5x_6$ = last two digits of year). The digits $x_7x_8x_9$ are a personal number and $x_{10}$ and $x_{11}$ are check digits defined by

$$x_{10} \equiv (8x_1 + 4x_2 + 5x_3 + 10x_4 + 3x_5 + 2x_6 + 7x_7 + 6x_8 + 9x_9)(\text{mod } 11)$$
$$x_{11} \equiv (6x_1 + 7x_2 + 8x_3 + 9x_4 + 4x_5 + 5x_6 + 6x_7 + 7x_8 + 8x_9 + 9x_{10})(\text{mod } 11)$$

(a) If your birthday is 1 July 1938 and your personal number is 943, what is your i.d. number?

(b) Determine the check equations, and hence find a class of double errors which are not detected.

### 4.4.3 Double-error-correcting code

The codes discussed up to now have been capable, at best, of correcting all single errors. It's now possible to describe a code which corrects all *double* errors, that is, errors in two digits of a codeword. Begin with the s.e.c. code in Section 4.4.2, and select those codewords which satisfy in addition to (4.54) and (4.55) two more check equations:

$$S_3 = \sum_{i=1}^{10} i^2 x_i \equiv 0(\text{mod } 11), \quad S_4 = \sum_{i=1}^{10} i^3 x_i \equiv 0(\text{mod } 11) \tag{4.58}$$

There are now *four* check digits $x_7$, $x_8$, $x_9$, $x_{10}$. Suppose a codeword $x_1x_2\ldots x_{10}$ is transmitted and two errors occur in positions $i$ and $j$ with magnitudes $e_1$ and $e_2$ respectively, so that the received word $r$ has $r_i = x_i + e_1$, $r_j = x_j + e_2$. From the expressions for the check sums it follows that in this case:

$$\left.\begin{array}{ll} S_1 = ie_1 + je_2, & S_2 = e_1 + e_2, \\ S_3 = i^2e_1 + j^2e_2, & S_4 = i^3e_1 + j^3e_2 \end{array}\right\} \tag{4.59}$$

The four equations in (4.59) are to be solved for the four unknowns $i, j, e_1, e_2$. Some appropriate algebra (see Problem 4.22) leads to the result that $i$ and $j$ (the locations of the errors) are the two roots of the quadratic equation

$$ax^2 + bx + c = 0 \tag{4.60}$$

where

$$a = S_1^2 - S_2 S_3, \quad b = S_2 S_4 - S_1 S_3, \quad c = S_3^2 - S_1 S_4 \qquad (4.61)$$

Once $i$ and $j$ have been found, it is easy to solve the first two equations in (4.59) (see Exercise 4.104) to obtain

$$e_2 = (iS_2 - S_1)(i - j)^{-1}, \quad e_1 = S_2 - e_2 \qquad (4.62)$$

Notice that if just one error occurs, say $e_1 \neq 0$, $e_2 = 0$ then in (4.59) $S_1 = ie_1$, $S_2 = e_1$, $S_3 = i^2 e_1$, $S_4 = i^3 e_1$, and substituting these values into (4.61) gives $a = 0$, $b = 0$, $c = 0$ (see Exercise 4.104).

Solving quadratic equations in $\mathbb{Z}_p$ was investigated in Example 2.42. The solution of (4.60) can be expressed as

$$i, j = \left[-b \pm \sqrt{(b^2 - 4ac)}\right] (2a)^{-1} \qquad (4.63)$$

using a formula given in Section 1.4.2. You were asked to show in Exercise 2.132, that square roots (where they exist) of members $q$ of $\mathbb{Z}_{11}$ are given by:

$$
\left.
\begin{array}{cccccc}
q & 1 & 3 & 4 & 5 & 9 \\
\sqrt{q} & 1 \text{ or } 10 & 5 \text{ or } 6 & 2 \text{ or } 9 & 4 \text{ or } 7 & 3 \text{ or } 8
\end{array}
\right\} \qquad (4.64)
$$

It was pointed out in Chapter 2 that although these square roots are not unique, only two different values are obtained from (4.63) (see Exercise 2.133).

The **decoding algorithm** for this code is as follows, where again all arithmetic is modulo 11:

(1) For a received word $r$, compute the syndrome

$$\mathbf{S} = H\mathbf{r} = \begin{bmatrix} S_2 \\ S_1 \\ S_3 \\ S_4 \end{bmatrix}$$

where

$$H = \begin{bmatrix} 1 & 1 & 1 & 1 & \cdots & 1 \\ 1 & 2 & 3 & 4 & \cdots & 10 \\ 1 & 2^2 & 3^2 & 4^2 & \cdots & 10^2 \\ 1 & 2^3 & 3^3 & 4^3 & \cdots & 10^3 \end{bmatrix} \qquad (4.65)$$

(2) If $\mathbf{S} = \mathbf{0}$ (that is, $S_1 = S_2 = S_3 = S_4 = 0$) then assume there is no error, and the codeword $r$ was transmitted.

(3) If $\mathbf{S} \neq \mathbf{0}$ and $a = b = c = 0$, then assume a single error occurred in digit $i = S_1 S_2^{-1}$, and the corrected digit is $r_i - S_2$ (this step is the same as (3) for the code in Section 4.4.2).

(4) If $S \neq 0$ and $a \neq 0$, $c \neq 0$, and $q = b^2 - 4ac$ has a square root in $\mathbb{Z}_{11}$ according to (4.64), then assume there are errors in positions $i$, $j$ given by (4.63), that is,

$$i, j = (-b \pm \sqrt{q})(2a)^{-1} \qquad (4.66)$$

The corrected digits are $r_i - e_1$, $r_j - e_2$ where $e_1$ and $e_2$ are given by (4.62).

(5) If none of (2), (3) or (4) holds, then at least three errors have been detected. Notice from (4.64) that this includes the case in (4) when $q$ takes one of the values 2, 6, 7, 8, since these have no square root in $\mathbb{Z}_{11}$.

It can be shown (see Problem 4.29) that the code has minimum distance 5, so by the theorem in Section 4.2 the code does indeed correct all double errors.

| Example 4.26 | **Applications of the decoding algorithm** |

(a) Suppose a received word is 3254571396. The sums $S_1$, $S_2$, $S_3$ and $S_4$ in (4.54), (4.55) and (4.56) are computed with modulo 11 arithmetic. To facilitate the computations it is convenient to set out the following array:

| $x_i$ | 3 | 2 | 5 | 4 | 5 | 7 | 1 | 3 | 9 | 6 |
|-------|---|---|---|---|---|---|---|---|---|---|
| $i$   | 1 | 2 | 3 | 4 | 5 | 6 | 7 | 8 | 9 | 10 |
| $i^2$ | 1 | 4 | 9 | 5 | 3 | 3 | 5 | 9 | 4 | 1 |
| $i^3$ | 1 | 8 | 5 | 9 | 4 | 7 | 2 | 6 | 3 | 10 |

Notice that the values of $i^2 \pmod{11}$ and $i^3 \pmod{11}$ have been used (see Exercise 4.99). Firstly, $S_2 = \sum x_i = 1$, and taking the scalar product modulo 11 of the first row of the table with each of the subsequent rows gives

$$\begin{aligned} S_1 &= 1 \times 3 + 2 \times 2 + 3 \times 5 + 4 \times 4 + 5 \times 5 + 6 \times 7 + 7 \times 1 \\ &\quad + 8 \times 3 + 9 \times 9 + 10 \times 6 \\ &= 2 \end{aligned}$$

and similarly

$$S_3 = \sum i^2 x_i = 10, \qquad S_4 = \sum i^3 x_i = 3$$

Next, from (4.61)

$$a = 4 - 1 \times 10 = -6 = 5, \quad b = 1 \times 3 - 2 \times 10 = -17 = 5,$$
$$c = 100 - 2 \times 3 = 6$$

so the quantity $q = b^2 - 4ac$ in step (2) is

$$q = 25 - 4 \times 5 \times 6 = -95 = 4$$

which has $\sqrt{q} = 2$ in (4.64). Hence the conditions in step (4) of the algorithm apply, so there are two transmission errors. From (4.66) these occurred in positions

$$i, j = (-5 \pm 2)10^{-1}$$

and since $10^{-1} = 10$ this gives

$$i = -3 \times 10 = -30 = 3, \quad j = -7 \times 10 = 7$$

The magnitudes of the errors are obtained from (4.62) as

$$e_2 = (3 \times 1 - 2)(3 - 7)^{-1} = 7^{-1} = 8$$

and $e_1 = 1 - 8 = 4$. Hence the corrected values of the third and seventh digits are $r_3 - 4 = 5 - 4 = 1$ and $r_7 - 8 = 1 - 8 = -7 = 4$. The received word is therefore decoded as 3214574396.

Notice that using the alternative value in (4.64) of $\sqrt{4} = 9$ gives

$$i, j = (-5 \pm 9)10$$
$$= -30, 40 = 3, 7$$

as before, confirming the earlier remark that the quadratic equation (4.60) has only two different solutions.

Notice also that the condition $c \neq 0$ in step (4) is necessary for the correction of two errors, for if $c = 0$ then (4.60) has a solution $i = 0$, which is invalid.

(b) Suppose a received word is 4063101012. Repeating the computations gives $S_1 = 9$, $S_2 = 7$, $S_3 = 10$, $S_4 = 2$ and $a = 0$, $b = 1$, $c = 5$. Because $a = 0$ it follows from step (5) of the algorithm that at least three errors have been detected (see Exercise 4.100), and retransmission is requested.

---

**Exercise**    **4.99** You were asked in Exercise 2.132, to determine the values of $i^2 \pmod{11}$ for $i = 1, 2, 3 \ldots, 10$. Compute the values of $i^3 \pmod{11}$, and hence write down a list of cube roots for members of $\mathbb{Z}_{11}$.

**Exercise**    **4.100** Verify that 4003101715 is a codeword for the code of this section, and that three transmission errors can cause the word in Example 4.26(b) to be received.

**Exercise**    **4.101** Using the code of this section, decode the received words:
(a) 2706028591    (b) 9082100863    (c) 0800731345.

**Exercise**    **4.102** Assuming the six information digits $x_1 x_2 x_3 x_4 x_5 x_6$ in the received word in part (b) of the preceding exercise are correct, use the result in Problem 4.21 to determine the check digits.

**Exercise**    **4.103** Solve the first two equations in (4.59) for $e_1$ and $e_2$.

**Exercise**    **4.104** Verify that when $e_1 \neq 0$, $e_2 = 0$ then (4.61) gives $a = b = c = 0$, and similarly when $e_1 = 0$, $e_2 \neq 0$.

## 4.5 BCH codes

An important class of binary codes which correct two (or more) errors is named after R.C. Bose, D.K. Ray-Chaudhuri and A. Hocquenghem, who introduced the codes around 1960. For short, the codes are referred to by the initials: **BCH**. The perfect binary Hamming code in Section 4.3.4 of length $n = 2^m - 1$ needed $m$ check bits to correct one error. In view of the fact that the decimal codes in Sections 4.4.2 and 4.4.3 needed respectively *two* check digits to correct one error, and *four* check digits to correct two errors, it seems reasonable to expect that generalizing binary Hamming codes so as to correct two errors will require double the number (that is, $2m$) of check bits.

The necessary mathematical structure is that of a **finite field**, the example $\mathbb{Z}_p$ (with $p$ a prime number) being developed in Section 2.4.4. The notation GF($N$) is often used for a finite field containing $N$ elements, being named after the French mathematician E. Galois who was killed in a duel in 1832 at the age of 21, but not before he had immortalized his name through what is now known as Galois Field Theory. Recall that such a field is defined as a finite set of elements (including the unit element 1) whose principal properties are as follows:

(1)  For any $a, b$ in GF($N$) then $a + b$, $a \times b$ also belong to GF($N$).

(2)  Every element has an additive inverse $-a$, that is, $a + (-a) = 0$.

(3)  Every non-zero element $a$ has a multiplicative inverse $a^{-1}$, that is, $a \times a^{-1} = 1$.

The commutative, associative and distributive laws set out in Section 2.4.4, also apply.

In this section the Galois Field GF(16) with elements $0, 1, \alpha, \alpha^2, \alpha^3, \ldots, \alpha^{14}$ will be used. No value is assigned to the parameter $\alpha$, which is called a **primitive element** for the field, since every non-zero member of GF(16) is a power of $\alpha$ (see Section 2.4.4 for an example of a primitive element for $\mathbb{Z}_5$). For technical reasons which will not be gone into in this book, $\alpha$ must satisfy a condition

$$\alpha^4 = 1 + \alpha \tag{4.67}$$

where all arithmetic is modulo 2. It then follows that every member of GF(16) can be expressed in terms of $1, \alpha, \alpha^2$ and $\alpha^3$ only. For example, using (4.67) shows that

$$\alpha^5 = \alpha.\alpha^4 = \alpha(1 + \alpha) = \alpha + \alpha^2$$
$$\alpha^6 = \alpha.\alpha^5 = \alpha(\alpha + \alpha^2) = \alpha^2 + \alpha^3$$
$$\alpha^7 = \alpha.\alpha^6 = \alpha^3 + \alpha^4 = 1 + \alpha + \alpha^3$$
$$\alpha^8 = \alpha.\alpha^7 = \alpha + \alpha^2 + \alpha^4 = \alpha + \alpha^2 + 1 + \alpha = 1 + \alpha^2$$

where in the expression for $\alpha^8$ the property $\alpha + \alpha \equiv 0 \pmod{2}$ has been used. The remaining powers of $\alpha$ can be computed similarly (see Exercise 4.105) and are listed in Table 4.2. Notice that (4.67) can be rewritten as $\alpha^4 + \alpha + 1 = 0$, and GF(16) is said to be **generated** by the polynomial $\alpha^4 + \alpha + 1$. An alternative

*Table 4.2*   GF(16) generated by $\alpha^4 + \alpha + 1$

| As a 4-bit word | As a polynomial | As a power of $\alpha$ |
|---|---|---|
| 0000 | 0 | 0 |
| 1000 | 1 | 1 |
| 0100 | $\alpha$ | $\alpha$ |
| 0010 | $\alpha^2$ | $\alpha^2$ |
| 0001 | $\alpha^3$ | $\alpha^3$ |
| 1100 | $1 + \alpha$ | $\alpha^4$ |
| 0110 | $\alpha + \alpha^2$ | $\alpha^5$ |
| 0011 | $\alpha^2 + \alpha^3$ | $\alpha^6$ |
| 1101 | $1 + \alpha + \alpha^3$ | $\alpha^7$ |
| 1010 | $1 + \alpha^2$ | $\alpha^8$ |
| 0101 | $\alpha + \alpha^3$ | $\alpha^9$ |
| 1110 | $1 + \alpha + \alpha^2$ | $\alpha^{10}$ |
| 0111 | $\alpha + \alpha^2 + \alpha^3$ | $\alpha^{11}$ |
| 1111 | $1 + \alpha + \alpha^2 + \alpha^3$ | $\alpha^{12}$ |
| 1011 | $1 + \alpha^2 + \alpha^3$ | $\alpha^{13}$ |
| 1001 | $1 + \alpha^3$ | $\alpha^{14}$ |

way of characterizing the members of GF(16) is simply in terms of the coefficients of $1, \alpha, \alpha^2, \alpha^3$, so for example

$$\alpha^5 = 0.1 + 1\alpha + 1\alpha^2 + 0\alpha^2 = 0110$$
$$\alpha^6 = 0.1 + 0\alpha + 1\alpha^2 + 1\alpha^3 = 0011$$

and the complete list of 4-bit words is given in Table 4.2. Notice also that

$$\alpha^{15} = \alpha.\alpha^{14} = \alpha(1 + \alpha^3)$$
$$= \alpha + \alpha^4 = \alpha + 1 + \alpha = 1$$

**Example 4.27**   **Arithmetic in GF(16)**

All arithmetical operations for the members of GF(16) are performed by referring to Table 4.2.

(a)  Addition can be done as follows:

$$\alpha^5 + \alpha^8 = (\alpha + \alpha^2) + (1 + \alpha^2)$$
$$= 1 + \alpha = \alpha^4$$

since $\alpha^2 + \alpha^2 \equiv 0 (\mathrm{mod}\ 2)$. As a simpler alternative, the 4-bit representations can be used with bit-by-bit addition modulo 2, as defined in Section 4.3.1. Thus for the same two elements in GF(16):

$$0110 + 1010 = 1100$$
$$\alpha^5 \qquad \alpha^8 \ = \ \alpha^4$$

Because the first column of Table 4.2 is the set of all possible 4-bit words, it follows that the sum of any two members remains in GF(16).

(b) The computation of additive inverses follows directly from $-1 = 1$, so for example $-\alpha^{10} = \alpha^{10}$, or equivalently $-1110 = 1110$.

(c) For multiplication it is preferable to use the third column in Table 4.2, together with the property $\alpha^{15} = 1$. For example:

$$\alpha^6 \times \alpha^{14} = \alpha^{6+14} = \alpha^{20} = \alpha^{15+5} = \alpha^{15} \times \alpha^5 = \alpha^5$$

or equivalently $(0011)(1001) = 0110$.

(d) Determining a multiplicative inverse is also easily done by using $\alpha^{15} = 1$. For example:

$$(\alpha^8)^{-1} = \alpha^{-8} = \alpha^{15}.\alpha^{-8} = \alpha^{15-8} = \alpha^7$$

or equivalently $(1010)^{-1} = 1101$.

Notice how the different representations in Table 4.2 for the members of GF(16) are used interchangeably.

---

The check matrix for a double-error-correcting binary BCH code of length 15 can now be defined as

$$H = \begin{bmatrix} 1 & \alpha & \alpha^2 & \alpha^3 & \alpha^4 & \alpha^5 & \alpha^6 & \alpha^7 & \alpha^8 & \alpha^9 & \alpha^{10} & \alpha^{11} & \alpha^{12} & \alpha^{13} & \alpha^{14} \\ 1 & \alpha^3 & \alpha^6 & \alpha^9 & \alpha^{12} & 1 & \alpha^3 & \alpha^6 & \alpha^9 & \alpha^{12} & 1 & \alpha^3 & \alpha^6 & \alpha^9 & \alpha^{12} \end{bmatrix}$$

$$(4.68)$$

To understand (4.68), two points need to be explained. Firstly, each element in the second row in (4.68) is the **cube** of the element above it, remembering that $\alpha^{15} = 1$, so for example

$$(\alpha^{13})^3 = \alpha^{39} = \alpha^{30}.\alpha^9 = (\alpha^{15})^2.\alpha^9 = \alpha^9$$

Secondly, the 4-bit words representing the powers of $\alpha$ in Table 4.2 can be written as **columns** (that is, the words are transposed). For example, in the third column in (4.68), $\alpha^2 = 0010$ and $\alpha^6 = 0011$, so the elements in the third column of $H$ are 0, 0, 1, 0, 0, 0, 1, 1 (reading from top to bottom). Thus the binary matrix form of $H$ in (4.68) has eight rows, so there are eight check bits $x_8$, $x_9, \ldots, x_{15}$, *twice* as many as for the perfect single-error-correcting Hamming code of length 15 in Section 4.3.4. The following scheme (see Exercises 4.114 and 4.115) shows that the code corrects all double errors.

## Decoding algorithm for BCH code

(1) For a received binary word $r$, compute the syndrome

$$S = Hr = \begin{bmatrix} u \\ v \end{bmatrix}$$

$$(4.69)$$

where $H$ is the matrix in (4.68) and $u$ and $v$ are members of GF(16).

(2) If $u = 0$, $v = 0$ then assume there is no error, and the codeword $r$ was transmitted.

(3) If $u \neq 0$ and $v = u^3$, then assume a single error in bit $r_{i+1}$, where $u = \alpha^i$.

(4) If $u \neq 0$, $v \neq u^3$ and the quadratic equation

$$X^2 + uX + (vu^{-1} + u^2) = 0 \qquad (4.70)$$

has two distinct roots $X = \alpha^i$, $X = \alpha^j$, then assume two errors have occurred in bits $r_{i+1}, r_{j+1}$.

(5) If (4.70) has no roots in GF(16), or if $u = 0$, $v \neq 0$, then at least three errors have been detected.

It is important to realize that $u$, $v$ and $X$ in (4.70) are all members of GF(16). Unfortunately the usual formula for solving quadratic equations does not work in GF(16). However, it's not difficult to obtain roots of (4.70) (if they exist) by trying to factorise the left side of (4.70) in the form

$$X^2 + uX + (vu^{-1} + u^2) = (X + \alpha^i)(X + \alpha^j)$$
$$= X^2 + (\alpha^i + \alpha^j)X + \alpha^{i+j}$$

Equating coefficients on either side of this identity gives

$$u = \alpha^i + \alpha^j, \quad vu^{-1} + u^2 = \alpha^{i+j} \qquad (4.71)$$

Substituting all possible combinations of $i$ and $j$ into (4.71) either determines the required solution of (4.70), or shows that none exists. This procedure is illustrated in the following examples.

| Example 4.28 | **Applications of the algorithm** |

(a) Suppose a received word produces a syndrome (4.69) with $u = \alpha^{14}$, $v = \alpha$. Since $v \neq u^3$, Step 3 does not apply. In Step 4 the last coefficient in the quadratic equation (4.70) is

$$\begin{aligned}
vu^{-1} + u^2 &= \alpha.\alpha^{-14} + \alpha^{28} \\
&= \alpha^{-13} + \alpha^{13+15} \\
&= \alpha^{15}.\alpha^{-13} + \alpha^{13}.\alpha^{15} \\
&= \alpha^2 + \alpha^{13} \\
&= 0010 + 1011 \\
&= 1001 = \alpha^{14}
\end{aligned}$$

To determine whether (4.70) has roots $\alpha^i$, $\alpha^j$ with $i \neq j$, the expressions (4.71) give

$$\alpha^{14} = \alpha^i + \alpha^j, \quad \alpha^{14} = \alpha^{i+j} \qquad (4.72)$$

The second identity in (4.72) shows that $i + j = 14$. To see whether the first identity in (4.72) can be satisfied, simply try all possible pairs of values of $i$ and $j$, with $i < j$. This can be set out as follows:

| $i$ | $j$ | $\alpha^i$ | $\alpha^j$ | $\alpha^i + \alpha^j$ |
|---|---|---|---|---|
| 1 | 13 | 0100 | 1011 | $1111 = \alpha^{12}$ |
| 2 | 12 | 0010 | 1111 | $1101 = \alpha^7$ |
| 3 | 11 | 0001 | 0111 | $0110 = \alpha^5$ |
| 4 | 10 | 1100 | 1110 | $0010 = \alpha^2$ |
| 5 | 9 | 0110 | 0101 | $0011 = \alpha^6$ |
| 6 | 8 | 0011 | 1010 | $1001 = \alpha^{14}$ |

The last entry in this table shows that $\alpha^6$ and $\alpha^8$ satisfy both conditions in (4.72), and so are the roots of (4.70). Hence by Step 4, two errors have occurred in positions 7 and 9 of the received word $r$, which is corrected to $r + 000000101000000$.

(b) Suppose that a received word is $r = 110010110100110$. Using $H$ in (4.68), the syndrome in (4.69) has

$$u = 1 + \alpha + \alpha^4 + \alpha^6 + \alpha^7 + \alpha^9 + \alpha^{12} + \alpha^{13} \qquad (4.73)$$

$$= 1000 + 0100 + 1100 + 0011 + 1101 + 0101 + 1111 + 1011 \qquad (4.74)$$

$$= 1111 = \alpha^{12}$$

Notice that the terms in (4.73) consist of those elements in the first row of $H$ in (4.68) which correspond to 1's in $r$; and that to add the terms in (4.74), simply count the numbers of 1's in each position – for example, there are five 1's in the first position, so their sum is 1. Similarly, using the second row of $H$ in (4.68) gives

$$v = 1 + \alpha^3 + \alpha^{12} + \alpha^3 + \alpha^6 + \alpha^{12} + \alpha^6 + \alpha^9$$

$$= 1 + \alpha^9$$

$$= 1000 + 0101$$

$$= 1101 = \alpha^7$$

The solution then continues as in part (a) (see Exercise 4.108).

(c) For a received word $r = 111110110110110$ the syndrome (4.69) is computed (see Exercise 4.109) to give $u = \alpha^2$, $v = \alpha^6$. Hence $v = u^3$, so Step 3 of the algorithm applies. A single error has occurred which is in $r_3$ (since $u = \alpha^2$), so the transmitted codeword was 110110110110110.

(d) Suppose a received word produces a syndrome (4.69) with $u = \alpha^7$, $v = \alpha^4$. The second part of (4.71) gives

$$\alpha^{i+j} = vu^{-1} + u^2 = \alpha^4 \alpha^{-7} + \alpha^{14}$$

$$= \alpha^{-3} + \alpha^{14}$$

$$= \alpha^{15-3} + \alpha^{14}$$

$$= 1111 + 1001 = 0110 = \alpha^5$$

so that $i + j = 5$. However, notice that since $\alpha^5 = \alpha^{20}$, an alternative condition is $i + j = 20$. The first identity in (4.71) requires that $\alpha^i + \alpha^j = \alpha^7$. As in part (a), trying some possible pairs of values of $i$ and $j$ produces the table:

| $i$ | $j$ | $\alpha^i$ | $\alpha^j$ | $\alpha^i + \alpha^j$ |
|---|---|---|---|---|
| 1 | 4 | 0100 | 1100 | $1000 = \alpha$ |
| 2 | 3 | 0010 | 0001 | $0011 = \alpha^6$ |
| 6 | 14 | 0011 | 1001 | $1010 = \alpha^8$ |

The reader should confirm that none of the remaining four possibilities with $i + j = 20$ produces $\alpha^7$ in the last column (see Exercise 4.111). In other words, equation (4.70) has no roots in GF(16), so from Step 5 of the algorithm it is concluded that at least three errors occurred in transmission, and retransmission is requested.

Nothing in the above development depends upon $N = 15$, and in fact a finite field GF($2^m$) can be used to obtain a double-error-correcting BCH code of length $2^m - 1$ with $2m$ check bits. Indeed, BCH codes can be constructed so as to correct any number of errors. For example, in telecommunications, codes of length 255 with 24 check bits are used to correct three errors, or to detect up to six errors. However, development of these codes relies on the concept of cyclic codes, which lie outside the scope of this book.

**Exercise** **4.105** Verify the entries for $\alpha^9$ to $\alpha^{14}$ in Table 4.2. Use it to compute square roots of the members of GF(16). For example, $(\alpha^5)^{1/2} = (\alpha^{20})^{1/2} = \alpha^{10}$.

**Exercise** **4.106** Construct a table like Table 4.2 for the finite field GF(8) generated by $\alpha^3 + \alpha + 1 = 0$. Verify that $\alpha^7 = 1$ and determine $(111)(011)$, $(110)^{-1}$.

**Exercise** **4.107** Construct a table like Table 4.2 if (4.67) is replaced by $\alpha^4 = 1 + \alpha^3$. Verify that $\alpha^{15} = 1$ and determine $(1111)(1011)$, $(0111)^{-1}$.

**Exercise** **4.108** Complete Example 4.28(b) by decoding the received word.

**Exercise** **4.109** Determine the syndrome (4.69) for the received word in Example 4.28(c).

**Exercise** **4.110** Decode received words which produce a syndrome (4.69) with:
(a) $u = \alpha^{14}$, $v = \alpha^5$ (b) $u = \alpha^2$, $v = \alpha^{11}$.

**Exercise** **4.111** Verify that in Example 4.28(d) none of the three remaining pairs of values of $i$ and $j$ gives roots of the quadratic equation (4.70).

**Exercise** **4.112** Verify that one possibility in Example 4.28(d) is that transmission errors occurred in positions 1, 2 and 4.

**Exercise**  **4.113**  Use the algorithm to decode the received words:

(a) 100111000000000   (b) 100011001010000

(c) 000111001000001   (d) 110111101011000.

**Exercise**  **4.114**  Use (4.68) to show that if a received word has a single error in the $i$th bit then the syndrome (4.69) has $u = \alpha^{i-1}$, $v = u^3$.

**Exercise**  **4.115**  Use (4.68) to show that if a received word has errors in bits $r_{i+1}$, $r_{j+1}$ then the syndrome (4.69) has $u = \alpha^i + \alpha^j$, $v = \alpha^{3i} + \alpha^{3j}$. Use the identity

$$a^3 + b^3 = (a+b)(a^2 + ab + b^2) \pmod 2$$

with $a = \alpha^i$, $b = \alpha^j$ to show that

$$\alpha^{i+j} = vu^{-1} + u^2$$

Hence deduce that the roots of (4.70) are $\alpha^i$, $\alpha^j$.

**4.6** **Cryptology**

The aim of error-correcting codes is to make messages clear, so that a receiver gets correct information. The objective of cryptology is the opposite: to conceal, or **encrypt**, a message so that it is unintelligible except to authorized receivers. This is the stuff of which spy novels and movies are made, but security of transmission of information in the financial and commercial worlds is just as important as military intelligence. **Cryptology** is the study of secrecy systems; the design of such systems is called **cryptography**, and the breaking of secrecy codes is called **cryptanalysis**. These names derive from the Greek 'kruptos', meaning 'hidden'. A very brief introduction to the subject is given in this section through some simple examples.

**Example 4.29**  **Scrambled television transmissions**

If you want to view most satellite or cable television programmes, you need to rent a 'decoder box' which unscrambles the transmissions. Your box is assigned a 'key', consisting of a binary word $k$ of length $n$, which enables it to 'unlock' the received information. In simplified terms, what happens is that each paying customer is issued with a different key, say $k_i$, $i = 1, 2, 3, \ldots$. The channel transmits a special binary word $p$, called the 'password' which is changed monthly, together with the sums $k_1 + p, k_2 + p, k_3 + p, \ldots$. It is assumed that no errors occur in transmission. Your decoder then adds your own key $k_j$ (say) to each of the received messages $r_i = k_i + p$, and compares each sum $k_i + k_j + p$ with the password. When $k_i + k_j + p = p$, the decoder unscrambles the television signal, since this means that $k_i + k_j = 0$, that is, $k_i = k_j$. Hence your key $k_j$ 'unlocks' the received word $r_j = k_j + p$ and no other. If you fail to renew your monthly payment, then the channel terminates your service by not transmitting

$k_j + p$ next month. One way of breaking the system would be to simply try all possible binary words of length $n$ as keys. However, there are $2^n$ such words, so if $n$ is large enough such a scheme is effectively doomed to failure (see Exercise 4.116).

**Example 4.30** **Character ciphers**

(a) **Caesar ciphers**: a simple way of encrypting (or **enciphering**) a string of letters, or 'characters', is to replace each letter by the one following it, so A is replaced by B, B by C, and so on, with Y replaced by Z, and Z by A. The message is called the **plaintext** and the encrypted string the **ciphertext**. For example, if the plaintext is BIG, then the ciphertext is CJH.

To express this mathematically, assign a numerical value to each letter of the Roman alphabet, starting with $A = 0$, $B = 1$ and ending with $Y = 24$, $Z = 25$. Let $p$ denote the numerical value of a plaintext letter, and $c$ the value of the corresponding ciphertext letter. Since each letter is shifted one place to the right, the scheme above is represented by

$$c \equiv (p + 1)(\mathrm{mod}\ 26) \qquad 0 \leqslant c \leqslant 25 \tag{4.75}$$

The congruence modulo 26 in (4.75) is necessary so that $Z = 25$ is encrypted into $25 + 1 \equiv 0$, which is A. A simple generalization of this encryption scheme is to shift each letter to the right by $k$ places. In other words, introduce a **key**, which is a single letter whose numerical value is $k$. This is added to each plaintext letter, so that (4.75) is replaced by

$$c \equiv (p + k)(\mathrm{mod}\ 26), \quad 0 \leqslant c \leqslant 25 \tag{4.76}$$

For example, if the key is the letter X which has numerical value $k = 23$, then BIG is encrypted as follows:

| Plaintext | B | I | G |
|---|---|---|---|
| Numerical value | 1 | 8 | 6 |
| +23(mod 26) | 24 | 5 | 3 |
| Ciphertext | Y | F | D |

To **decrypt** (or **decipher**) a received message the process is carried out in reverse, so that (4.76) is solved for each plaintext letter according to $p \equiv (c - k)(\mathrm{mod}\ 26)$, with $0 \leqslant p \leqslant 25$. For example, with the same key $k = 23$, if QLM is received then the deciphering procedure is as follows:

| Ciphertext | Q | L | M |
|---|---|---|---|
| Numerical value | 16 | 11 | 12 |
| −23(mod 26) | 19 | 14 | 15 |
| Plaintext | T | O | P |

Ciphers of the type described above are sometimes named after Julius Caesar, who is said to have used them. A further extension is to use a key consisting of more than one letter, repeated in sequence if necessary. For example, if the key is CAT and a plaintext message is CIRCUS, then the encryption procedure is:

| | | | | | | |
|---|---|---|---|---|---|---|
| *CIRCUS* | 2 | 8 | 17 | 2 | 20 | 18 |
| *CATCAT* | 2 | 0 | 19 | 2 | 0 | 19 |
| *Sum (mod 26)* | 4 | 8 | 10 | 4 | 20 | 11 |
| *Ciphertext* | E | I | K | E | U | L |

To decipher messages, the receiver subtracts the key (repeated if necessary). For example, if the ciphertext is FOZIY then deciphering produces:

| | | | | | |
|---|---|---|---|---|---|
| *FOZIY* | 5 | 14 | 25 | 8 | 24 |
| *CATCA* | 2 | 0 | 19 | 2 | 0 |
| *Difference (mod 26)* | 3 | 14 | 6 | 6 | 24 |
| *Plaintext* | D | O | G | G | Y |

(b) **Linear transformation**: Here (4.76) is replaced by a more general linear congruence

$$c \equiv (ap + k)(\text{mod } 26), \qquad 0 \leqslant c \leqslant 25 \tag{4.77}$$

where $k$ is the key, and $a$ is a positive integer such that $(a, 26) = 1$. This condition is necessary so that as $p$ runs through a complete system of residues modulo 26, then $c$ does also. To decipher, solve (4.77) for $p$, giving

$$p \equiv a^{-1}(c - k)(\text{mod } 26), \qquad 0 \leqslant p \leqslant 25 \tag{4.78}$$

For example, suppose $a = 3$ and $k = 5$ so that (4.77) becomes $c \equiv (3p + 5)(\text{mod } 26)$. A few letters of the alphabet are encrypted as follows:

| *Plaintext* | A | B | C | D | X | Y | Z |
|---|---|---|---|---|---|---|---|
| *Numerical value p* | 0 | 1 | 2 | 3 | 23 | 24 | 25 |
| *(3p + 5)(mod 26)* | 5 | 8 | 11 | 14 | 22 | 25 | 2 |
| *Ciphertext* | F | I | L | O | W | Z | C |

In (4.78), $a^{-1} = 3^{-1} \equiv 9(\text{mod } 26)$, so that

$$p = 9(c - 5) = 9c - 45 \equiv (9c + 7)(\text{mod } 26)$$

For example, the ciphertext letter L which has value $c = 11$ is deciphered as

$$p \equiv (9 \times 11 + 7)(\text{mod } 26) \equiv 106(\text{mod } 26) \equiv 2(\text{mod } 26)$$

which corresponds to the plaintext letter C, agreeing with the table above.

An important encryption scheme is called **RSA** from the initials of R.L. Rivest, A. Shamir and L.M. Adleman who published the idea in 1978. It is an example of a so-called **public-key** cryptosystem, where the key is not secret, but deciphering requires the key to be decomposed into two prime factors. If the key has at least 230 decimal digits then an impossibly large amount of computer time is needed to break the system. Discussion of the scheme is outside the scope of this book, but you are invited to investigate it in Problem 4.30.

**Exercise**  **4.116**  Show that if $n = 56$ in the encryption scheme in Example 4.29 then there are about $7.2 \times 10^{16}$ keys. Show also that a computer hacker attempting to break the system would need to try about $2.7 \times 10^{10}$ keys per second for a month. Hence even if tens of millions of keys are issued, the probability of finding a correct one by trial is extremely small.

**Exercise**  **4.117**  If $k = 4$ in (4.76), encrypt the plaintext CLOWN, and decipher the message VMRK.

**Exercise**  **4.118**  Using the key RST with the extended Caesar cipher, encrypt the plaintext LION, and decipher the message KAZVJ.

**Exercise**  **4.119**  Complete the encryption of the alphabet for the code in Example 4.30(b).

**Exercise**  **4.120**  A message is known to have been encrypted using a Caesar cipher (4.76), but the key is lost. Decipher EHOX FTMAXFTMBVL by shifting the letters to the right until an intelligible message appears.

**Exercise**  **4.121**  Suppose that in the linear transformation (4.77) $a = 5$ and $k = 6$. Encipher the plaintext FINISH, and decipher the message JGSX ARANQUSA.

## Miscellaneous problems

**4.1**  A transmitted binary codeword $x_1 x_2 \ldots x_n$ is received as $y_1 y_2 \ldots y_n$, where the last bit is the even parity check bit. Prove that if

$$\text{parity}(y_1 y_2 \ldots y_{n-1}) \neq \text{parity}(y_n) \tag{4.79}$$

then there must be an odd number of transmission errors. (*Hint*: consider separately the cases $y_n = 0$ and $y_n = 1$.) Similarly, if equality holds in (4.79) show that the total number of errors must be even (including zero errors). Apply these results to the three received words in Exercise 4.13.

**4.2**  Consider a $p$-ary code of length $n$, where $x_n$ is the parity check digit chosen so that

$$\sum_{i=1}^{n} x_i \equiv 0 (\text{mod } p)$$

and $x_i \in \mathbb{Z}_p$. If a received word is $y_1 y_2 \dots y_n$ and

$$\sum_{i=1}^{n} y_i \equiv a(\bmod\ p)$$

determine the significance of the cases $a = 0$ and $a \neq 0$. Use $p = 16$ (see Section 2.2.4) and assume that at most one transmission error occurs. Determine the information message when the received word is $4B8FA$.

**4.3** Investigate the situation for the 13-digit EAN in Example 4.5 when two non-adjacent digits are interchanged.

**4.4** The identity number on some machine-readable passports is a 7-digit codeword. The first six digits are the date of birth in the form $x_1 x_2 = $ day, $x_3 x_4 = $ month, $x_5 x_6 = $ last two digits of year. The check digit $x_7$ is determined by

$$7(x_1 + x_4) + 3(x_2 + x_5) + x_3 + x_6 + x_7 \equiv 0(\bmod\ 10)$$

Prove that:

(a) all single errors are detected

(b) all transposition errors involving adjacent digits are detected (notice that $x_4 \leftrightarrow x_5$ and $x_5 \leftrightarrow x_6$ require visual inspection of the passport holder – why?).

**4.5** Consider the US money order code in Example 4.6. Show:

(a) there are $99 \times 10^{10}$ words containing an error in a single digit

(b) there are $2 \times 10^{10}$ single-digit errors which go undetected

(c) the percentage of undetected single errors is 2.02.

**4.6** For the ZIP code in Example 4.7, what happens if:

(a) there is a single error in each of two different 5-bit blocks?

(b) there are three errors in the same 5-bit block?

**4.7** (a) Suppose that the congruence (4.7) in Example 4.6 is replaced by

$$\sum_{i=1}^{10} x_i \equiv x_{11}(\bmod\ 10)$$

Show that this code detects all single errors, and also detects all transpositions of $x_{10}$ and $x_{11}$ except when $|x_{10} - x_{11}| = 5$.

(b) Suppose that instead (4.7) is replaced by

$$x_1 - x_2 + x_3 - x_4 + x_5 - x_6 + x_7 - x_8 + x_9 - x_{10} \equiv x_{11}(\bmod\ 10)$$

Show that this code also detects all single errors, and in addition detects most errors involving transposition of adjacent digits (excluding $x_{11}$). Deduce that 8/9 of such transpositions are detected.

**4.8** Show that for two binary codewords $x = x_1 x_2 \dots x_n$ and $y = y_1 y_2 \dots y_n$ then

$$d(x, y) = \sum_{i=1}^{n} |x_i - y_i|$$

Use this expression, together with the standard inequality $|a + b| \leqslant |a| + |b|$ to obtain an alternative proof of the triangle inequality (4.9) for a binary code.

**4.9** Let $x$, $y$ be any two codewords in a linear binary code $C$. By setting $z = \mathbf{0}$ in the triangle inequality (4.9), and using (4.15) and (4.16), show that $w(x + y) \leqslant w(x) + w(y)$.

**4.10** Show that if $x$ is a word of length $n$, then the number of words of length $n$ which are distance at most $t$ from $x$ is $\sum_{i=0}^{t} C(n, i)$.

How many words of length ten are distance at most 3 from the zero word?

**4.11** It can be shown that there exists a binary linear code of length $n$, dimension $k$ and minimum distance $\delta$ if

$$\sum_{i=0}^{\delta-2} C(n-1, i) < 2^{n-k}$$

Does there exist a binary linear code of length 12, dimension 2, and minimum distance 5?

**4.12** Suppose that $C$ is a linear binary code of length $n$. Construct a new code $\hat{C}$ of length $n + 1$ by appending an even parity check bit to each word of $C$ (as in Section 4.1.2). Show that $\hat{C}$ is also linear. Use the result in Exercise 4.53 to prove that if the minimum distance $\delta$ for $C$ is odd, then $\hat{C}$ has minimum distance $\delta + 1$.

**4.13** The **extended** $[n + 1, k]$ binary Hamming code is obtained from the $[n, k]$ binary Hamming code by appending an extra even parity check bit $x_{n+1}$ to each codeword, as in the preceding problem. Show that the extended code detects all double errors.

**4.14** Starting with the check matrix $H$ in (4.37) for the perfect $[7, 4]$ binary Hamming code, construct the check matrix $\hat{H}$ for the $[8, 4]$ extended binary Hamming code defined in the preceding problem. Do this by taking the first check equation to be $\sum_{i=1}^{8} x_i = 0$.

Show that for this extended code, the first element of a syndrome is 1 if there is a single transmission error, and 0 if there are two errors.

Hence encode the information message 1001, and decode the received words 11010110, 11010100.

**4.15** For the $p$-ary Hamming code defined in Section 4.3.5, obtain expressions for the first three check digits in terms of the information digits. Hence show that $(p - 1)(p - 1)100 \dots 0$ is a codeword, and deduce that the minimum distance for the code is 3.

**4.16** The **rate** $R$ of an $[n, k]$ code is defined as $k/n$, and is a measure of the efficiency of a code, since $n$ digits are required to send $k$ information digits. Show that for a perfect binary Hamming code with $m$ check bits, $R = 1 - m/(2^m - 1)$, and deduce that $R \longrightarrow 1$ as $m \longrightarrow \infty$.

**4.17** Show that the number of ternary words of length ten containing an even number of 1's is

$$\sum_{i=0}^{5} C(10, 2i) 2^{10-2i}$$

**4.18** Suppose a linear binary code $C$ of length $n$ is being used only for error detection. Errors will only go undetected if a codeword $x$ is sent and a different codeword $y$ is received. Since the code is linear, the error $e = y - x$ is also a codeword.

(a) Show that the probability of $e$ occurring with weight $i$ is $p^i(1 - p)^{n-i}$, where $p$ is the probability of an error in a single bit.

(b) If there are $a_i$ codewords of weight $i$ in $C$, show that the probability of an incorrect received message going undetected is

$$P_1 = \sum_{i=1}^{n} a_i p^i(1 - p)^{n-i}.$$

(c) When an error is detected, retransmission of the message is requested. Show that the retransmission probability is $1 - (1 - p)^n - P_1$.

(d) A linear binary code has check matrix

$$\begin{bmatrix} 1 & 0 & 0 & 1 & 1 & 1 & 0 \\ 0 & 1 & 0 & 0 & 1 & 1 & 1 \\ 0 & 0 & 1 & 1 & 0 & 1 & 1 \end{bmatrix}$$

If $p = 0.01$, show that the probability of an incorrect word being undetected is $6.79 \times 10^{-6}$, and that about 7% of received words have to be retransmitted.

**4.19** Use the idea in Problem 1.8, to show that for a decimal number $s = s_1 s_2 \ldots s_n$ then

$$s \equiv (s_n - s_{n-1} + s_{n-2} - s_{n+3} + \ldots) \mod 11$$

Hence obtain (4.51) from (4.50).

**4.20** For the decimal code in Section 4.4.2, solve the check equations (4.54) and (4.55) to obtain (modulo 11):

$$x_9 = \sum_{i=1}^{8} (i + 1)x_i, \quad x_{10} = \sum_{i=1}^{8} (9 - i)x_i$$

**4.21** For the decimal code in Section 4.4.3, let

$$a = \sum_{i=1}^{6} x_i, \quad b = \sum_{i=1}^{6} i x_i, \quad c = \sum_{i=1}^{6} i^2 x_i, \quad d = \sum_{i=1}^{6} i^3 x_i$$

Solve the four equations represented by (4.54), (4.55) and (4.58) for the four check digits $x_7, x_8, x_9, x_{10}$, using modulo 11 arithmetic, to obtain

$$x_7 = a + c + 2d, \quad x_8 = 7a + 4b + 2c + 5d,$$
$$x_9 = 6a + 4b + 4c + 6d, \quad x_{10} = 7a + 3b + 4c + 9d$$

Hence show that

$$x_7 = 4x_1 + 10x_2 + 9x_3 + \phantom{0}2x_4 + \phantom{0}x_5 + 7x_6$$
$$x_8 = 7x_1 + \phantom{0}8x_2 + 7x_3 + \phantom{00}x_4 + 9x_5 + 6x_6$$
$$x_9 = 9x_1 + \phantom{00}x_2 + 7x_3 + \phantom{0}8x_4 + 7x_5 + 7x_6$$
$$x_{10} = \phantom{0}x_1 + \phantom{0}2x_2 + 9x_3 + 10x_4 + 4x_5 + x_6$$

**4.22** Consider the equations in (4.59), numbered as follows:

$$S_2 = e_1 + e_2 \text{ (1)}, \quad S_1 = ie_1 + je_2 \text{ (2)}$$
$$S_3 = i^2 e_1 + j^2 e_2 \text{ (3)}, \quad S_4 = i^3 e_1 + j^3 e_2 \text{ (4)}$$

Apply the following operations to eliminate $e_2$:

$$j \times (1) - (2), \quad j \times (2) - (3), \quad j \times (3) - (4)$$

and denote the resulting equations by (5), (6) and (7) respectively. Then eliminate $e_1$ and $i$ by applying $(6)^2 - (5) \times (7)$, and verify that you obtain

$$aj^2 + bj + c = 0$$

where $a$, $b$ and $c$ are defined in (4.61).

Similarly, by eliminating first $e_1$, and then $e_2$ and $j$, verify that $i$ satisfies the same quadratic equation.

**4.23** The double-error-correcting code in Section 4.4.3 can be extended to correct more than two errors. For example, to correct three errors two more check equations, and hence two more check digits, are required. The check equations are $S_1 = 0, S_2 = 0, \ldots, S_6 = 0 \pmod{11}$ where

$$S_{j+1} = \sum_{i=1}^{10} i^j x_i, \quad j = 0, 1, 2, 3, 4, 5$$

If during transmission of a codeword three errors of magnitudes $e_1, e_2, e_3$ occur respectively in positions $k_1, k_2, k_3$ then these position numbers are the roots of the cubic equation

$$x^3 + a_1 x^2 + a_2 x + a_3 = 0$$

where the coefficients $a_i$ are the solutions of the linear equations

$$S_4 + a_1 S_3 + a_2 S_2 + a_3 S_1 = 0$$
$$S_5 + a_1 S_4 + a_2 S_3 + a_3 S_2 = 0$$
$$S_6 + a_1 S_5 + a_2 S_4 + a_3 S_3 = 0$$

Since arithmetic is performed modulo 11, the roots of the cubic equation are found simply by trying the values $x = 0, 1, 2, \ldots, 10$. The magnitudes of the errors are then obtained by solving the equations

$$e_1 + e_2 + e_3 = S_1$$
$$k_1 e_1 + k_2 e_2 + k_3 e_3 = S_2$$
$$k_1^2 e_1 + k_2^2 e_2 + k_3^2 e_3 = S_3$$

Suppose that a word $r$ is received for which it is computed that $S_1 = 2$, $S_2 = 8$, $S_3 = 4$, $S_4 = 5$, $S_5 = 3$, $S_6 = 2$. Assuming that three transmission errors have occurred, find the corrections for the appropriate digits of $r$.

**4.24** Consider GF(32) generated by $\alpha^5 + \alpha^2 + 1 = 0$. Express $\alpha^i$ for $i = 5, 6, 7, \ldots, 18$ in terms of $1, \alpha. \alpha^2, \alpha^3, \alpha^4$, and show that $\alpha^{31} = 1$.

**4.25** A linear binary code $C$ of length $n$ is called **cyclic** if whenever $x_1 x_2 x_3 \ldots x_n$ is a codeword then so is $x_n x_1 x_2 x_3 \ldots x_{n-1}$ (that is, every bit is shifted one place to the right). Prove that if $C$ contains a word having odd weight, then it contains the codeword all of whose bits are 1.

**4.26** Use the result in Exercise 2.92 with $a = k$, $b = 15$ to show that if $\alpha$ is a primitive element for GF(16), then so is $\alpha^k$ for any positive integer $k < 16$ such that $(k, 15) = 1$.

Verify that $\alpha^2$, $\alpha^8$ and $\alpha^{14}$ are each primitive elements for GF(16).

**Student project**

**4.27** For readers familiar with linear algebra, investigate the concepts of linear independence and linear dependence of sets of column vectors, and the rank of a matrix, using appropriate textbooks.

Prove the following results:

(1) If $H$ is the check matrix for a linear binary code $C$ of length $n$, then the dimension of the code is $n - \text{rank } H$.

(2) The code has minimum distance $\delta$ if and only if *every* set of $\delta - 1$ columns of $H$ is linearly *independent*, but there is at least *one* set of $\delta$ columns of $H$ which is linearly *dependent*.

Apply this result to prove:

(a) when

$$H = \begin{bmatrix} 0 & 1 & 1 & 1 & 1 & 0 & 0 & 0 \\ 1 & 1 & 1 & 0 & 0 & 1 & 0 & 0 \\ 1 & 1 & 0 & 1 & 0 & 0 & 1 & 0 \\ 1 & 0 & 1 & 1 & 0 & 0 & 0 & 1 \end{bmatrix}$$

then $C$ has $\delta = 4$.

(b) Perfect binary Hamming codes have $\delta = 3$. (*Hint*: find three columns of $H$ which are linearly dependent.)

(c) The code with check matrix $H$ in (4.57) has $\delta = 3$.

(3) The minimum distance satisfies $\delta \leqslant n - k + 1$.

**Student project**

**4.28** Some knowledge of linear algebra is again required. Let $H$ be the $r \times n$ check matrix for a linear binary code $C$, and denote the rows of $H$ by $\mathbf{h}_1, \mathbf{h}_2, \ldots, \mathbf{h}_r$.

(a) Suppose the row vector $\mathbf{v} = [1,1,\ldots,1,1]$, with $n$ elements each equal to 1, is linearly dependent on the rows of $H$. Prove that in this case all the codewords in $C$ have even weight.

(b) Suppose, alternatively, that $\mathbf{v}$ is linearly independent of the rows of $H$. Consider the extended code $\hat{C}$ whose check matrix is obtained from $H$ by adding an extra top row of 1's, as in Problem 4.14. Prove that in this case exactly half the codewords in $C$ have even weight.

Taken together, these results establish that for a linear binary code, either all the codewords have even weight, or half have even weight and half have odd weight.

**Student project**

**4.29** Some knowledge of matrices and linear algebra is also required for this problem.

(a) Define the $n \times n$ **Vandermonde determinant**

$$\det V_n = \begin{vmatrix} 1 & 1 & \cdots & 1 \\ v_1 & v_2 & \cdots & v_n \\ v_1^2 & v_2^2 & \cdots & v_n^2 \\ \vdots & \vdots & \cdots & \vdots \\ v_1^{n-1} & v_2^{n-1} & \cdots & v_n^{n-1} \end{vmatrix}$$

For example, when $n = 2$

$$\det V_2 = \begin{vmatrix} 1 & 1 \\ v_1 & v_2 \end{vmatrix} = v_2 - v_1$$

Show that

$$\det V_3 = (v_3 - v_2)(v_3 - v_1)(v_2 - v_1)$$

and prove by induction that

$$\det V_n = \prod_{i>j} (v_i - v_j)$$

where $\prod$ means the product of all terms $(v_i - v_j)$ for which $i > j$, as illustrated for $\det V_3$.

(b) A result in linear algebra states that the columns of any $n \times n$ matrix $A$ are linearly independent if and only if $\det A \neq 0$.

Apply the result in part (a) to the check matrix in (4.65), together with the result (2) in Problem 4.27, to deduce that the code in Section 4.4.3 has minimum distance 5.

## Student project

**4.30**  Write a detailed essay on the RSA cryptosystem, using appropriate textbooks, including those listed in Further Reading.

## Further reading

Section 4.1    Backhouse J.K. (1983). Retail article numbering and bar codes. *Bulletin Institute of Mathematics and its Applications*, **19**, 17–18

Bernard J. (1986). Compact discs bit-by-bit. *Radio Electron*, August, 62–63, 85

Berry J., Burghes D. and Huntley, I. (1986) *Decision Mathematics*. Chichester: Ellis Horwood, Chapter 13

Blocksma M. (1989). *Reading the Numbers*. New York: Penguin, pp.7, 208

Comap (1994). *For All Practical Purposes*. New York: Freeman, Chapter 9

Connor S. (1984). The invisible border guard. *New Scientist*, 5 January, 9–14

Gallian, J.A. (1986). The Zip code bar code. *UMAP Journal*, **7**, 191–5

Gallian J.A. and Winters S. (1988). Modular arithmetic in the marketplace. *American Math. Monthly*, **95**, 548–51

Hill R. (1989). Error-correcting codes I. *Math. Spectrum*, **22**(3), 94–103

Pohlmann K.C. (1989). *Principles of Digital Audio*, 2nd edn. Indianapolis: Howard W. Sams

Price N. (Ed.) (1994). *Discrete Mathematics*, 2nd edn. Oxford: Heinemann, Chapter 8

Rae I.D. (1984). Machine readable codes. *New Zealand Mathematics Magazine*, **21**, 109–13

Savir D. and Laurer G.J. (1975). The characteristics and decodability of the Universal Product Code symbol. *IBM Systems Journal*, **14**(1), 16–34

Selmer, E.S. (1967). Registration numbers in Norway: Some applied number theory and psychology. *Journal Royal Statistical Society* (Ser. A), **130**, 225–31

Section 4.2    Hill R. (1986). *A First Course in Coding Theory*. Oxford: Oxford University Press, Chapter 1

Section 4.3    Comap (1994). *op. cit.*, Chapter 10

Hill R. (1986). *op. cit.*, Chapters 5–8

Hoffman D.G., Leonard D.A., Lindner C.C., Phelps K.T., Rodger C.A. and Wall J.R. (1991). *Coding Theory, The Essentials*. New York: Marcel Dekker, Chapter 2

Section 4.4    Hill R. (1986). *op. cit.*, pp.36, 76

Hill R. (1990). Error-correcting codes II. *Math. Spectrum*, **23**(1), 14–23

Tuchinsky P.M. (1985). International Standard Book Numbers. *UMAP Journal*, **5**, 41–54

Section 4.5    Hill R. (1986). *op. cit.*, Chapter 12

Hoffman D.G. et al. (1991). *op. cit.*, Chapter 5

Lidl, R. and Niederreiter H. (1986). *Introduction to Finite Fields and their Applications*. Cambridge: Cambridge University Press, Chapter 8

MacWilliams F.J. and Sloane N.J.A. (1975). *The Theory of Error-Correcting Codes*. Amsterdam: North-Holland, Chapter 3

Section 4.6    Albertson M.O. and Hutchinson J.P. (1988) *Discrete Mathematics with Algorithms*. New York: Wiley, p.222

Comap (1994). *op. cit.*, Chapter 10

Rosen K.H. (1993). *Elementary Number Theory and its Applications*, 3rd edn. Reading MA: Addison-Wesley, Chapter 7

Schroeder, M.R. (1986). *Number Theory in Science and Communication*, 2nd enlarged edn. Berlin: Springer-Verlag, Part IV

Welsh D. (1988). *Codes and Cryptography*. Oxford: Clarendon Press, Chapters 7, 11

# 5 Recurrence Relations

When you press the square root button on your calculator the result comes up on the display almost at once. But the electronic chip which constitutes the calculator's 'brain' has actually done a series of calculations extremely rapidly using the scheme described in Exercise 1.24. The same operation is done repeatedly, and the formula involved is called a 'recurrence relation', from the word 'recur' which has the dictionary meaning 'to happen again, especially at regular intervals'. Because of its repetitive nature, the mathematical process is called 'iteration'. Digital computers and pocket calculators are ideally suited for such work, and iterative methods are widely used to solve equations.

Recurrence relations also provide a way of constructing 'mathematical models'. This idea arises from scale models of buildings, civil engineering projects, motor vehicles, aircraft, and so on, which are built to assist the process of design. In the same way, setting up a mathematical description which incorporates the rules obeyed by some real-life situation can help to investigate what is going on. An essential feature of mathematical models is their ability to describe how things evolve as time passes – the so-called 'dynamic' behaviour. Before the advent of high-speed digital computation, the language of calculus and differential equations was the main tool of mathematical modelling of dynamic situations, and a range of analytical techniques was developed. However, many differential equations can only be solved by casting them into a form which can be solved numerically on a computer. It therefore makes sense to see if the model itself can be set up from the beginning using a digital description, involving recurrence relations (or 'difference equations') instead of differential equations. There is a second compelling reason for using difference equations: their essential feature is that the variables involved only exist at distinct or 'discrete' intervals of time. This mirrors how time is used and measured in the real world. For example, an athlete will be timed to hundredths of seconds when running a race; the index of prices on the Stock Market is published in newspapers on a daily basis; the rate of inflation of retail prices is announced monthly; interest on savings accounts is usually added annually; the population of the UK is determined by census every ten years. Thus in practice time is not regarded as a continuous commodity, but as a string of finite 'packets', which may be hundredths of seconds, minutes, hours, days, months, years or whatever you wish. To analyse behaviour, difference equations relate the future to times past. For example, after observing how weather behaves in your locality, you could invent a scheme which forecasts the weather in an hour's time from a knowledge of weather conditions now and one hour ago. You could then use the same scheme to forecast the weather in two hours' time based on the conditions predicted for an hour ahead together with present conditions. The use of values at specified intervals of time (here one hour) and the repetitive nature of the calculations exemplify the methods and applications of this chapter.

## 5.1 Iteration

A scheme for computing $\sqrt{3}$ was outlined in Exercise 1.26. This can be formalized in the following way. Suppose that $x_0$ is an initial guess (taken as 1.5) for $\sqrt{3}$. Then a better approximation $x_1$ is given by

$$x_1 = \frac{1}{2}\left(x_0 + \frac{3}{x_0}\right)$$

This new value $x_1$ is then used to give a better approximation

$$x_2 = \frac{1}{2}\left(x_1 + \frac{3}{x_1}\right)$$

and so on. If $x_{n+1}$ denotes the value obtained after $n+1$ computations then

$$x_{n+1} = \frac{1}{2}\left(x_n + \frac{3}{x_n}\right) \tag{5.1}$$

where $n$ takes the values 0, 1, 2, 3, .... . The same calculation is done repeatedly using the iterative formula (5.1). In Exercise 1.26 the values $x_1 = 1.75$, $x_2 = 1.7321$, $x_3 = 1.72320508$ were given, and one further repetition gives $x_4 = 1.73205081$, so the value of $\sqrt{3}$ correct to seven decimal places is 1.7320508, since $x_3$ and $x_4$ agree to this accuracy. This statement is only true provided it can be assumed that the procedure is 'convergent', which means that by taking the value of $n$ large enough you can make $x_n$ as close as you like to the exact value of $\sqrt{3}$. It can be proved that this is indeed the case. Moreover, an extension to determine $\sqrt{a}$ for any positive number $a$ is given by the formula

$$x_{n+1} = \frac{1}{2}\left(x_n + \frac{a}{x_n}\right), \qquad n = 0, 1, 2, 3, \ldots$$

You were asked in Exercise 1.26 to use this formula (without its being explicitly stated) to compute $\sqrt{5}$. You keep on repeating the calculation for successive value of $n$ until the difference $x_{n+1} - x_n$, or more precisely its numerical value ignoring sign, denoted by $|x_{n+1} - x_n|$, is less than some specified amount, depending upon the accuracy required. The square root button on a calculator does the whole process automatically, much quicker than you can! A similar formula is used by the cube root button, if your calculator has one (see Exercise 5.1). Both these formulae are examples of what is known as the **Newton–Raphson** iterative method. This states that to find a root $x^*$ of an equation $f(x) = 0$, starting with an initial estimate $x_0$ near to $x^*$, then successively better approximations are given by

$$x_{n+1} = x_n - \frac{f(x_n)}{f'(x_n)}, \qquad n = 0, 1, 2, 3, \ldots \tag{5.2}$$

where (for readers familiar with calculus) $f'(x)$ is the **derivative** of the function $f(x)$ with respect to $x$. It can be shown that if $x_n$ is close to $x^*$, then the error in $x_{n+1}$ is proportional to the *square* of the error in $x_n$, that is,

$$x_{n+1} - x^* \approx k(x_n - x^*)^2$$

where $k$ is a constant. For example, if the error at one stage of the calculation is 0.01, then the error at the next stage will be proportional to $(0.01)^2 = 0.0001$. This shows that the process rapidly converges to the desired value.

## Example 5.1    Solution of quadratic equation

The Newton–Raphson method is now used to find the roots of the quadratic equation $x^2 - 7x + 3 = 0$ correct to five decimal places.

Here $f(x) = x^2 - 7x + 3$ so $f'(x) = 2x - 7$ and the formula (5.2) gives

$$x_{n+1} = x_n - \frac{(x_n^2 - 7x_n + 3)}{2x_n - 7}$$

$$= \frac{x_n^2 - 3}{2x_n - 7}, \qquad n = 0, 1, 2, 3, \ldots \tag{5.3}$$

Since $f(0) = 3$, $f(1) = 1 - 7 + 3 = -3$, this means that if you plot a graph of $f(x)$ it must cross the axis somewhere between 0 and 1, that is, there is a root in this region. This is especially easy to see if you use a graphics calculator. It therefore makes sense to take $x_0 = 0.5$, which when substituted into (5.3) with $n = 0$ gives $x_1 = 0.423077$. You repeat the calculation with this new value to find $x_2$, and continue until $x_n$ and $x_{n+1}$ agree to five decimal places. This gives $x_2 = 0.458413$, $x_3 = 0.458619$, $x_4 = 0.458619$ so the required root correct to five decimal places is 0.45862.

The second root is found in the same way. Since $f(6) = 36 - 42 + 3 = -3$ and $f(7) = 49 - 49 + 3 = 3$, a suitable starting value is $x_0 = 6.5$. Using (5.3) gives $x_1 = 6.541667$, $x_2 = 6.541381$, $x_3 = 6.541381$ so the required root is 6.54138. In both cases you can see that the convergence is very rapid.

You can check that these computed roots are correct by using the explicit formula for the roots of a quadratic equation given in Section 1.4.2, page 19.

In some awkward cases the sequence of successive calculations may not converge to a definite value, due to inherent difficulties in the problem, not the Newton–Raphson method. It is then necessary to increase the number of decimal places in the intermediate calculations so as to obtain the required accuracy for the root (see Problem 5.3 for an illustration of this). Other difficulties concerning convergence are discussed in textbooks on numerical analysis.

---

**Exercise**  **5.1**  The iteration

$$x_{n+1} = \frac{1}{3}\left(2x_n + \frac{a}{x_n^2}\right), \qquad n = 0, 1, 2, 3, \ldots$$

produces successively closer approximations to the cube root of a positive number $a$, starting with an initial value $x_0$. Use it to compute $6^{1/3}$, correct to five decimal places.

**Exercise**  **5.2**  Use the Newton–Raphson iteration to obtain the roots of

$$2x^2 + 7x - 10 = 0$$

correct to four decimal places.

**Exercise**  **5.3**  Use the Newton–Raphson iteration to find a positive root of the cubic equation

$$x^3 + 2x - 1 = 0$$

correct to six decimal places.

## 5.2  First order linear difference equations

In all the examples in the previous section there was a relationship between the value $x_{n+1}$ and the previously computed value $x_n$. Starting with an initial value $x_0$ you were able to compute $x_1$ via some formula; this *same* formula was then used to compute $x_2$ from $x_1$, and the process was then repeated until some required value was reached. The suffix gives the position of the variable in the sequence

$$x_0, x_1, x_2, x_3, \ldots, x_n, x_{n+1}, \ldots$$

The recurrence formulae involved only two neighbouring values $x_n$ and $x_{n+1}$ in this sequence. Because these suffices differ only by 1, the equations are known as **first order**. The term 'difference equation' is also in common use as an alternative to 'recurrence relation', because you can think of finding the difference between $x_n$ and $x_{n+1}$. In many applications, but not all, the variable $n$ represents time, which is measured in distinct finite amounts. If the difference equation involves $x_{n+2}$ and $x_n$ (and also possibly $x_{n+1}$) then since the suffices differ by 2, the equation is called **second order**. This type will be dealt with in Section 5.3.

### 5.2.1 Example: compound interest

When you deposit a sum of money in a savings account, you do so because money, called 'interest' will be added to it. This is one of the foundations of the international economy, and explains why 'the rich get richer'. There are two different ways in which interest is 'earned'. In the first, a *fixed* percentage of the sum deposited is added at predetermined intervals. For example, if you deposit £1000 and the interest is at a fixed annual percentage rate (APR) of 5, then after one year an amount

$$£1000 \times \tfrac{5}{100} = £50$$

will be added. The *same* amount will be added every year, provided the APR remains the same. This type of interest is called **simple**, since the fixed interest is merely added to the original deposit, and does not itself gain interest. Thus after two years the amount in the account will be $£1000 + 2 \times £50 = £1100$. In general, if $£x_n$ denotes the balance of the account at the end of the $n$th year, then with an APR of 5

$$x_n = 1000 + 50n, \qquad n = 0, 1, 2, 3, \ldots$$

This is an example of **linear** growth, since if you plot the values of $x_n$ against $n$, as in Figure 5.1, you will obtain a series of points which lie on a straight line.

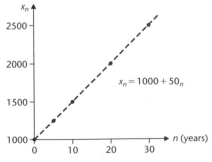

*Figure 5.1*   Simple interest (APR $= 5$)

A snag with an account which only pays interest annually is that you lose out if you close the account early, say after 18 months when you would only get the first year's interest, giving a total withdrawal of £1050. You might therefore prefer an account where the interest is added more often, say at the end of every month. If the APR remains at 5, the *monthly* interest rate will be (5/12)%. At the end of each month the interest paid will be

$$£1000 \times \frac{5/12}{100} = £4\tfrac{1}{6}$$

After one year the total interest will be £$4\tfrac{1}{6} \times 12 = £50$, the same as before. With simple interest, it doesn't make any difference to the annual amount of interest you receive if the interest is paid more frequently. However, when interest is paid monthly then after 18 months you would get £$4\tfrac{1}{6} \times 18 = £75$ interest, so if you closed your account then the total withdrawal would be £1075.

The second type of interest is called **compound**, and is the one usually paid on bank and savings accounts. After interest is added to the account, the new *total* sum earns interest.

---

| Example 5.2 | **Savings account with compound interest** |

Suppose you decide to invest £1000 in an account which has an APR of 5 as before, but which pays compound interest instead of simple interest. At the end of the first year there will be no change, the interest gained will be £50, giving a new balance of £1050. However, compounding means that the *whole* of this amount earns interest, which at the end of the second year is

$$£1050 \times \tfrac{5}{100} = £52.50$$

The total balance at the end of the two-year period is therefore

$$£1050 + £52.50 = £1102.50$$

It's now instructive to again write $x_n$ for the amount (in pounds) in the account at the end of year $n$, so that $x_0 = 1000$. After one year you get

$$x_1 = \underset{\text{deposit}}{1000} + \underset{\text{interest}}{1000 \times \tfrac{5}{100}}$$
$$= 1000\left(1 + \tfrac{5}{100}\right)$$
$$= 1000(1.05)$$

and after two years in an exactly similar way you get

$$x_2 = x_1 + x_1 \times \tfrac{5}{100}$$
$$= x_1(1.05)$$

Another year produces

$$x_3 = x_2(1.05)$$

and in general you get

$$x_{n+1} = 1.05x_n, \qquad n = 0, 1, 2, 3, \ldots \tag{5.4}$$

This is the recurrence relation, or difference equation, which expresses the amount in the account at the end of year $n + 1$ in terms of the account at the end of year $n$. You can see that like the recurrence relations in Section 5.1, it has first order. However, because (5.4) is a linear equation it's possible to actually obtain an explicit expression for the solution of (5.4). First combine together the expressions for $x_2$ and $x_1$ to obtain

$$\begin{aligned} x_2 &= x_1(1.05) \\ &= 1000(1.05)^2 \end{aligned}$$

Similarly,

$$x_3 = x_2(1.05) = 1000(1.05)^3$$

and you can see that in general

$$x_n = 1000(1.05)^n, \qquad n = 1, 2, 3, \ldots \tag{5.5}$$

which is the required solution of (5.4). The expression (5.5) reveals that as $n$ increases the value of $x_n$ increases more rapidly than with simple interest at the same APR (see Figure 5.2).

The value of $x_n$ can be made as large as you like simply by taking $n$ large enough. For example, how long would it take for your balance to reach a total of £100,000? You can see from (5.5) that to achieve this $n$ must be such that $(1.05)^n = 100$. You can experiment with the $x^y$ button on your calculator, for example

$$(1.05)^{70} = 30.4, \quad (1.05)^{100} = 131.5$$

and you soon find that $(1.05)^{95} = 103.0$ so your balance will exceed £100,000 after 95 years – unfortunately, it's highly unlikely you'll be around by then!

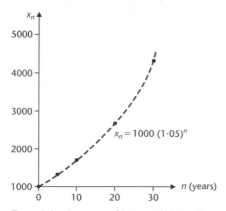

*Figure 5.2*  Compound interest (APR $= 5$)

In the difference equation (5.4) the suffix $n$ represents time in years. Suppose that interest is compounded every six months (that is, twice per year) instead of annually. The APR is 5, so the rate of interest per six-month period is $5/2 = 2.5\%$. The formula (5.4) is replaced by

$$x_{n+1} = 1.025x_n, \qquad n = 0, 1, 2, 3, \ldots$$

where now the suffix $n$ represents the new time period of six months. The initial deposit remains at £1000, so after six months the total amount in the account is

$$x_1 = 1.025x_0 = £1025$$

After a further six months (that is, after one year altogether) the account contains

$$x_2 = 1.025x_1 = £1050.625$$

which is slightly more than the first situation when the interest was compounded annually.

To be more general, suppose the interest is compounded $k$ times per year, so the interest rate for each time period (of length $1/k$ years) is $(5/k)\%$. The case just considered had $k = 2$. The difference equation this time is

$$x_{n+1} = \left(1 + \frac{5}{100k}\right)x_n, \qquad n = 0, 1, 2, 3, \ldots \tag{5.6}$$

instead of (5.4), where now $x_n$ is the amount in the account after $n$ time intervals of length $1/k$ years have elapsed. By analogy with (5.5) the solution of (5.6) is

$$x_n = 1000\left(1 + \frac{5}{100k}\right)^n, \qquad n = 1, 2, 3, \ldots \tag{5.7}$$

For example, if interest is compounded monthly, that is, 12 times per year so $k = 12$, then after a year (= 12 months) has elapsed the amount in the account is obtained by setting $n = 12$ in (5.7) to give

$$x_{12} = 1000\left(1 + \frac{5}{100 \times 12}\right)^{12} \tag{5.8}$$
$$= £1051.16$$

This is a bit more than when the interest was compounded twice a year, and that in turn was a little more than when the interest was compounded annually. Does this mean that if the interest is compounded more and more frequently, then the total amount in the account at the end of a year keeps on getting bigger and bigger? This amount is $x_k$, and its value is obtained by setting $n = k$ in (5.7) to give

$$£1000\left(1 + \frac{5}{100k}\right)^k \tag{5.9}$$

The case $k = 12$ (monthly compounding) is displayed in (5.8). The question now is: what happens to the expression in (5.9) as $k$ gets larger and larger? The answer is provided in Problem 1.21, where a version of the compound interest problem using slightly more convenient numbers was discussed. It turns out that the sum in (5.9) does *not* go on increasing indefinitely. As $k$ gets larger and larger, this sum actually gets ever closer to, but does not exceed, the amount

$$£1000e^{5/100} = £1051.27$$

where e is the exponential function, making a surprise appearance. It is no doubt very comforting to the world's bankers to know that increasing the frequency at which interest is compounded does *not* mean ever-increasing payouts!

---

Now bring together the facts about compound interest. If the APR is $r$, and interest is compounded $k$ times per year, then the amount $x_n$ in the account after $n$ time periods (each of length $1/k$ years) have elapsed is

$$x_n = \left(1 + \frac{r}{100k}\right)^n x_0, \qquad n = 0, 1, 2, 3, \ldots \tag{5.10}$$

where $x_0$ is the initial deposit. The value of $x_n$ for $n > 1$ will always be greater than if simple interest is paid at the same APR. The difference equation relating $x_{n+1}$ and $x_n$ is

$$x_{n+1} = \left(1 + \frac{r}{100k}\right) x_n, \qquad n = 0, 1, 2, 3, \ldots \tag{5.11}$$

This is a first order equation which is said to be **linear** because it contains no terms $x_n^2$, $x_{n+1}^2$, or higher powers of $x_n$ or $x_{n+1}$. However, there can be other such terms, for example the equation

$$x_{n+1} = (n+1)^3 x_n, \qquad n = 0, 1, 2, 3, \ldots$$

is first order and linear.

If $n = k$ in (5.10) and $k$ gets larger and larger, that is, interest is added more and more frequently at the same APR, say every month, or day, or even second, then the amount at the end of one year gets ever closer to $x_0 e^{r/100}$.

| Example 5.3 | **Inflation** |

Economic inflation is the name given to the situation where money depreciates in value. The percentage increase in the Retail Prices Index over a year is called the annual rate of inflation. For example, if the rate of inflation is 3% per year, this means that you need

$$£\left(1 + \frac{3}{100}\right) = £1.03$$

to buy now items which cost £1 a year ago. If the inflation rate is constant from year to year, then the calculations involved are the same as those in the compound interest case. For example, after 20 years the amount you would need to buy something which originally cost £1000 is

$$£1000\left(1 + \tfrac{3}{100}\right)^{20} = £1806.11$$

The formula used is (5.10) with $r = 3$, $k = 1$, $n = 20$. Although this inflation rate is quite low, over a long period of time the value of money goes down significantly. For example, still assuming a constant annual rate of inflation of 3%, how long will it take for money to halve in value? To answer this, you need to find the value of $n$ such that $(1.03)^n = 2$, since this means that after $n$ years you will need £2 to buy what cost £1 at the beginning of the period, so prices have doubled. You can use your calculator to obtain $(1.03)^{24} = 2.03$, so the answer to the above question is that after 24 years the pound will buy a little less than half of the goods you could get at the start of the period.

| Example 5.4 | **Instant access account** |
| --- | --- |

In Example 5.2 you opened an account with an initial deposit which you then left alone to accrue interest. In practice you may well prefer an account of the 'instant access' type, where you can make withdrawals or additional deposits whenever you please. Suppose again you make an initial deposit of £1000, and that this time the APR is 6, paid monthly (at a rate of $6/12 = 0.5$). The net amount which is in the account at the *start* of a month gains interest at the start of the *following* month. For example, if you make your initial deposit $x_0$ (= £1000) on 1 January, and do nothing with the account during the rest of January, then on 1 February the interest added is £$(0.5/100)1000 = £5$. Suppose during February you deposit £200, but later withdraw £300 to pay a bill, so the net deposit is −£100 (that is, a net withdrawal of £100). On 1 March you gain interest on the sum of £1005 which was in the account on 1 February, that is, £$(0.5/100)1005 = £5.025$, so the net sum in the account is

$$£1005 + £5.025 − £100 = £910.025$$

In general, if £$x_n$ denotes the amount in your account at the start of month $n$, then this gains interest of £$(0.5/100) \times x_n$ at the start of month $n + 1$. If £$D_n$ is the net deposit during month $n$ (for example, $D_1 = −100$) then at the start of month $n + 1$ you have in total

$$x_{n+1} = x_n + \frac{0.5}{100} x_n + D_n$$

$$= 1.005 x_n + D_n, \qquad n = 0, 1, 2, 3, \ldots \tag{5.12}$$

For example, when $n = 0$ this gives

$$x_1 = 1.005x_0 + D_0$$
$$= 1.005 \times 1000 + 0$$
$$= 1005$$

and when $n = 1$,

$$x_2 = 1.005x_1 + D_1$$
$$= 1.005 \times 1005 - 100$$
$$= 910.025$$

as before. The difference equation (5.12) is again first order and linear, but now contains the extra term $D_n$. Knowing the values of the $D$'s, you can obtain the values of $x_3, x_4, x_5, \ldots$ sequentially using the difference equation (5.12).

In certain cases, for example if $D_n$ is the same every month (you might perhaps debit a fixed amount to your electricity supplier each month, or put in a fixed monthly savings deposit) then it's possible to find an explicit expression for the solution of equations like (5.12), just as (5.10) is the solution of the equation (5.11). Details will be given in the next section.

**Exercise**

**5.4** You decide to invest £1000 in an account which pays simple interest at a fixed APR. At the end of six years the total in the account is £1330. What is the APR?

**Exercise**

**5.5** Suppose in the previous exercise the initial sum had instead been invested in an account which paid interest compounded annually at a fixed rate. Determine this APR if at the end of the six years the total amount received is again £1330.

**Exercise**

**5.6** Suppose you invest £1000 in an account paying interest compounded annually at a fixed APR of 5 for six years. How much will you get? If instead the interest was compounded quarterly, what would the total return be at the end of the six years?

**Exercise**

**5.7** You have £1000 to invest and wish to double this sum as quickly as possible. You are offered a choice of three fixed interest accounts which pay as follows:

(a) APR of $7\frac{1}{2}$ compounded annually

(b) APR of $7\frac{1}{4}$ compounded semi-annually

(c) APR of 7 compounded monthly.

Which option should you select?

**Exercise** **5.8** You have been nominated as an executor for your uncle's will. He has left his estate to be divided equally amongst his three surviving children, his wife having died some years previously. One of the items is Government stock which earns interest at $7\frac{3}{4}\%$, compounded annually. The stock was purchased three years ago and is currently worth £15011.81. How much did your uncle originally invest? When the stock matures in seven years' time, how much will each of your cousins receive?

**Exercise** **5.9** You invest £2000 in an account which pays interest compounded annually at a fixed APR of 4 for six years. A feature of this account is that you are allowed to withdraw up to £300 without penalty at the end of each year. You decide to use this facility to its full extent. Compute the amount in the account at the end of each year, and hence determine the total sum you will receive at the end of the period (net of the £1800 withdrawn).

Let $x_n$ denote the amount in the account at the end of year $n$, with $x_0 = 2000$. What is the difference equation relating $x_{n+1}$ and $x_n$?

**Exercise** **5.10** In the 'instant access' account described in Example 5.4 you make deposits of £125, £70 and £95 respectively during each of the first three months. What will be the amount in the account at the start of the fourth month?

You then arrange for hire purchase payments on a washing machine to be debited to the account in nine monthly instalments of £40 each. If you make no other deposits or withdrawals, how much will be left in the account at the end of the year?

### 5.2.2 General solution

The difference equations (5.4) and (5.11) in Section 5.2.1 are examples of a general first order linear equation

$$x_{n+1} = c\,x_n, \qquad n = 0, 1, 2, 3, \ldots \tag{5.13}$$

where $c$ is a *constant*. In most applications you are likely to encounter $c$ will be a real number, but in principle it could be a complex number. The general solution of (5.13) is

$$x_n = c^n x_0, \qquad n = 1, 2, 3, \ldots \tag{5.14}$$

where $x_0$ is the initial value of $x_n$. To check this is correct, write $n + 1$ in place of $n$ in (5.14) to get

$$x_{n+1} = c^{n+1} x_0 = c(c^n x_0) = c x_n$$

In applications it is often of interest to know what happens to $x_n$ when the value of $n$ becomes large. The notation $n \longrightarrow \infty$ is used to mean that the positive integer $n$ increases indefinitely. Clearly if $c$ is a positive real number less than one, then $c^n$ gets smaller and smaller as $n \to \infty$, and this is written $c^n \to 0$. We say that '$c^n$ tends to zero as $n$ tends to infinity', and this means that $c^n$ can be made as close to zero as you like simply by taking $n$ large enough. For example, if $c = 0.5$ then you can use your calculator to find that $(0.5)^{34} = 5.8 \times 10^{-11}$, so you can see that if $n > 33$ then $c^n < 10^{-10}$. A useful notation is $|c|$, the **modulus** of $c$, which denotes the numerical value of a real number $c$, so $|c| < 1$ means that $c$ lies between $-1$ and $1$, that is, $-1 < c < 1$. If $c$ is negative, that is, $-1 < c < 0$, then $c^n$ will oscillate in sign as $n$ increases, but $|c|^n$ will tend to zero as $n \to \infty$. If $c$ is a complex number with real part $a$ and imaginary part $b$ so that $c = a + ib$, then the modulus of $c$ is defined by

$$|c| = (a^2 + b^2)^{1/2}$$

which is the distance from the origin to the point represented by $c$ on an Argand diagram (see Section 1.4.2). The condition for $|c|^n \to 0$ as $n \to \infty$ remains the same, that is, $|c| < 1$. If $c = 1$ then it is trivial to see from (5.14) that $x_n = x_0$ for all values of $n$. If $|c| > 1$ (in both the real and complex cases) then it's clear that $|c|^n$ gets larger and larger as $n \to \infty$, that is, $|c|^n \to \infty$. To summarize: the solution $x_n$ in (5.14) of the difference equation (5.13) has the property as $n \to \infty$ that $|x_n| \to \infty$ if $|c| > 1$, and $|x_n| \to 0$ if $|c| < 1$. The former type of behaviour is called **exponential growth**, as illustrated in Figure 5.2. This name seems confusing, since the exponential function is not explicitly involved in the formula (5.14) for $x_n$. However, the underlying reason for the name is best explained through Figure 5.2. The points marked are given by the formula (5.5), namely $x_n = 1000(1.05)^n$. The curve shown in the figure by a dashed line has the equation $x(t) = 1000(1.05)^t$, where $t$ is a continuous variable representing time, measured along the horizontal axis. The two formulae coincide when $t$ is a positive integer (that is, at the discrete intervals of time). However, the expression for $x(t)$ can be written as

$$x(t) = 1000e^{kt}$$

where $e^k = 1.05$ (so $k = \ln 1.05$) and this is an exponential expression. In a similar way, when $|c| < 1$ the behaviour of $x_n$ in (5.14) is called **exponential decay**.

In Example 5.4 describing an instant access savings account, the difference equation (5.12) was an example of an equation having the general form

$$x_{n+1} = cx_n + D_n, \qquad n = 0, 1, 2, 3, \dots \tag{5.15}$$

where $c$ is a constant, as before. To solve this, simply substitute successive values of $n$ in (5.15) to obtain

$$
\begin{aligned}
(n = 1) \quad & x_1 = cx_0 + D_0 \\
(n = 2) \quad & x_2 = cx_1 + D_1 = c(cx_0 + D_0) + D_1 \\
& \qquad\qquad\qquad\quad = c^2 x_0 + cD_0 + D_1 \\
(n = 3) \quad & x_3 = cx_2 + D_2 = c^3 x_0 + c^2 D_0 + cD_1 + D_2
\end{aligned}
$$

It's easy to spot that the pattern in general is

$$x_n = c^n x_0 + c^{n-1} D_0 + c^{n-2} D_1 + c^{n-3} D_2 + \ldots + D_{n-1} \qquad (5.16)$$

Notice that the first term in this expression is just (5.14), the solution when all the $D$'s are zero. The other terms form the **particular solution**, so-called because it depends upon the particular values of the $D$'s. When all the $D$'s are equal to some constant $d$, say, then the particular solution in (5.16) becomes

$$(c^{n-1} + c^{n-2} + c^{n-3} + \ldots + c + 1)d$$

The expression inside the brackets, namely

$$S = 1 + c + c^2 + \ldots + c^{n-2} + c^{n-1} \qquad (5.17)$$

is called a **geometric series**. Each term is $c$ times the preceding one, beginning with 1. To find a simpler expression for $S$, notice that

$$Sc = c + c^2 + c^3 + \ldots + c^{n-1} + c^n$$
$$= S - 1 + c^n$$

so that

$$S(c - 1) = c^n - 1$$

Hence, provided $c \neq 1$ then

$$S = \frac{c^n - 1}{c - 1}$$

which is a well-known formula for the sum of $n$ terms of a geometric series. If $c = 1$ then it is obvious that in (5.17) there are $n$ terms each equal to 1, so that $S = n$. The solution (5.16) of the difference equation

$$x_{n+1} = c x_n + d, \qquad n = 0, 1, 2, 3, \ldots \qquad (5.18)$$

can therefore be written as

$$x_n = c^n x_0 + \begin{cases} \dfrac{c^n - 1}{c - 1} d, & c \neq 1 \\ nd, & c = 1 \end{cases} \qquad (5.19)$$

In particular if $|c| < 1$ then since $|c|^n \longrightarrow 0$ as $n \longrightarrow \infty$, the terms $|c^n x_0|$ and $|c^n d|$ in (5.19) also tend to zero, and the expression for $x_n$ in (5.19) reduces to

$$x_n \longrightarrow \frac{d}{1 - c} \text{ as } n \longrightarrow \infty \qquad (5.20)$$

| Example 5.5 | **Savings account with withdrawals** |

Return to the savings account described in Example 5.4, where you deposited £1000 to gain interest compounded monthly at an APR of 6. Suppose that you arrange for a monthly direct debit of £50 to be paid to your electricity supplier.

How much will remain in the account after one year has passed? The monthly interest rate is $6/12 = 0.5\%$. Assume interest is paid on the 26th of each month, and that the debit is made on the penultimate day of each month. If $£x_n$ is the amount in the account at the end of the $n$th month then initially $x_0 = 1000$. After one month has elapsed the balance in the account is

$$x_1 = 1.005x_0 - 50$$

In general the difference equation describing the account is

$$x_{n+1} = 1.005x_n - 50, \qquad n = 0, 1, 2, 3, \ldots$$

This has the form of (5.18), and the solution (5.19) has $c = 1.005$ and $d = -50$. Setting $n = 12$ gives the amount in the account after one year has elapsed, namely

$$x_{12} = (1.005)^{12}1000 - \frac{(1.005)^{12} - 1}{1.005 - 1} \times 50$$

$$= £444.90$$

**Example 5.6**    **Loan repayment**

The expression (5.19) can also be used to determine repayments on a loan. For example, suppose that you borrow £2000 from a bank to help pay for your studies. The loan carries interest of 6% compounded annually, and you want to pay off the loan by equal monthly instalments over a period of three years. Let $£m$ denote the amount of your monthly repayment, and $£x_n$ the amount of the loan after $n$ years have elapsed, so that $x_0 = 2000$. The interest is charged on the balance at the beginning of the year, so that after one year the amount outstanding is

$$x_1 = \underset{\substack{\text{initial} \\ \text{loan}}}{2000} + \underset{\text{interest}}{2000 \times \tfrac{6}{100}} - \underset{\substack{\text{one year's} \\ \text{payment}}}{12m}$$

$$= \left(1 + \tfrac{6}{100}\right)2000 - 12m$$

In general, after $n + 1$ years

$$\underset{\substack{\text{balance after} \\ n + 1 \text{ years}}}{x_{n+1}} = \underset{\substack{\text{balance plus} \\ \text{interest after} \\ n \text{ years}}}{1.06x_n} - \underset{\substack{\text{payments made} \\ \text{during year}}}{12m}$$

This equation has the form (5.18) with $c = 1.06$ and $d = -12m$, and the solution is given by (5.19). Repayment of the loan after three years means that $x_3 = 0$, so substituting $n = 3$ into the expression (5.19) with $x_0 = 2000$ gives

$$x_3 = 0 = 2000(1.06)^3 - 12m\frac{(1.06)^3 - 1}{1.06 - 1}$$

Rearranging to solve for $m$ produces

$$m = \frac{2000(1.06)^3 \times 0.06}{12[(1.06)^3 - 1]}$$
$$= 62.35$$

and this is the monthly repayment.

Another special case of (5.15) whose general solution can be found algebraically is when $D_n = n$ and $c = 1$. Substituting these values into the expression (5.16) shows that the solution of

$$x_{n+1} = x_n + n, \qquad n = 0, 1, 2, 3, \ldots \tag{5.21}$$

is

$$x_n = x_0 + 1 + 2 + 3 + \ldots + (n - 1)$$
$$= x_0 + \tfrac{1}{2}n(n - 1) \tag{5.22}$$

where a standard formula for the sum of the integers from 1 to $n - 1$ has been used (see Exercise 5.15). The term $\tfrac{1}{2}n(n - 1)$ in (5.22) is the particular solution of the equation (5.21) corresponding to the term $n$ on the right-hand side.

If $D_n$ in (5.15) consists of a *sum* of terms, for example

$$x_{n+1} = cx_n + 4d + 5n \qquad (c \neq 1) \tag{5.23}$$

then because the equation is *linear*, the particular solution is the *sum* of the solutions corresponding to each of the terms. Thus in (5.23) the particular solution for $4d$ is 4 times that in (5.19), and the particular solution for $5n$ is 5 times that in (5.22). Overall, the complete solution of (5.23) is therefore

$$x_n = c^n x_0 + \frac{4d(c^n - 1)}{c - 1} + \frac{5n(n - 1)}{2}$$

This property of linear equations makes them easier to solve than non-linear equations. Textbooks usually concentrate on linear equations because, as you've seen, it's possible to obtain neat algebraic expressions for the solution such as those in (5.19) and (5.21). However, it is important to realize that difference equations of the form

$$x_{n+1} = f(x_n, D_n), \qquad x_0 \text{ given}$$

where $f$ is some given function, can always be solved iteratively. For example, to solve the non-linear first order equation

$$x_{n+1} = 2x_n^2 - 4x_n + 5, \qquad x_0 = 1$$

simply compute

$$x_1 = 2x_0^2 - 4x_0 + 5 = 2 - 4 + 5 = 3$$
$$x_2 = 2x_1^2 - 4x_1 + 5 = 18 - 12 + 5 = 11$$
$$x_3 = 2x_2^2 - 4x_2 + 5 = 242 - 44 + 5 = 203$$

and so on. It's interesting that first order non-linear *differential* equations usually have to be solved numerically by first converting them into difference equation form.

**Exercise**
5.11 Determine the solution of the following equations valid for $n = 0, 1, 2, 3, \ldots$, subject to the given condition:

(a) $x_{n+1} - 3x_n = 7, \qquad x_0 = -1$

(b) $x_{n+1} + 0.5x_n = -4, \qquad x_0 = 3$

(c) $x_{n+1} - x_n = 2, \qquad x_0 = 4$

(d) $x_{n+1} - x_n = 2 + 5n, \qquad x_0 = 4$

**Exercise**
5.12 Obtain the general solution of the difference equation you found in Exercise 5.9, and use it to recompute $x_6$.

**Exercise**
5.13 Investigate what happens to the solution of equation (5.13) when $c = -1$, $i$ or $-i$.

**Exercise**
5.14

(a) If $x_{n+1} = 4x_n + 2$ and $x_0 = -1$, determine $x_8$.

(b) If $x_{n+1} + x_n = -2$ and $x_{10} = -11$, determine $x_0$.

(c) If $2x_{n+1} + 6x_n = d$ and $x_0 = 4$, $x_8 = 6564$, determine $d$.

**Exercise**
5.15 Let $S_n$ denote the sum of the integers from 1 to $n - 1$, that is, $S = 1 + 2 + 3 + \ldots + (n - 1)$. Write the terms in reverse order, and hence show that $S = (1/2)n(n - 1)$.

**Exercise**
5.16 It is known that a certain process is described by the equation

$$x_{n+1} = ax_n + b, \qquad n = 0, 1, 2, 3, \ldots$$

where $a$ and $b$ are constants. Because of delays in setting up measuring equipment $x_0$ is unknown, but it is found that $x_2 = 2$, $x_3 = 3$, $x_4 = 5$. Determine $x_0$ and $x_{20}$.

**Exercise**
5.17 Show that the solution of equation (5.13) satisfying a given condition $x_k = X$ at $n = k$ instead of $n = 0$ is

$$x_n = c^{n-k}X, \qquad n = 0, 1, 2, 3, \ldots$$

Hence show that the solution of the equation (5.18) subject to $x_k = X$ is

$$x_n = c^{n-k}X + \begin{cases} \dfrac{c^{n-k} - 1}{c - 1}d, & c \neq 1 \\ (n - k)d, & c = 1 \end{cases}$$

**Exercise**    **5.18**   Use the result in the preceding exercise to solve the equation in part (a) of Exercise 5.11 subject to the condition $x_5 = 4$ instead of $x_0 = -1$.

**Exercise**    **5.19**   You borrow £5000 to buy a car. The loan is to be paid off by equal quarterly instalments over a period of four years. The interest is compounded annually on the amount outstanding at the beginning of each year at an APR of 12. What will be your quarterly payment?

**Exercise**    **5.20**   Suppose that instead of the compound interest charged on the loan of £5000 in the preceding exercise, you are offered a loan bearing *simple* interest at an APR of 10.75, charged on the total sum borrowed. Would you be better off choosing this option?

**Exercise**    **5.21**   Show that for $n$ sufficiently large the solution of

$$x_{n+1} + 0.4x_n = 28, \qquad n = 0, 1, 2, 3, \ldots$$

subject to $x_0 = 1000$, is approximately equal to 20. Compare this with the value of $x_{15}$. How large does $n$ have to be so that $|x_n - 20| \leq 0.1$?

**Exercise**    **5.22**   Consider equation (5.15) with $D_n = ak^n$, where $k$ is a non-zero constant. Show that the solution of the equation in this case is

$$x_n = c^n x_0 + \begin{cases} \dfrac{a(k^n - c^n)}{k - c}, & k \neq c \\ anc^{n-1}, & k = c \end{cases}$$

Hence determine the solution of:
(a) $x_{n+1} - 2x_n = 7(5^n), \qquad x_0 = 4$
(b) $x_{n+1} - 2x_n = 3(2^n), \qquad x_0 = 4$
(c) $x_{n+1} - 2x_n = 7(5^n) + 3(2^n), \qquad x_0 = 4.$

**Exercise**    **5.23**   If

$$x_{n+1} = \frac{x_n}{1 + x_n^2} + 5, \qquad n = 0, 1, 2, 3, \ldots$$

and $x_0 = 1$, compute $x_6$. Try to interpret what this result represents.

### 5.2.3 More applications

Example 5.7 **Non-renewable resources**

It is estimated that at the present level of consumption, Britain's reserves of oil under the North Sea will run out in 50 years' time. If instead of remaining constant, consumption were to increase at an annual rate of 5%, how long would the oil reserves last?

Suppose the current annual level of usage of oil is $A$ units, so that the total reserves are $50A$ units. If $x_n$ is the amount of oil consumed during the $n$th year, then $x_1 = A$, and with a 5% annual growth rate

$$x_{n+1} = 1.05x_n, \qquad n = 1, 2, 3, \ldots$$

This equation has the form (5.13), with solution given by (5.14) as

$$x_n = (1.05)^{n-1}A, \qquad n = 1, 2, 3, \ldots$$

The index $n - 1$ is used because the given condition is that consumption in the first year is $x_1 = A$ (see Exercise 5.17). The total usage over a period of $N$ years is

$$x_1 + x_2 + x_3 + \ldots + x_N = A + 1.05A + \ldots + (1.05)^{N-1}A$$
$$= [1 + 1.05 + (1.05)^2 + \ldots + (1.05)^{N-1}]A$$

The series inside the square brackets has the same form as $S$ in (5.17), and so can be simplified to

$$\frac{(1.05)^N - 1}{1.05 - 1}A = \frac{(1.05)^N - 1}{0.05}A$$

In order to find how long the reserves of oil will last, it is therefore necessary to determine the value of $N$ such that the total consumption is equal to $50A$, that is,

$$\frac{(1.05)^N - 1}{0.05}A = 50A$$

This reduces to

$$(1.05)^N = 50 \times 0.05 + 1 = 3.5$$

and it is easy to use your calculator to find $(1.05)^{25} = 3.38$, $(1.05)^{26} = 3.55$. This shows that the reserves would run out in a little over 25 years. This is only about half the time if consumption was kept constant, even though the growth rate of 5% seems quite modest. This illustrates the importance of curbing the growth in usage of non-renewable resources such as oil, coal and natural gas. In practice, as supplies are depleted prices tend to rise, and this has a dampening effect on consumption. Nevertheless, such 'market forces' cannot be relied on to preserve resources which once used up are gone forever.

Example 5.8  **Radioactive decay**

Radioactive materials emit particles and thereby decrease in quantity at a predictable rate. The length of time in which an amount of radioactive substance decays to half its original value is called its **half-life**. Let $x_n$ be the amount of radioactive material after $n$ years. If the annual rate of decay is $r$ then

$$x_{n+1} = (1 - r)x_n, \qquad n = 0, 1, 2, 3, \ldots$$

where instead of using percentages the number $r$ lies between 0 and 1. This equation again has the usual form (5.13), so according to (5.14) the solution is

$$x_n = (1 - r)^n x_0, \qquad n = 1, 2, 3, \ldots \tag{5.24}$$

where $x_0$ is the initial amount of radioactive material. If the half-life is $H$ years then $x_H = (1/2)x_0$, so

$$\tfrac{1}{2}x_0 = (1 - r)^H x_0$$

which can be solved to give

$$1 - r = \left(\tfrac{1}{2}\right)^{1/H}$$

Substituting this back into (5.24) produces

$$x_n = (0.5)^{n/H} x_0 \tag{5.25}$$

which is the amount of material remaining after $n$ years.

A useful application of the formula (5.25) is the application to what is called **carbon dating**. When a fossil or other remains of an animal or plant is found, measurement of the content of the radioactive isotope carbon-14 enables the age of the relic to be computed accurately. This is because the proportion of carbon-14 in the tissues of an animal or plant is constant during life, but starts decaying after death. The half-life of carbon-14 is approximately 5730 years. For example, a few years ago the body of a man was found preserved in a glacier in the Austrian Alps. Measurement of the carbon-14 content revealed that this was about 54% of the original amount of carbon-14 present. In order to determine the age of the body, use (5.25) with $x_n = 0.54x_0$ and $H = 5730$. This gives

$$0.54x_0 = (0.5)^{n/5730} x_0$$

and it is easy to use your calculator to find that $(0.5)^{0.89} \approx 0.54$, from which $n/5730 = 0.89$, that is, $n \approx 5100$ – the man died about 5100 years ago.

Incidentally, if you're familiar with logarithms, then you can solve the equation $(0.5)^m = 0.54$ directly with your calculator, rather than trying out different values of $m$ on the '$x^y$' button. Taking logs of both sides gives

$$m \log 0.5 = \log 0.54$$

and $m = \log 0.54 / \log 0.5 \approx 0.89$.

| Example 5.9 | **Roll of kitchen paper** |

A standard roll of kitchen paper consists of a cardboard tube of diameter 4 cm with the paper wound around it. A typical roll contains 60 sheets of paper each 25 cm long, so the total length of paper is 1500 cm. Imagine the paper being wound onto the inner tube. After each complete winding the total diameter of the roll increases by an amount $2t$, where $t$ cm is the thickness of the paper. Let $x_n$ cm denote the total length of paper when it is wrapped $n$ times around the inner tube, so that $x_0 = 0$.

At this stage the outer diameter of the roll is $(4 + 2tn)$ cm. When one more complete winding is carried out, the length of paper needed to do this is $\pi(4 + 2tn)$ cm. This is the increase in length of paper used in going from $n$ wrappings to $n + 1$, which by definition is $x_{n+1} - x_n$, so

$$x_{n+1} - x_n = \pi(4 + 2tn)$$

Rewrite this as

$$x_{n+1} = x_n + 4\pi + 2\pi tn, \qquad n = 0, 1, 2, 3, \ldots \tag{5.26}$$

which is a difference equation in the form (5.15) with $c = 1$ and $D_n = 2\pi(2 + tn)$. Knowing the paper thickness $t$ you can solve (5.26) to obtain the length of paper needed for a given number of windings.

Because $x_0 = 0$, the expression (5.16) shows that the solution of (5.26) consists of the sum of the particular solutions corresponding to the terms $4\pi$ and $2\pi tn$ on the right-hand side. Since $c = 1$ the particular solution corresponding to $4\pi$ is obtained from (5.19) as $4\pi n$. From (5.22) the particular solution corresponding to $2\pi tn$ is $2\pi t \times (1/2)n(n - 1)$. Overall therefore the solution of (5.26) is

$$x_n = 4\pi n + \pi tn(n - 1), \qquad n = 0, 1, 2, \ldots \tag{5.27}$$

For example, suppose $t = 0.07$ and $n = 70$, then

$$x_n = 4\pi \times 70 + \pi \times 0.07 \times 70 \times 69$$
$$= 1941.8$$

and this is the length of paper needed to produce 70 windings.

Alternatively, if you know the total length of paper, and count the number of windings (by unrolling and then rewinding) you can use (5.27) to estimate the thickness of the paper (see Exercises 5.28 and 5.29).

| Exercise | **5.24** Suppose that in Example 5.7 the continued introduction of more fuel-efficient vehicles leads to an annual *reduction* of 1% in oil consumption. If this rate of reduction is maintained, how long would Britain's oil reserves last? |

**Exercise** **5.25** In Exercise 5.24 use your calculator to estimate the annual rate of reduction in oil consumption which would have to be maintained in order to make the oil reserves last for 100 years.

**Exercise** **5.26** Suppose that one kilogram of a certain radioactive substance decays to 0.983 kg after one year. What is its half-life?

**Exercise** **5.27** You participate in an archaeological dig on the site of a Roman encampment, and unearth a piece of wood which is found to contain 80.4% of its original content of carbon-14. What is the approximate age of the wood?

**Exercise** **5.28** You unwind a roll of kitchen paper, and then find that it goes round 61 times to rewind. The roll consists of 60 sheets each 25 cm long. Use equation (5.27) to estimate the thickness of the paper.

In practice the paper is soft and easily compressed, so it's not possible to make an accurate measurement of its thickness.

**Exercise** **5.29** A roll of aluminium cooking foil is wound round a cardboard tube whose diameter is 5 cm. The total length of foil is 10.25 m and there are 60 windings. Determine the approximate thickness of the foil.

Another roll of the same foil has an outer diameter of 5.42 cm. What length of foil does it contain?

**Exercise** **5.30** If the length of foil in the preceding exercise is 6 m, what is the outer diameter of the roll?

**Exercise** **5.31** The total population of the world in 1995 was approximately 5.6 billion, and was increasing at the rate of 1.7% per year. Assuming that growth continues at this rate, what will be the world's population in the year 2015?

A more accurate representation of the way the population of the world is increasing is to group together countries having similar growth rates, as follows:

| Group | Population 1995 (billions) | Growth rate (%) |
| --- | --- | --- |
| Developed countries | 1.24 | 0.5 |
| Less developed countries, excluding China | 3.16 | 2.3 |
| China | 1.20 | 1.4 |

Assuming the growth rates remain constant, what estimate for the world's population in 2015 does this model produce?

**Exercise** **5.32** Because of misgivings about the size of the first prize in the National Lottery, it is decided that in future this prize will be paid out in ten equal annual instalments. Suppose you are lucky enough to be the sole winner of £8 million. Assuming that the annual rate of inflation is maintained at 3% over the next ten years, what will be the value of your prize in real terms?

**5.3**   **Second order linear difference equations with constant coefficients**

In the previous section the recurrence formulae contained only two consecutive values of the variable $x_n$ and $x_{n+1}$. The next step up is when the equations involve variables $x_{n+2}$ and $x_n$ whose suffices differ by 2, hence the description **second order**. There is usually, but not always, a term involving $x_{n+1}$ as well.

**5.3.1**   **General solution**

By analogy with the first order equation in (5.13), the simplest second order linear equation has the form

$$x_{n+2} = cx_n, \qquad n = 0, 1, 2, 3, \ldots \tag{5.28}$$

where $c$ is a constant. Substitute successive values of $n$, starting at zero, into (5.28) to obtain

$$x_2 = cx_0, \quad x_3 = cx_1,$$
$$x_4 = cx_2 = c^2 x_0, \quad x_5 = cx_3 = c^2 x_1$$

and if you continue this you will see that when the suffix $n$ is even $x_n$ is a multiple of $x_0$, whereas when $n$ is odd $x_n$ is a multiple of $x_1$. Specifically, you can write

$$x_{2m} = c^m x_0, \quad x_{2m+1} = c^m x_1, \qquad m = 1, 2, 3, \ldots \tag{5.29}$$

This general solution of (5.28) is reminiscent of the solution of (5.14) of (5.13), but in this case depends upon *two* given values, namely $x_1$ and $x_0$. This fact carries over to the general form of second order linear equation with constant coefficients containing $x_{n+1}$ as well as $x_n$ and $x_{n+2}$. This can be written as

$$x_{n+2} = bx_{n+1} + cx_n, \qquad n = 0, 1, 2, 3, \ldots \tag{5.30}$$

where $b$ and $c$ are constants. If $b = 0$ then (5.30) reduces to the case already considered in (5.28). Substituting successive values of $n$ into (5.29) gives

$$(n = 0) \quad x_2 = bx_1 + cx_0$$
$$(n = 1) \quad x_3 = bx_2 + cx_1 = b(bx_1 + cx_0) + cx_1$$
$$\qquad\qquad = (b^2 + c)x_1 + bcx_0$$
$$(n = 2) \quad x_4 = bx_3 + cx_2$$
$$\qquad\qquad = (b^3 + bc)x_1 + b^2 cx_0 + bcx_1 + c^2 x_0$$
$$\qquad\qquad = (b^3 + 2bc)x_1 + (b^2 c + c^2)x_0$$

and you can continue with this to convince yourself that $x_n$ can be expressed in terms of $x_1$ and $x_0$. Unfortunately, however, no pattern can be discerned from the algebraic expressions for $x_2$, $x_3$, $x_4$, $x_5$, ..., so to find a convenient expression for the general solution of (5.30) it's necessary to look for a different approach. The expressions (5.13) and (5.29) suggest that it may be worth trying $x_n = \lambda^n a$ where $\lambda$ and $a$ are some numbers yet to be found. However, it can be assumed that $a \neq 0$, $\lambda \neq 0$ since otherwise $x_n = 0$, which is useless! Since $x_{n+1} = \lambda^{n+1}a$, $x_{n+2} = \lambda^{n+2}a$ then if this guess is to work, substituting into (5.30) gives

$$\lambda^{n+2}a = b\lambda^{n+1}a + c\lambda^n a$$

or

$$\lambda^n a(\lambda^2 - b\lambda - c) = 0 \qquad (5.31)$$

The equation (5.31) will be satisfied if $\lambda$ is such that it satisfies the quadratic equation

$$\lambda^2 - b\lambda - c = 0 \qquad (5.32)$$

because the initial factor $\lambda^n a$ in (5.31) is non-zero. The equation (5.32) will in general have two roots $\lambda_1$ and $\lambda_2$. These will either be both real, or a complex conjugate pair $a \pm i\beta$. The difference equation (5.30) therefore has two different solutions, $\lambda_1^n a$ and $\lambda_2^n a$. It follows that because the equation is *linear*, the general solution of (5.30) is the *sum* of these two separate solutions, namely

$$x_n = a_1\lambda_1^n + a_2\lambda_2^n \qquad (5.33)$$

where $a_1$ and $a_2$ are constants whose values are determined from the given values $x_0$ and $x_1$.

**Example 5.10** **Second order equation**

Consider the equation

$$x_{n+2} = 3x_{n+1} + 28x_n, \qquad n = 0, 1, 2, 3, \ldots \qquad (5.34)$$

subject to $x_0 = 7$, $x_1 = -6$. The so-called **characteristic equation** in (5.32) is

$$\lambda^2 - 3\lambda - 28 = 0$$

which factorizes into

$$(\lambda - 7)(\lambda + 4) = 0$$

so the roots are $\lambda_1 = -4$, $\lambda_2 = 7$ and the general solution (5.33) is

$$x_n = a_1(-4)^n + a_2 7^n \qquad (5.35)$$

Setting $n = 0$ and $n = 1$ in (5.35) gives the two equations

$$7 = a_1 + a_2, \quad -6 = -4a_1 + 7a_2$$

and you should solve these equations to get $a_1 = 5$, $a_2 = 2$. The solution of (5.34) is therefore

$$x_n = 5(-4)^n + 2(7)^n, \qquad n = 0, 1, 2, 3, \ldots$$

If the two specified values had been other than the initial two $x_0$ and $x_1$, then the method of solution is unaltered. Suppose the conditions to be satisfied change to $x_1 = 5$, $x_2 = -9$. Substituting $n = 1, 2$ into (5.35) gives

$$5 = -4a_1 + 7a_2, \quad -9 = 16a_1 + 49a_2$$

and the solution of these equations is $a_1 = -1$, $a_2 = 1/7$, so in this case the solution of (5.34) is

$$x_n = -(-4)^n + \left(\tfrac{1}{7}\right)7^n = 7^{n-1} - (-4)^n$$

---

It was remarked earlier that it can be assumed that $\lambda \neq 0$. In fact, $\lambda = 0$ is a root of the characteristic equation (5.32) only when $c = 0$. In this case (5.30) reduces to

$$x_{n+2} = bx_{n+1}$$

which is simply a first order equation, in a slightly disguised form, since $x_{n+2}$ and $x_{n+1}$ are consecutive values. The solution in this case is therefore just $x_n = b^n x_0$, as in Section 5.2.2.

The case when (5.32) has two *equal* roots is not so easily dismissed. In this case if $\lambda_1 = \lambda_2$ then the expression (5.33) becomes

$$x_n = a_1 \lambda_1^n + a_2 \lambda_1^n = (a_1 + a_2)\lambda_1^n$$

or simply $x_n = a_3 \lambda_1^n$, where $a_3$ is some other constant. This can't be the complete solution to the second order equation, since there must be *two* constants dependent on the *two* given conditions. In fact the second part of the solution is $x_n = n\lambda_1^n$, as can be verified by substitution into (5.30). This involves doing the following straightforward but slightly tedious algebra. The first step is to find the implications of the fact that $\lambda = \lambda_1$ is a repeated root of the characteristic equation (5.32). This means that $\lambda^2 - b\lambda - c$ is a perfect square, that is,

$$\begin{aligned} \lambda^2 - b\lambda - c &\equiv (\lambda - \lambda_1)^2 \\ &\equiv \lambda^2 - 2\lambda_1 \lambda + \lambda_1^2 \end{aligned}$$

The two sides are identical, so in particular the coefficients of $\lambda$ are the same, that is, $b = 2\lambda_1$. Now substitute $n\lambda_1^n$ into (5.30), giving the expression

$$\begin{aligned} x_{n+2} - bx_{n+1} - cx_n &= (n+2)\lambda_1^{n+2} - b(n+1)\lambda_1^{n+1} - cn\lambda_1^n \\ &= n\lambda_1^n \underbrace{(\lambda_1^2 - b\lambda_1 - c)}_{\substack{=0, \text{ since } \lambda_1 \text{ is} \\ \text{a root of (5.32)}}} + \lambda_1^{n+1} \underbrace{(2\lambda_1 - b)}_{\substack{=0, \text{ as} \\ \text{above}}} \end{aligned}$$

This confirms that $n\lambda_1^n$ satisfies (5.30), so once again the general solution of the difference equation is the sum of the two parts, that is,

$$\begin{aligned} x_n &= a_3 \lambda_1^n + a_4 n\lambda_1^n \\ &= (a_3 + a_4 n)\lambda_1^n, \qquad n = 0, 1, 2, 3, \ldots \end{aligned} \tag{5.36}$$

where $a_3$ and $a_4$ are constants determined by the given conditions.

**Example 5.11**  **Equal roots**

The equation

$$x_{n+2} = 10x_{n+1} - 25x_n, \qquad n = 0, 1, 2, 3, \ldots$$

has the characteristic equation (5.32)

$$0 = \lambda^2 - 10\lambda + 25 = (\lambda - 5)^2$$

so the general solution in (5.36) is

$$x_n = (a_3 + a_4 n)5^n, \qquad n = 0, 1, 2, 3, \ldots$$

If the given conditions are $x_0 = -1$, $x_1 = 2$ then substituting $n = 0$ and $n = 1$ gives

$$-1 = a_3, \quad 2 = (a_3 + a_4)5$$

so that $a_4 = 7/5$ and the solution is therefore

$$x_n = (-1 + \tfrac{7}{5}n)5^n$$

---

The discussion in Section 5.2.2 about what happens to the solution when $n \longrightarrow \infty$ is easily extended. The solution (5.33) has the properties:

$|x_n| \longrightarrow \infty$ if either $|\lambda_1| > 1$ or $|\lambda_2| > 1$, or both

$|x_n| \longrightarrow 0$ if both $|\lambda_1| < 1$ and $|\lambda_2| < 1$

As for first order equations, the next increase in complexity is when there is an additional constant term $d (\neq 0)$ in the equation, so (5.30) becomes

$$x_{n+2} = bx_{n+1} + cx_n + d, \qquad n = 0, 1, 2, 3, \ldots \tag{5.37}$$

In the first order case the solution in (5.19) is a multiple of $d$, so it is reasonable to investigate whether the particular solution for the term $d$ in (5.37) also has this form. Trying $x_n = kd$ in (5.37), where $k$ is a constant, gives

$$kd = bkd + ckd + d$$

or

$$kd(1 - b - c) = d$$

so that $k = 1/(1 - b - c)$ provided $1 - b - c \neq 0$. The complete solution of (5.37) is therefore the sum of the solution when $d = 0$ (that is, the solution of (5.30), using either (5.33) or (5.36) as appropriate) together with the particular solution $d/(1 - b - c)$.

**Example 5.12**  **Equation (5.37)**

The general solution of the equation

$$x_{n+2} = 3x_{n+1} + 28x_n + 60, \qquad n = 0, 1, 2, \ldots$$

has two parts. The particular solution is $60/(1 - 3 - 28) = 60/(-30) = -2$. The solution when there is no constant term was found in Example 5.10. The complete solution is therefore the sum of (5.35) and $-2$, that is,

$$x_n = a_1(-4)^n + a_2 7^n - 2$$

If two given conditions, say $x_0 = 3$, $x_1 = 0$ are to be satisfied then substituting $n = 0, n = 1$ gives the equations

$$3 = a_1 + a_2 - 2, \quad 0 = -4a_1 + 7a_2 - 2$$

which have solution $a_1 = 3$, $a_2 = 2$.

---

When $1 - b - c = 0$ the method used to find the particular solution for (5.37) must be modified, and this is explored in Problems 5.11 and 5.12.

As in the case of first order equations, you should realize that even when a nice algebraic formula cannot be derived, any second order equation in the form

$$x_{n+2} = f(x_{n+1}, x_n, D_n) \qquad x_0, x_1 \text{ given}$$

where $f$ is some function, can always be solved iteratively. However, general linear difference equations with non-constant coefficients, or general non-linear equations, are much more difficult to solve.

**Example 5.13**  **Analyzing the economy**

One very simple mathematical description of a national economy begins with the assumption that in year $n$ the national income $I_n$ is equal to the sum of three components. These are consumer expenditure $C_n$ (for example, on food, domestic appliances, cars); private investment $P_n$ (for example, on new factories); and government expenditure $G_n$ (for example, on the National Health Service or education).

This means that

$$I_n = C_n + P_n + G_n, \qquad n = 1, 2, 3, \ldots \tag{5.38}$$

The various types of expenditure are not independent of each other, and by analyzing past economic data it is suggested that the following assumptions are reasonable.

(1) If the national income increases in one year, then consumers will spend more the following year. Suppose that in year $n + 1$ consumers spend half of the previous year's income, that is, $C_{n+1} = (1/2)I_n$.

(2) An increase in consumer spending from one year to the next results in an increase in private investment, that is, $P_{n+1} = C_{n+1} - C_n$. Substituting the expression for $C_{n+1}$ from (1) gives $P_{n+1} = (1/2)(I_n - I_{n-1})$.

Suppose that the government decides to keep its expenditure constant, that is, $G_n = G$ for all values of $n$. Replacing $n$ by $n+2$ in equation (5.38) gives, using the expressions for $C_{n+1}$ in (1) and $P_{n+1}$ in (2):

$$I_{n+2} = C_{n+2} + P_{n+2} + G_{n+2}$$
$$= \tfrac{1}{2}I_{n+1} + \tfrac{1}{2}(I_{n+1} - I_n) + G$$
$$= I_{n+1} - \tfrac{1}{2}I_n + G \qquad\qquad (5.39)$$

This is a second order linear difference equation relating the national income in year $n+2$ to that in the previous two years, as well as the level of government expenditure. Suppose that $I_1 = G$ and $I_2 = 2G$. It is easy to compute successive values of $G$ from (5.39), for example:

$(n = 1)\quad I_3 = I_2 - 0.5I_1 + G = 2G - 0.5G + G = 2.5G$
$(n = 2)\quad I_4 = I_3 - 0.5I_2 + G = 2.5G - G + G = 2.5G$

You should use your calculator to compute the following table:

| $n$ | 5 | 6 | 7 | 8 | 9 | 10 | 11 | 12 | 13 | 14 | 15 |
|---|---|---|---|---|---|---|---|---|---|---|---|
| $I_n$ | 2.25G | 2G | 1.88G | 1.88G | 1.94G | 2G | 2.03G | 2.03G | 2.02G | 2G | 1.99G |

The value of the national income seems to be stabilizing at a value around twice government expenditure. You can confirm that $I_n \to 2G$ as $n \to \infty$ by substituting $I_{n+2} = 2G$, $I_{n+1} = 2G$, $I_n = 2G$ into (5.39) and verifying that the equation is indeed satisfied. Hence when $n$ is sufficiently large, the constant government expenditure $G$ produces a fairly healthy economy with national income at a steady stable level of $2G$.

Unfortunately, a major problem in any kind of economic modelling in mathematical terms is the unreliability of the data used. By the time economic statistics have been collected they are out of date, and because of the sheer volume of data available they are bound to be incomplete – for example, there is no way that the expenditure of every individual citizen can be monitored. Moreover the assumptions used, such as (1) and (2) above, are often based upon personal opinions and preferences. For these and similar reasons, it is no wonder that economists and their predictions seldom agree!

Some effects of changing the assumptions (1) and (2) are investigated in Exercise 5.37 and Problems 5.15 and 5.16.

**Exercise**    **5.33**    Determine the solution of the following equations valid for $n = 0, 1,$ $2, 3, \ldots$ subject to the given conditions:

(a) $x_{n+2} = 12x_n - x_{n+1}, \qquad x_0 = 8, \quad x_1 = 3$

(b) $x_{n+2} = 6x_{n+1} - 10x_n, \qquad x_0 = 4, \quad x_1 = 12 - 6i$

(c) $x_{n+2} = 6x_{n+1} - 9x_n, \qquad x_1 = 33, \quad x_2 = 162.$

**Exercise**    **5.34**    If $x_{n+2} = 5x_{n+1} - 6x_n + 2,$ and $x_0 = 2, x_1 = 5,$ determine the solution and hence compute $x_{10}$.

**Exercise**    **5.35**    If $x_{n+2} - 7x_{n+1} + 12x_n = d,$ and $x_0 = -1, x_1 = 3, x_3 = 99,$ determine $d$.

**Exercise**    **5.36**    Show that for $n$ sufficiently large the solution of

$$10x_{n+2} = 3x_{n+1} + x_n + 18, \qquad x_0 = 5, \quad x_1 = 5.4$$

is approximately equal to 3. Compare this with the value of $x_5$. How large does $n$ have to be so that $|x_n - 3| \le 0.05$?

**Exercise**    **5.37**    In Example 5.13, if you have certain political attitudes you might think that private investment should be *double* the increase in public spending. This would replace assumption (2) by $P_{n+1} = 2(C_{n+1} - C_n),$ with all the other conditions remaining the same as before. Obtain the new difference equation satisfied by $I_n$.

Compute the values of $I_n$ for $n = 3, 4, \ldots, 9$. What can you say about the effects of this choice of policy? What happens for $n > 9$?

**Exercise**    **5.38**    If

$$x_{n+2} = \frac{x_n + 1}{1 + x_n^2} - \frac{2}{x_{n+1}} + 3, \qquad n = 0, 1, 2, 3, \ldots$$

and $x_0 = 1, x_1 = 3,$ compute $x_{10}$. Try to interpret what this result represents.

**Exercise**    **5.39**    A starship is travelling through hyperspace in a straight line with phenomenal acceleration. The navigational computer reports that the distance travelled during any hour is 2.5 times the distance travelled in the previous hour. If $x_n$ denotes the distance (in light years) from Earth $n$ hours after lift-off, show that

$$x_{n+2} - 3.5x_{n+1} + 2.5x_n = 0$$

If $x_1 = 6,$ determine a general expression for $x_n$. How long will it take to reach a galaxy 10,000 light years from Earth?

**Exercise**    **5.40**    Show that the general solution of

$$x_{n+2} = bx_{n+1} + cx_n + k^n, \qquad n = 0, 1, 2, 3, \ldots$$

where $k$ is a non-zero constant, is $x_n = a_1\lambda_1^n + a_2\lambda_2^n + P_n$ with

$$P_n = \begin{cases} \dfrac{k^n}{k^2 - bk - c}, & k \neq \lambda_1 \text{ or } \lambda_2 \\[2mm] \dfrac{nk^{n-1}}{2k - b}, & k = \lambda_1 \text{ or } \lambda_2 \end{cases}$$

where $\lambda_1$, $\lambda_2$ are the two roots (assumed different) of $\lambda^2 - b\lambda - c = 0$, and $a_1$, $a_2$ are constants depending upon the values of $x_0$ and $x_1$.

Hence solve the equations:

(a) $x_{n+2} = 12x_n - x_{n+1} + 2(5^n)$,     $n = 0, 1, 2, 3, \ldots$, subject to
$x_0 = 10/9$,   $x_1 = 95/9$

(b) $x_{n+2} = 12x_n - x_{n+1} - 3^n$,     $n = 0, 1, 2, 3, \ldots$, subject to
$x_0 = -7$,   $x_1 = -(1/7)$

See Problem 5.14 for the case when $\lambda_1 = \lambda_2$.

## 5.3.2   Fibonacci numbers

In the year 1202 an Italian mathematician usually known as Fibonacci (this means 'son of Bonaccio', his real name was Leonardo of Pisa) published an influential book called *Liber Abaci* which means 'Book of the Abacus', but is actually about arithmetic and algebra. It included the following problem, which has perpetuated Fibonacci's name:

> How many pairs of rabbits will be produced in a year, beginning with a single pair, if in every month each pair produces one new pair which begins to bear offspring two months after its own birth?

In order to translate Fibonacci's problem into mathematical terms, you need to make two further assumptions:

(1) once paired, rabbits remain faithful to each other

(2) rabbits do not die.

Of course, all the assumptions are idealized and cannot be achieved in practice. For example, (2) requires an unlimited food supply and no predators, so that no deaths occur due to starvation, overcrowding or killing. You might think that a very large fertile, empty island is a rabbit's idea of paradise! In fact, the introduction of rabbits into Australia in the nineteenth century is not too different from this situation, and everybody knows about the subsequent catastrophic growth in the rabbit population. However, what is important is not this rather artificial model of a rabbit population, but the sequence of **Fibonacci numbers** which it generates. Although these numbers were not studied until the nineteenth century, they have an amazing range of applications in the world of nature, and elsewhere. There is a regularly published journal as well as several books devoted to Fibonacci numbers, their properties and applications.

The rule which generates Fibonacci numbers is simplicity itself: each number in the Fibonacci sequence is the *sum* of the two preceding numbers, starting with 1,1. Thus if $F_n$ denotes the $n$th number, with $F_1 = 1$, $F_2 = 1$, then

$$F_3 = F_2 + F_1 = 1 + 1 = 2, \quad F_4 = F_3 + F_2 = 2 + 1 = 3,$$
$$F_5 = F_4 + F_3 = 3 + 2 = 5, \quad F_6 = F_5 + F_4 = 5 + 3 = 8$$

and in general the numbers satisfy the second order difference equation

$$F_{n+2} = F_{n+1} + F_n, \qquad n = 1, 2, 3, 4, \ldots \tag{5.40}$$

The first 15 Fibonacci numbers are:

| 1 | 1 | 2 | 3 | 5 | 8 | 13 | 21 | 34 | 55 | 89 | 144 | 233 | 377 | 610 |
|---|---|---|---|---|---|----|----|----|----|----|-----|-----|-----|-----|
| $F_1$ | $F_2$ | $F_3$ | $F_4$ | $F_5$ | $F_6$ | $F_7$ | $F_8$ | $F_9$ | $F_{10}$ | $F_{11}$ | $F_{12}$ | $F_{13}$ | $F_{14}$ | $F_{15}$ |

Some authors begin with $F_0 = 0$, $F_1 = 1$ but this makes no difference except that in (5.40) the values of $n$ start at zero. The sequence was briefly introduced in Section 2.3.2 – see equation (2.16), page 92. It was shown there that any two consecutive Fibonacci numbers are relatively prime, and some other properties were investigated in Exercises 2.97, 2.98 and 2.99.

Before going into details of the rabbit population problem and other applications, the equation (5.40) can be solved using the method of the previous section. The characteristic equation (5.32) is

$$\lambda^2 - \lambda - 1 = 0 \tag{5.41}$$

and this has roots

$$\frac{1 \pm \sqrt{(1+4)}}{2} = \frac{1 \pm \sqrt{5}}{2}$$

Write for convenience $\lambda_1 = (1 + \sqrt{5})/2$, $\lambda_2 = (1 - \sqrt{5})/2$. The general solution in (5.33) is

$$F_n = a_1 \lambda_1^n + a_2 \lambda_2^n, \qquad n = 0, 1, 2, 3, \ldots$$

where the constants $a_1$ and $a_2$ are determined from $F_0 = 0$, $F_1 = 1$ (this is easier than using $F_1 = 1$, $F_2 = 1$). This gives

$$0 = a_1 + a_2, \quad 1 = a_1 \lambda_1 + a_2 \lambda_2$$

so that $a_2 = -a_1$ and

$$a_1(\lambda_1 - \lambda_2) = 1$$

Since $\lambda_1 - \lambda_2 = \sqrt{5}$, this reduces to $a_1 = 1/\sqrt{5}$, $a_2 = -1\sqrt{5}$ so overall

$$F_n = \frac{1}{\sqrt{5}} \left[ \left( \frac{1 + \sqrt{5}}{2} \right)^n - \left( \frac{1 - \sqrt{5}}{2} \right)^n \right], \qquad n = 0, 1, 2, 3, \ldots \tag{5.42}$$

$$= \frac{1}{2^n \sqrt{5}} [(1 + \sqrt{5})^n - (1 - \sqrt{5})^n] \tag{5.43}$$

The appearance of $\sqrt{5}$ in the formulae (5.42) and (5.43) is puzzling at first sight, since all the $F$'s are integers, but what happens is that the terms involving $\sqrt{5}$ cancel out (see Exercise 5.41). For example, substituting $n = 4$ into (5.43) and using the binomial expansion (see Section 3.2) gives

$$F_4 = \frac{1}{16\sqrt{5}}[(1 + \sqrt{5})^4 - (1 - \sqrt{5}^4)]$$

$$= \frac{1}{16\sqrt{5}}[1 + 4\sqrt{5} + 6(\sqrt{5})^2 + 4(\sqrt{5})^3 + (\sqrt{5})^4 - (1 - 4\sqrt{5} + 6(\sqrt{5})^2$$
$$-4(\sqrt{5})^3 + (\sqrt{5})^4)]$$

$$= \frac{1}{16\sqrt{5}}(8\sqrt{5} + 4 \times 5\sqrt{5} + 4 \times 5\sqrt{5})$$

$$= 48\sqrt{5}/16\sqrt{5} = 3$$

You can of course use your calculator to determine $F_n$ from the formula (5.42) or (5.43).

Having established this explicit formula for $F_n$, it's now appropriate to look at some applications of Fibonacci numbers, beginning with the idea which started it all.

---

**Example 5.14**  **Rabbit population**

Suppose that on 1 January there is one pair of baby rabbits ($B$). On 1 February this pair is mature ($M$) and on 1 March produces a pair of baby rabbits, so there is then a total of *two* pairs. On 1 April the original pair produces another pair, giving a total of *three* pairs. This process is shown in Figure 5.3, where the continuation of the birth table is shown for the next couple of months.

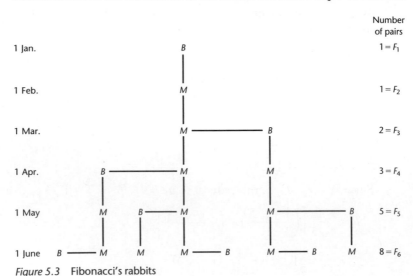

*Figure 5.3*  Fibonacci's rabbits

In general, at the start of the $n$th month there is a total of $F_n$ pairs of rabbits. To see why this is so, look for example at the fifth row (for 1 May) in Figure 5.3. All three ($= F_4$) pairs in the previous row are now mature ($M$); the two ($= F_3$) pairs in the row above that (1 March) have produced baby pairs ($B$) on 1 May. Hence the total for the fifth row is $F_4 + F_3$, that is, $F_5$.

It is easy to extend this argument to show that the total in row $n$ is $F_{n-1} + F_{n-2}$, that is, $F_n$. For example, after one year has elapsed (that is, at the beginning of the thirteenth month) there will be $F_{13} = 233$ pairs.

You can use (5.41) to see what happens as time passes. Notice that because $\lambda_2 \approx -0.618$ the second term within the square brackets in (5.42) rapidly becomes very small; for example, after three years $[(1 - \sqrt{5})/2]^{37} \approx -1.85 \times 10^{-8}$, so the number of rabbit pairs is then very nearly

$$\frac{1}{\sqrt{5}} \left( \frac{1 + \sqrt{5}}{2} \right)^{37} \approx 24 \text{ million}$$

After 30 years the number is approximately $1.25 \times 10^{75}$ which is absurdly large – more than the number of subatomic particles in the known universe! Fortunately, in the real world, rabbits do not live forever.

Another population model which is often used to illustrate how Fibonacci numbers arise is that of bees in a hive. You are asked to develop this in Problem 5.20.

**Example 5.15** **Hydrogen gas**

The behaviour of an electron in an atom of hydrogen gas can be described by the following rules:

(1) An electron can be in one of three states, labelled 0, 1, 2, depending upon its energy level. It alternately gains and loses energy in succession.

(2) When the gas gains energy all the electrons in state 1 rise to state 2; half of those in state 0 rise to state 1, and half to state 2.

(3) When the gas loses energy, all the electrons in state 1 fall to state 0; half those in state 2 fall to state 1, and half to state 0.

The history of an electron which starts off in state 0 is represented in Figure 5.4.

After the initial energy gain there are *two* possible outcomes, either state 0 followed by state 1, or state 0 followed by state 2. It's convenient to record these events with the notation 01 and 02. There is then an energy loss, and the *three* possibilities shown in Figure 5.4 are 010, 021, 020, since according to rule (3) 01 goes to 010 and 02 divides equally into 021 and 020. You can see from Figure 5.4 that the numbers of histories follow the Fibonacci pattern. You should extend the diagram for the next two cases of energy gain and loss, and

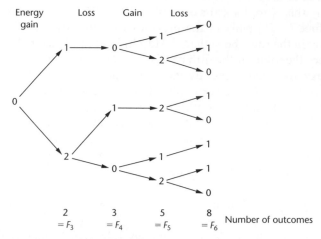

*Figure 5.4* Hydrogen gas electron

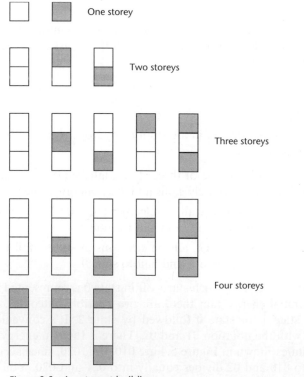

*Figure 5.5* Apartment buildings

also try to prove that in general after $n$ energy gains and losses there are $F_{n+2}$ possible histories.

**Example 5.16** | **Painting buildings**

A holiday complex is to be built consisting of apartment buildings of different heights. The architect has decided for aesthetic reasons that the buildings are to be painted in horizontal bands, storey by storey, in white or green with no two consecutive floor levels both green.

In Figure 5.5 you can see that there are *two* possible colour schemes for 1-storey buildings, *three* for two storeys, *five* for three storeys and *eight* for four storeys. Once again the Fibonacci numbers are appearing. To see how this works, look at the 4-storey buildings. The top floor can either be white or green. Since no two consecutive floors can be green, the top floor can be *white* when added to each of the five ($= F_5$) 3-storey buildings in the previous row. The top floor of a 4-storey building can be *green* when added to a 3-storey building whose top floor is white – there are *three* ($= F_4$) of these corresponding to the number of 2-storey buildings. In other words, the total number of differently painted 4-storey buildings is $F_5 + F_4 = F_6$. The argument is easily extended to show that the number of differently painted buildings with $n$ stories is $F_{n+1} + F_n = F_{n+2}$.

**Example 5.17** | **The golden rectangle**

With your calculator compute the ratios of successive Fibonacci numbers:

$$\frac{F_3}{F_2} = 2, \quad \frac{F_4}{F_3} = 1.5, \quad \frac{F_5}{F_4} = 1.67, \quad \frac{F_6}{F_5} = 1.6,$$

$$\frac{F_7}{F_6} = 1.625, \quad \frac{F_8}{F_7} = 1.615, \quad \frac{F_9}{F_8} = 1.619, \quad \frac{F_{10}}{F_9} = 1.618$$

As $n$ increases the values of the ratio alternately increase and decrease, but as $n \longrightarrow \infty$ the ratio gets closer and closer to the number

$$\phi = \tfrac{1}{2}(\sqrt{5} + 1) = 1.61803398\ldots \tag{5.44}$$

which is the positive root (denoted previously by $\lambda_1$) of equation (5.41). Since the time of the ancient Greeks, a rectangle whose sides are in the ratio $\phi : 1$ has been called 'golden', since it has a visually pleasing shape. The number $\phi$ is known as the 'golden ratio' or 'divine ratio', the latter because it was believed to express God-given beauty. To construct a golden rectangle, first construct the square $ABCD$ shown in Figure 5.6(a), with sides 2 units in length.

Let $E$ be the mid-point of $DC$. Then by Pythagoras's theorem

$$EB^2 = EC^2 + CB^2 = 1 + 4 = 5$$

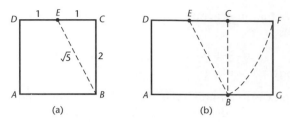

*Figure 5.6*  Golden rectangle

so *EB* has length $\sqrt{5}$. With centre *E* and radius *EB* draw a circle to meet the line through *DC* at the point *F*. The required golden rectangle is *AGFD* in Figure 5.6(b), since *EF* has length $\sqrt{5}$ so the length of *DF* is $1 + \sqrt{5}$, and $DF/FG = (1 + \sqrt{5})/2 = \phi$.

The nearest most of us get to gold is that indispensable item the credit card – so it's perhaps appropriate that if you measure one you'll find that its dimensions are (more or less) those of a golden rectangle.

Since $\phi$ is by definition the positive root of the equation (5.41), it satisfies

$$\phi^2 - \phi - 1 = 0 \qquad\qquad (5.45)$$

Rearranging (5.45) gives $\phi - 1 = 1/\phi$, so that $1/(\phi - 1) = \phi$. From this it follows that if a golden rectangle is divided into a square and another rectangle, as shown in Figure 5.7, then the smaller rectangle is also golden, because the ratio of the longer to the shorter side is

$$\frac{a}{(\phi - 1)a} = \frac{\phi}{1}$$

A further consequence of (5.45) is that in Figure 5.8 the point *B* divides up the line *AC* into what is called the 'golden section' or 'golden cut' of *AC*. The ratio of *AB* to *BC* is $\phi : 1$. Since *AC* has length $\phi + 1$ the ratio of *AC* to *AB* is

$$\frac{AC}{AB} = \frac{\phi + 1}{\phi} = \frac{\phi^2}{\phi} = \phi$$

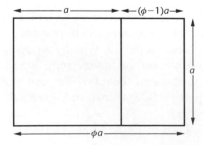

*Figure 5.7*  Subdivision of golden rectangle

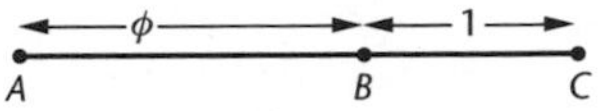

Figure 5.8 Golden section

using $\phi^2 = \phi + 1$ from (5.45). Hence $B$ divides $AC$ so that

$$\frac{AB}{BC} = \frac{AC}{AB}$$

This property of the point $B$ has also been used to create aesthetically pleasing shapes.

**Example 5.18** **Continued fraction**

A different and intriguing way of generating the golden ratio $\phi$ is by considering a so-called 'continued fraction', constructed in the following way:

$$c_0 = 1, \quad c_1 = 1 + \frac{1}{1}, \quad c_2 = 1 + \cfrac{1}{1 + \cfrac{1}{1}},$$

$$c_3 = 1 + \cfrac{1}{1 + \cfrac{1}{1 + \cfrac{1}{1}}}, \ldots$$

You can see that

$$c_1 = 1 + \frac{1}{c_0}, \quad c_2 = 1 + \frac{1}{c_1}, \quad c_3 = 1 + \frac{1}{c_2},$$

and in general the sequence of fractions *continues* indefinitely according to the formula

$$c_{n+1} = 1 + \frac{1}{c_n}, \qquad n = 0, 1, 2, 3, \ldots \tag{5.46}$$

Compare (5.46) with the relationship $\phi = 1 + 1/\phi$, obtained above. Simple arithmetic produces

$$c_1 = 2 = \frac{F_3}{F_2}, \quad c_2 = 1 + \frac{1}{c_1} = 1 + \frac{1}{2} = \frac{3}{2} = \frac{F_4}{F_3}$$

and it looks as though $c_n = F_{n+2}/F_{n+1}$. This can be proved by the method of induction (see the Interlude for details): The result is certainly valid for $n = 1$ and $n = 2$, so Step 1 of the induction procedure is established. For Step 2, assume the result is true for $n$. Then from (5.46)

$$c_{n+1} = 1 + \frac{1}{c_n}$$

$$= 1 + \frac{1}{(F_{n+2}/F_{n+1})}, \qquad \text{by assumption}$$

$$= 1 + \frac{F_{n+1}}{F_{n+2}}$$

$$= \frac{F_{n+2} + F_{n+1}}{F_{n+2}}$$

$$= \frac{F_{n+3}}{F_{n+2}}, \qquad \text{using (5.40) with } n \text{ replaced by } n+1$$

Hence if the formula is correct for $c_n$ it is also correct for $c_{n+1}$. By Step 3 of the induction procedure it therefore must be true for all values of $n$.

Since $c_n = F_{n+2}/F_{n+1} \longrightarrow \phi$ as $n \longrightarrow \infty$, the recurrence relation (5.46) gives an easy way of obtaining an approximation to $\phi$ using only the '$1/x$' button on your calculator. Starting with $n = 2$ and $c_2 = 1.5$, see how many iterations you need to get a value for $\phi$ correct to five decimal places.

---

There is a large literature on continued fractions in general. The sequence $c_0$, $c_1$, $c_2$, $c_3$, ... is an appealing one because each fraction is the simplest continued fraction which can be constructed, consisting only of 1's, yet it generates $\phi$ and the Fibonacci numbers.

As well as geometrical applications in art and architecture, Fibonacci numbers arise in a wide variety of other situations. Some of these are now listed.

(1) The numbers of petals of many flowers are Fibonacci numbers. For example, most daisies have 13, 21 or 34 petals, and some other common plants are shown in the table.

| Number of petals | Plant |
|---|---|
| 3 | lily, iris |
| 5 | buttercup, larkspur |
| 8 | cosmos, delphinium |
| 13 | ragwort |
| 21 | aster |
| 34 | pyrethrum |
| 55, 89 | Michaelmas daisies |

(2) Spirals on the outer surface of pine cones contain 3, 5, 8 or 13 scales; similarly, the spirals on pineapples contain 8, 13 or 21 scales.

(3) Spirals of seeds on sunflower heads contain 34 or 55 seeds, with 89, 144 or 233 on giant sunflowers.

(4) The spiralling arrangement of leaves on the branches of trees and shrubs follows a pattern involving the ratios of Fibonacci numbers.

(5) Suppose a function $f(x)$ has a single maximum value on a fixed interval $a \leq x \leq b$. An efficient algorithm 'searches' for this maximum value by successively computing values of $f(x)$. The locations of the points on the $x$-axis at which $f(x)$ is evaluated are determined from ratios of Fibonacci numbers.

More applications of Fibonacci numbers are introduced in Exercises 5.43 and 5.44 and Problems 5.20 and 5.21.

**Example 5.19** | **Properties of Fibonacci numbers**

Fibonacci numbers have all sorts of interesting properties. For example, if you look at the list of the first 15 numbers given above you should spot that every third number (that is, $F_3$, $F_6$, $F_9$, $F_{12}$, $F_{15}$) is an even number. The method of induction can be used to prove that $F_{3p}$ is an even number for *any* positive integer $p$. Step 1 of the procedure (again refer to the Interlude) has already been verified for $p = 1, 2, 3, 4, 5$. For Step 2 of the method it is required to prove that if $F_{3p}$ is assumed to be even for $p$, then $F_{3(p+1)}$ is also even. To do this, the recurrence formula (5.40) applied twice gives

$$
\begin{aligned}
F_{3(p+1)} = F_{3p+3} &= F_{3p+2} + F_{3p+1} \\
&= (F_{3p+1} + F_{3p}) + F_{3p+1} \\
&= 2F_{3p+1} + F_{3p}
\end{aligned}
$$

showing that $F_{3p+3}$ is also even because it is the sum of two even numbers $2F_{3p+1}$ and $F_{3p}$. Hence by Step 3 of the induction procedure the result is true for all values of $p$, that is, $F_n$ is even for any value of $n$ which is a multiple of 3.

A more complicated property can be identified by looking at sums of squares of Fibonacci numbers:

$$
F_1^2 = 1 = F_1 F_2, \quad F_1^2 + F_2^2 = 1 + 1 = 2 = F_2 F_3
$$
$$
F_1^2 + F_2^2 + F_3^2 = 1 + 1 + 4 = 6 = F_3 F_4
$$

It seems a reasonable guess that

$$
F_1^2 + F_2^2 + F_3^2 + \ldots + F_n^2 = F_n F_{n+1}
$$

Again the method of induction can be used to prove that this is true for *all* values of $n = 1, 2, 3, \ldots$. This follows because for Step 2 of the induction procedure

$$
\begin{aligned}
F_1^2 + F_2^2 + \ldots F_n^2 + F_{n+1}^2 &= F_n F_{n+1} + F_{n+1}^2, \qquad \text{by assumption} \\
&= F_{n+1}(F_n + F_{n+1}) \\
&= F_{n+1} F_{n+2}, \qquad \text{using (5.40)}
\end{aligned}
$$

Several other similar properties of Fibonacci numbers are given in Exercise 5.46 and Problems 5.22 and 5.23.

**Exercise**   **5.41**   Show that (5.43) can be simplified to

$$F_n = \frac{1}{2^{n-1}}[C(n,1) + C(n,3)5 + C(n,5)5^2 + C(n,7)5^3 + \ldots]$$

where $C(n, r)$ is the binomial coefficient defined in Section 3.1.3.

**Exercise**   **5.42**   Suppose as a result of genetic experiments which get out of hand, a new breed of 'super rabbits' escapes into the countryside. These obey the same assumptions as Fibonacci's for reproduction except that at each new birth there are *two* new pairs. If $G_n$ denotes the total number of pairs at the start of the $n$th month show that

$$G_{n+2} = G_{n+1} + 2G_n, \qquad n = 1, 2, 3, \ldots$$

with $G_1 = 1$, $G_2 = 1$ as before.

   How many pairs will there be after a year has elapsed? How long will it take to exceed 10,000 pairs?

**Exercise**   **5.43**   Suppose that you climb a flight of steps by taking either one step or two steps at a time. Let $x_n$ be the number of different ways of climbing $n$ steps. Obviously $x_1 = 1$ since there is only one way of going up one step; and when $n = 2$ you can either take two single steps or one double step, so $x_2 = 2$. Prove that in general

$$x_{n+2} = x_{n+1} + x_n$$

so the numbers follow the Fibonacci pattern 1, 2, 3, 5, 8, .... .

**Exercise**   **5.44**   Suppose a child is playing with a set of building blocks, all of the same size. Because its hands are small the child can only pick up either one block or two blocks at a time. The child decides to move a pile of blocks from the floor onto a table. Let $x_n$ be the number of different ways in which the child can pick up $n$ blocks. Show that the numbers $x_1, x_2, x_3, \ldots$ are the Fibonacci numbers, using an argument like that in the preceding exercise.

**Exercise**   **5.45**   Use the method of induction to prove that every fourth number in the Fibonacci sequence is divisible by 3, and every fifth number is divisible by 5.

**Exercise**   **5.46**   Prove the following identities, each valid for $n = 1, 2, 3, \ldots$, by the method of induction, or otherwise.

(a) $F_1 + F_2 + F_3 + \ldots F_n = F_{n+2} - 1$

(b) $F_2 + F_4 + F_6 + \ldots + F_{2n} = F_{2n+1} - 1$

(c) $F_1 + F_3 + F_5 + \ldots + F_{2n-1} = F_{2n}$

(d) $F_n F_{n+2} - F_{n+1}^2 = (-1)^{n+1}$

(e) $F_n^2 + F_{n+1}^2 = F_{2n+1}$.

## 5.4 Extensions

Some more complicated problems involving linear difference equations are described in this section, including an introduction to the use of matrix methods.

### Example 5.20 Car rental

Suppose you are the area manager of a car rental company in charge of two offices where cars can be rented, one in the city centre and one at the airport. You have a total of 105 cars in the fleet, and want to know how many cars you should keep at each location so as to satisfy demand. After studying the records of rentals for the past few years you decide that each month you can expect 70% of cars rented at the city office to be returned there, and the other 30% to be dropped off at the airport. For cars rented from the airport office, 25% are returned there whereas 75% are taken to the city. Let $x_n$, $y_n$ be the numbers of cars kept at the city centre and airport offices respectively at the start of month $n$. After one month has elapsed, the number $x_{n+1}$ of cars at the city depot consists of those returned there during the previous month, namely $0.7x_n$, together with those on a one-way rental from the airport, namely $0.75y_n$. The difference equation is therefore

$$x_{n+1} = \underset{\substack{\text{cars returned} \\ \text{to city centre}}}{0.7x_n} + \underset{\substack{\text{cars from} \\ \text{airport}}}{0.75y_n}, \qquad n = 1, 2, 3, \ldots \tag{5.47}$$

Similarly the number $y_{n+1}$ of cars at the airport office at the start of month $n+1$ is

$$y_{n+1} = \underset{\substack{\text{cars from} \\ \text{city centre}}}{0.3x_n} + \underset{\substack{\text{cars returned} \\ \text{to airport}}}{0.25y_n}, \qquad n = 1, 2, 3, \ldots \tag{5.48}$$

The two equations (5.47) and (5.48) comprise a linked pair. Each equation is first order (there are no terms in $x_{n+2}$ or $y_{n+2}$) and linear (there are no terms like $x_n^2$ or $y_n^2$). Suppose that you begin a new year on 1 January with $x_1 = 60$ cars at the city depot and $y_1 = 45$ cars at the airport. Substituting $n = 1$ into (5.47) gives the number of cars at the city centre on 1 February as

$$x_2 = 0.7 \times 60 + 0.75 \times 45 = 75.75$$

To compute $y_2$ you can use (5.48) with $n = 1$, but since there are 105 cars in total, then $y_2 = 105 - 75.75 = 29.25$. You should compute the entries in the table below, using (5.47).

| $n$ | $x_n$ | $y_n$ |
| --- | --- | --- |
| 3 | 74.96 | 30.04 |
| 4 | 75.00 | 30.00 |
| 5 | 75.00 | 30.00 |

You can see that very rapidly the numbers settle down to a 'steady state' with 75 cars at the city centre and 30 at the airport. This means that if you keep these numbers at the two locations at the beginning of one month, then they will stay the same at the beginning of every subsequent month. To check this, simply substitute $x_n = 75$, $y_n = 30$ into (5.47) to get

$$x_{n+1} = 0.7 \times 75 + 0.75 \times 30 = 75$$

and

$$y_{n+1} = 105 - x_{n+1} = 30$$

In fact, you can compute these steady state values directly from (5.47). You are looking for a number $c$ such that $x_n = c$, $y_n = 105 - c$ for all values of $n$. Substituting these values into (5.47) gives

$$c = 0.7c + 0.75(105 - c)$$

so that $1.05c = 0.75 \times 105$, that is, $c = 75$ as found above. Notice also that the pair of equations (5.47) and (5.48) can actually be combined together into a single equation. Simply write $y_n = 105 - x_n$ in (5.47) to obtain

$$\begin{aligned} x_{n+1} &= 0.7x_n + 0.75(105 - x_n) \\ &= -0.05x_n + 78.75 \end{aligned} \tag{5.49}$$

You should verify that writing $y_{n+1} = 105 - x_{n+1}$ in (5.48) produces the *same* equation (5.49). Equation (5.49) has exactly the form of equation (5.18), so by (5.19) with $c = -0.05$ and $d = 78.75$ the solution is

$$x_n = (-0.05)^n x_0 + \frac{(-0.05)^n - 1}{-0.05 - 1} \times 78.75, \qquad n = 1, 2, 3, \ldots$$

As $n$ increases the terms $(-0.05)^n$ rapidly become very small, so as $n \longrightarrow \infty$ then $x_n \longrightarrow 78.75/1.05 = 75$, agreeing with what was found above.

This problem is looked at again in Exercise 5.47.

**Example 5.21**  **Use of matrices**

If you are familiar with the basic concepts of matrix algebra, then it is interesting to tackle the previous example using these ideas. First notice that equations (5.47) and (5.48) can be written in the combined form

$$\begin{bmatrix} x_{n+1} \\ y_{n+1} \end{bmatrix} = \begin{bmatrix} 0.7 & 0.75 \\ 0.3 & 0.25 \end{bmatrix} \begin{bmatrix} x_n \\ y_n \end{bmatrix}$$

or

$$X_{n+1} = CX_n, \qquad n = 1, 2, 3, \ldots \tag{5.50}$$

where $C$ is the $2 \times 2$ matrix

$$C = \begin{bmatrix} 0.7 & 0.75 \\ 0.3 & 0.25 \end{bmatrix} \tag{5.51}$$

and $X_n$ is the $2 \times 1$ matrix (that is, the column vector)

$$X_n = \begin{bmatrix} x_n \\ y_n \end{bmatrix}$$

The advantage of expressing the problem in the matrix-vector form (5.50) is that both variables $x_n$ and $y_n$ can be handled simultaneously. For example, substituting $n = 1$ into (5.50) gives $X_2 = CX_1$, where

$$X_1 = \begin{bmatrix} x_1 \\ y_1 \end{bmatrix} = \begin{bmatrix} 60 \\ 45 \end{bmatrix}$$

and similarly

$$X_3 = CX_2 = C(CX_1) = C^2 X_1$$
$$X_4 = CX_3 = C^3 X_1$$

In general the solution of (5.50) is

$$X_{n+1} = C^n X_1, \qquad n = 1, 2, 3, \ldots \tag{5.52}$$

where $C^n$ is the matrix $C$ multiplied by itself $n$ times. Evaluation of $C^n$ is easy with a suitable graphics calculator. For example, with $C$ in (5.51)

$$C^2 = \begin{bmatrix} 0.715 & 0.7125 \\ 0.285 & 0.2875 \end{bmatrix}$$

Hence from (5.52) with $n = 2$, using the usual rule for matrix multiplication,

$$X_3 = C^2 X_1 = C^2 \begin{bmatrix} 60 \\ 45 \end{bmatrix}$$
$$= \begin{bmatrix} 74.96 \\ 30.04 \end{bmatrix} = \begin{bmatrix} x_3 \\ y_3 \end{bmatrix}$$

agreeing with the values for $x_3$ and $y_3$ in the table in Example 5.20.

The expression (5.52) is a generalization of (5.20), when $C$ is a number instead of a matrix (starting at $n = 1$ instead of $n = 0$ is of no significance). An explicit expression for $C^n$ can be developed using a little matrix theory. Textbooks on the subject show that

$$C^n = \lambda_1^n Z_1 + \lambda_2^n Z_2, \qquad n = 0, 1, 2, 3, \ldots \tag{5.53}$$

where $Z_1$ and $Z_2$ are constant $2 \times 2$ matrices to be determined shortly, and $\lambda_1$ and $\lambda_2$ are the **eigenvalues** of $C$ (it is assumed $\lambda_1 \neq \lambda_2$). This requires some explanation. Let $C$ be an arbitrary $2 \times 2$ matrix with elements as shown:

$$C = \begin{bmatrix} c_1 & c_2 \\ c_3 & c_4 \end{bmatrix}$$

The eigenvalues of $C$ are defined as the roots of the quadratic equation

$$(\lambda - c_1)(\lambda - c_4) - c_2 c_3 = 0 \tag{5.54}$$

In order to determine the matrices $Z_1$ and $Z_2$ in (5.53), set $n = 0$ and $n = 1$ to obtain two equations

$$\left.\begin{array}{l} I = Z_1 + Z_2 \\ C = \lambda_1 Z_1 + \lambda_2 Z_2 \end{array}\right\} \tag{5.55}$$

where $C^0 = I$ is the $2 \times 2$ **unit matrix** defined by

$$I = \begin{bmatrix} 1 & 0 \\ 0 & 1 \end{bmatrix}$$

Although the simultaneous equations (5.55) involve matrices they can be solved in the usual way: multiply the first equation by $\lambda_2$ and subtract the second from it to obtain

$$\lambda_2 I - C = (\lambda_2 - \lambda_1) Z_1$$

so that

$$Z_1 = \frac{1}{\lambda_2 - \lambda_1} (\lambda_2 I - C) \tag{5.56}$$

By the usual rules of matrix algebra

$$\lambda_2 I - C = \begin{bmatrix} \lambda_2 & 0 \\ 0 & \lambda_2 \end{bmatrix} - \begin{bmatrix} c_1 & c_2 \\ c_3 & c_4 \end{bmatrix}$$

$$= \begin{bmatrix} \lambda_2 - c_1 & -c_2 \\ -c_3 & \lambda_2 - c_4 \end{bmatrix} \tag{5.57}$$

Similarly, by eliminating $Z_1$ between the two equations in (5.55) you get

$$\lambda_1 I - C = (\lambda_1 - \lambda_2) Z_2$$

so that

$$Z_2 = \frac{1}{\lambda_1 - \lambda_2} (\lambda_1 I - C) \tag{5.58}$$

where the matrix $\lambda_1 I - C$ has the form (5.57) with $\lambda_2$ replaced by $\lambda_1$. Now apply these formulae when $C$ is the matrix in (5.51). The quadratic equation (5.54) becomes

$$(\lambda - 0.7)(\lambda - 0.25) - 0.75 \times 0.3 = 0$$

that is,

$$\lambda^2 - 0.95\lambda - 0.05 = 0$$

which factorizes into

$$(\lambda - 1)(\lambda + 0.05) = 0$$

so $\lambda_1 = 1$, $\lambda_2 = -0.05$. The matrix (5.57) becomes

$$\lambda_2 I - C = \begin{bmatrix} (-0.05 - 0.7) & -0.75 \\ -0.3 & (-0.05 - 0.25) \end{bmatrix}$$

$$= \begin{bmatrix} -0.75 & -0.75 \\ -0.3 & -0.3 \end{bmatrix}$$

and similarly

$$\lambda_1 I - C = \begin{bmatrix} 0.3 & -0.75 \\ -0.3 & 0.75 \end{bmatrix}$$

Finally, since $\lambda_1 - \lambda_2 = -1.05$, the expression (5.53) together with those for $Z_1$ in (5.56) and $Z_2$ in (5.58) gives

$$C^n = (1)^n Z_1 + (-0.05)^n Z_2 \tag{5.59}$$

$$= \frac{1}{-1.05}(\lambda_2 I - C) + \frac{(-0.05)^n}{1.05}(\lambda_1 I - C)$$

$$= \frac{1}{1.05}\begin{bmatrix} 0.75 & 0.75 \\ 0.3 & 0.3 \end{bmatrix} + \frac{(-0.05)^n}{1.05}\begin{bmatrix} 0.3 & -0.75 \\ -0.3 & 0.75 \end{bmatrix}$$

$$= \frac{1}{1.05}\begin{bmatrix} 0.75 + 0.3(-0.05)^n & 0.75 - 0.75(-0.05)^n \\ 0.3 - 0.3(-0.05)^n & 0.3 + 0.75(-0.05)^n \end{bmatrix}$$

You should verify that setting $n = 2$ in this expression produces the same values for the entries of $C^2$ as before. For example, the entry in the top left corner is

$$\frac{0.75 + 0.3(-0.05)^2}{1.05} = 0.715$$

The term $(-0.05)^n$ in (5.59) rapidly becomes very small as $n$ increases, so that as $n \longrightarrow \infty$ then $C^n \longrightarrow Z_1$. Hence by (5.52) the solution $X_n$ of the equation (5.50) becomes

$$X_n \longrightarrow Z_1 X_1, \text{ as } n \longrightarrow \infty$$

$$= \frac{1}{1.05}\begin{bmatrix} 0.75 & 0.75 \\ 0.3 & 0.3 \end{bmatrix}\begin{bmatrix} 60 \\ 45 \end{bmatrix}$$

$$= \frac{1}{1.05}\begin{bmatrix} 78.75 \\ 31.5 \end{bmatrix} = \begin{bmatrix} 75 \\ 30 \end{bmatrix}$$

and these are the steady-state values of $x_n$ and $y_n$ found in Example 5.20.

For further details on the use of matrix methods to solve difference equations (including the case when $\lambda_1 \neq \lambda_2$) you should consult books on matrix algebra.

| Example 5.22 | **Insect population** |

An entomologist observes a certain species of insect over a period of time, and suggests the following rules to describe the breeding and survival patterns of the female population.

(1) Half of the insects survive into their second year.

(2) Of these, two-thirds live into a third year.

(3) None of the original insects lives longer than three years.

(4) On average, a female produces no new insects in the first year of her life, two new insects in her second year, and three new insects in her final year.

Let $x_n$, $y_n$, $z_n$ denote the numbers of new-born, one-year-old and two-year-old insects respectively, at the start of year $n$. Because of rule (4), $2y_n + 3z_n$ insects are born in year $n$, so

$$x_{n+1} = 2y_n + 3z_z, \qquad n = 1, 2, 3, \ldots$$

Rule (1) states that half of the new-born insects survive past their first birthday, so that

$$y_{n+1} = \tfrac{1}{2}x_n, \qquad n = 1, 2, 3, \ldots$$

Finally, according to (2) only two-thirds of the one-year-olds survive past their second birthday, so that

$$z_{n+1} = \tfrac{2}{3}y_n, \qquad n = 1, 2, 3, \ldots$$

Suppose that initially there are 6000 new-born insects, 2000 one-year-olds and 1500 two-year-olds. Using your calculator you should compute the values in the following table (given in thousands) from the above difference equations. For example, with $n = 1$ you get

$$x_2 = 2y_1 + 3z_1 = 8.5, \quad y_2 = 0.5x_1 = 3.0, \quad z_2 = 0.67y_1 = 1.34$$

| $n$ | 1 | 2 | 3 | 4 | 5 | 6 | 7 | 8 | 9 | 10 |
|-----|-----|------|-------|-------|-------|-------|-------|-------|-------|-------|
| $x_n$ | 6.0 | 8.5 | 10.02 | 14.50 | 18.57 | 24.58 | 33.14 | 43.24 | 57.83 | 76.54 |
| $y_n$ | 2.0 | 3.0 | 4.25 | 5.01 | 7.25 | 9.28 | 12.29 | 16.57 | 21.62 | 28.91 |
| $z_n$ | 1.5 | 1.34 | 2.0 | 2.85 | 3.36 | 4.86 | 6.22 | 8.23 | 11.10 | 14.49 |

This table suggests that the numbers of insects in each group continue to grow, producing a 'population explosion'. To try to be more specific, look at the annual percentage increase in the values of $x_n$. For example, after one year the increase in $x_1$ is

$$\frac{x_2 - x_1}{x_1} \times 100 = \frac{8.5 - 6.0}{6.0} \times 100 = 41\%$$

Similarly, for subsequent values of $n$ you should compute the table:

| $n$ | 2 | 3 | 4 | 5 | 6 | 7 | 8 | 9 |
|---|---|---|---|---|---|---|---|---|
| Percentage increase in $x_n$ | 17.9 | 44.7 | 28.1 | 32.4 | 34.8 | 30.4 | 33.7 | 32.3 |

It looks as though the annual increase is settling down to a value somewhere around 32%. To investigate this, first combine together the three difference equations into a single one. To do this, replace $n$ by $n+2$ in the first equation to get

$$x_{n+3} = 2y_{n+2} + 3z_{n+2}$$

However, from the second equation

$$y_{n+2} = \tfrac{1}{2}x_{n+1}$$

and from the third equation

$$z_{n+2} = \tfrac{2}{3}y_{n+1} = \tfrac{2}{3}\left(\tfrac{1}{2}x_n\right) = \tfrac{1}{3}x_n$$

Hence overall the equation becomes

$$\begin{aligned}x_{n+3} &= 2\left(\tfrac{1}{2}x_{n+1}\right) + 3\left(\tfrac{1}{3}x_n\right)\\ &= x_{n+1} + x_n\end{aligned} \tag{5.60}$$

which is a **third** order linear difference equation because it involves terms in $x_{n+3}$ and $x_n$, with a difference of 3 in the suffices. If there is a solution of (5.60) in which as $n \longrightarrow \infty$ the value of $x_n$ increases by a constant percentage $p$, this means that

$$\frac{100(x_{n+1} - x_n)}{x_n} = p$$

so that

$$\begin{aligned}x_{n+1} &= \frac{px_n}{100} + x_n\\ &= \left(1 + \frac{p}{100}\right)x_n\\ &= kx_n, \quad \text{say}\end{aligned}$$

This implies that

$$x_{n+2} = kx_{n+1} = k^2 x_n, \quad x_{n+3} = kx_{n+2} = k^3 x_n$$

and substituting into (5.60) gives

$$k^3 x_n = kx_n + x_n$$

or

$$(k^3 - k - 1)x_n = 0$$

This will hold for all $x_n$ (with $n$ sufficiently large) provided

$$k^3 - k - 1 = 0$$

You can check with your calculator that $k = 1.325$ is a close approximation to a root of this cubic equation. Since $k = 1 + p/100$, this gives $p = 32.5$, which agrees with what was observed numerically in the table. You should calculate a few more values of $x_n$ to confirm that the annual percentage increase gets closer and closer to 32.5% as $n$ increases. The long-term increases in $y_n$ and $z_n$ are the same.

**Exercise**

**5.47**  In the car rental problem described in Example 5.20, after a couple of years you find things begin to go wrong. The number of cars at the city depot at the beginning of the month falls short of the target. On analyzing rental data you find that 40% of people collecting cars from the city centre are now dropping them off at the airport. Write down the new difference equations.

You decide to buy an extra ten cars in view of increased overall demand. If the new trend is maintained, how should you allocate the extra cars in the fleet?

**Exercise**

**5.48**  The solution of the pair of equations

$$x_{n+1} = ax_n + by_n, \quad y_{n+1} = cx_n + dy_n, \quad n = 0, 1, 2, 3, \ldots$$

when $x_0 = 1$, $y_0 = 1$ is

$$x_n = 5(-3)^n - 4(-4)^n, \quad y_n = -15(-3)^n + 16(-4)^n$$

Determine the values of $a$, $b$, $c$, $d$. Hence find the solution of the equations when the initial conditions are $x_0 = 2$, $y_0 = 1$.

**Exercise**

**5.49**  Determine the eigenvalues of the matrix

$$C = \begin{bmatrix} 1 & 3 \\ 3 & 1 \end{bmatrix}$$

Use (5.53) to obtain an expression for $C^n$. Check your result for $n = 4$ by computing $C^4$ directly.

Use your expression for $C^n$ to obtain the solution of the equations

$$\left. \begin{array}{l} x_{n+1} = x_n + 3y_n \\ y_{n+1} = 3x_n + y_n \end{array} \right\} \quad n = 0, 1, 2, 3, \ldots$$

subject to $x_0 = 3$, $y_0 = 1$.

**Exercise**

**5.50**  Two cable television channels run competing news programmes in the same time slot every evening. Audience research amongst those who always watch the news shows that the viewers are equally divided. That's to say, if a viewer watches Channel A one evening then there is a 50% chance they will watch Channel B the next evening, and conversely.

In an attempt to gain a larger proportion of the audience, Channel A redesigns its programme. As a result of this, subsequent audience research shows that if a viewer watches Channel A news there is a 55% chance they will watch again the following evening, and only a 45% chance that they will watch Channel B. However, the probability of switching from Channel B to Channel A remains at 50%. Let $a_n$, $b_n$ be the probabilities that a viewer is watching Channel A or B respectively on evening $n$. Show that

$$a_{n+1} = 0.55a_n + 0.5b_n, \quad b_{n+1} = 0.45a_n + 0.5b_n, \quad n = 1, 2, 3, \ldots$$

Using the fact that $a_n + b_n = 1$ (since the group of viewers under consideration always watches one or other of the two news programmes) show that in the long term the audience for Channel A increases to 52.6% of the total.

**Exercise**

**5.51**  Suppose that in Example 5.22 a different species of insect obeys similar breeding and survival rules except for the following modifications:

(2) two-thirds is replaced by one-fifth

(4) females do not produce new insects until their third year, when on average ten new insects are born.

Suppose initially there are 1000 insects in each age group. Compute the numbers in each group for the next five years. Prove that the numbers follow a cyclic pattern which repeats every three years.

## Miscellaneous problems

**5.1**  Show that applying the Newton–Raphson iterative method to a general quadratic equation $ax^2 + bx + c = 0$ produces the formula

$$x_{n+1} = \frac{ax_n^2 - c}{2ax_n + b}, \quad n = 0, 1, 2, 3, \ldots$$

Prove that if $n$ is large enough so that $x_n \approx x^*$, where $x^*$ is an exact root of the equation, then

$$x_{n+1} - x^* \approx k(x_n - x^*)^2$$

where $k = a/(2ax^* + b)$.

**5.2** The Newton–Raphson method can also be applied to non-algebraic equations. Use it to find a positive solution of the equation

$$2\cos x = x$$

correct to five decimal places.

**5.3** Use the standard formula (see Section 1.4.2) to show that the roots of the quadratic equation

$$x^2 - 1.064x + 0.283 = 0$$

correct to three decimal places are 0.527, 0.537.

Apply the Newton–Raphson iteration with $x_0 = 0.55$, working to four decimal places. Show that $x_{10} = x_2 = 0.5361$, so that further iterations will simply reproduce the values from $x_3$ to $x_{10}$. Then work to five decimal places and show that this leads to a correct approximation of the larger root.

**5.4** A solicitor informs you that an old will has been just discovered, dating back 60 years. In it your great-grandfather left an amount to be given to his first great-grandchild (you!). The sum was £100 (enough to buy a small new car back then) and has been earning interest of 3% compounded annually. How much will you receive now?

Suppose that you can now buy a small new car for £5000.

(a) How much would your ancestor have needed to leave you in his will so that you would have enough to buy a car?

(b) Over the 60-year period, what *constant* annual rate of inflation does the increase in car prices represent?

**5.5** The European Union establishes a new Administration Capital in the year 2010, containing initially 170,000 bureaucrats. Ten years later this number has grown to 245,000. Assuming that the percentage increase in the number of bureaucrats is the same every year, how many will there be in the year 2050?

**5.6** The EuroSceptic Party (ESP) finally takes Britain out of the European Union in the year 2010, and establishes the NUK (New United Kingdom). One of their first acts as a government is to pass a law fixing the rate of inflation at 3.5% per year.

(a) If a kilogram of sugar costs five New Pounds in 2010, how much will it cost 10 years later?

(b) After the ESP has been in office for a decade the New Pound has fallen from parity with the European Union Currency Unit to a value of 0.60 units. Has the inflation law been successful?

**5.7** You take out a mortgage of £50,000 to buy a small house. There is a fixed APR of 12.5, compounded annually on the balance outstanding at the beginning of each year. What is the monthly cost of repaying the loan over a period of 20 years? How much will you still owe halfway through the term?

**5.8** One way to solve the difference equation

$$x_{n+1} = x_n + (n+1)^2, \qquad n = 0, 1, 2, 3, \dots$$

subject to the condition $x_0 = 0$ is to try as a solution $x_n = an^3 + bn^2 + cn$, and equate coefficients of powers of $n$ on either side of the resulting identity. Show that this produces three equations for the three unknown constants $a$, $b$, $c$ and solve these equations to get $a = 1/3$, $b = 1/2$, $c = 1/6$.

Notice from the original equation that

$$x_1 = x_0 + 1^2, \quad x_2 = x_1 + 2^2 = 1^2 + 2^2,$$
$$x_3 = x_2 + 3^2 = 1^2 + 2^2 + 3^2$$

and so on. Hence deduce that

$$1^2 + 2^2 + \dots + n^2 = \tfrac{1}{6} n(n+1)(2n+1)$$

**5.9** Use the method of the preceding problem to obtain the solution of the equation

$$x_{n+1} = x_n + (2n+1)^2, \qquad n = 0, 1, 2, 3, \dots$$

subject to $x_0 = 0$. Hence prove that

$$1^2 + 3^2 + 5^2 + \dots + (2n-1)^2 = \tfrac{1}{3} n(4n^2 - 1)$$

**5.10** The growth of the population of a certain animal in a wildlife reserve is described by

$$x_{n+1} = 1.25 x_n, \qquad n = 0, 1, 2, 3, \dots$$

where $x_n$ is the number in year $n$. If $x_0 = 200$, determine how many years it will take for the population to exceed 2000.

Because of its ever-increasing numbers the animal becomes a pest and it is necessary to cull the population by an annual amount $2^n$, so that

$$x_{n+1} = 1.25 x_n - 2^n$$

Obtain the general expression for $x_n$ using an appropriate formula from Exercise 5.22. What will be the size of the population in 10 years, assuming that the culling begins when the population size is 2000? Assuming the culling policy is unaltered, how long would it take for the animal to become extinct in the reserve?

**5.11** Show that when $1 - b - c = 0$ then the characteristic equation (5.32) has a root $\lambda = 1$. Using the solution (5.33), explain why the particular solution for (5.37) in this case cannot be a constant.

**5.12** Verify by direct substitution that the general solution of (5.37) when $1 - b - c = 0$ is

$$x_n = a_1 + a_2(b-1)^n + \frac{dn}{2-b}, \qquad b \neq 2$$
$$= a_1 + a_2 n + \tfrac{1}{2} dn^2, \qquad b = 2$$

where $a_1$ and $a_2$ are constants determined by two given conditions.

**5.13** Use the results in the preceding problem to determine the solution of the following equations valid for $n = 0, 1, 2, 3, \ldots$ subject to the given conditions:

(a) $x_{n+2} = 3x_{n+1} - 2x_n + 5$, $\quad x_0 = 6$, $\quad x_1 = 0$

(b) $x_{n+2} = 2x_{n+1} - x_n - 1$, $\quad x_2 = -5$, $\quad x_4 = -17$.

**5.14** Verify that the general solution of the equation

$$x_{n+2} = 2cx_{n+1} - c^2 x_n + k^n$$

where $c \neq 0$, is

$$x_n = (a_1 + a_2 n)c^n + \begin{cases} \dfrac{k^n}{k^2 - 2ck + c^2}, & k \neq c \\ \frac{1}{2}n(n-1)c^{n-2}, & k = c \end{cases}$$

where $a_1$ and $a_2$ are constants.

Hence solve the equation

$$x_{n+2} = 6x_{n+1} - 9x_n + 2(3^n) - 4^n$$

subject to $x_0 = 5$, $x_1 = 8$.

**5.15** In the economic model described in Example 5.13, the strategy suggested in Exercise 5.37 proves disappointing, as the economy goes through extremes of 'bust' and 'boom'. In an attempt to improve matters you decide to increase investment to *three* times consumer spending, that is,

$$P_{n+1} = 3(C_{n+1} - C_n)$$

Compute the national income $I_n$ and hence determine whether this policy is a success.

**5.16** For the economic model described in Example 5.13, investigate what happens when you vary the assumptions (1) and (2), whilst keeping $I_1 = G$, $I_2 = 2G$, $G_n = G$, in the following ways:

(a) $C_{n+1} = 0.8I_n$, $\quad P_{n+1} = 0.5(C_{n+1} - C_n)$

(b) $C_{n+1} = 0.3I_n$, $\quad P_{n+1} = 5(C_{n+1} - C_n)$.

Also, use different values of $I_1$ and $I_2$ (for example $I_1 = G$, $I_2 = 1.5G$) to see if this affects the long-term behaviour of the economy in these two cases.

**5.17** A certain process is known to be described by the equation

$$x_{n+2} = ax_{n+1} + bx_n + c, \qquad n = 0, 1, 2, 3, \ldots$$

where $a$, $b$ and $c$ are constants. It is found that $x_1 = 4$, $x_2 = 12$, $x_3 = 23$, $x_4 = 37$, $x_5 = 54$. Determine $x_0$ and $x_{10}$. Also, obtain a general expression for $x_n$.

**5.18** Wild flowers propagate by self-seeding. A botanist has observed a certain species in a nature reserve, and has suggested the following rules:

(a) A plant flowers and produces seeds either one year or two years after it has germinated from seed.

(b) After flowering all plants die.

(c) A plant flowering after one year produces (on average) 264 seeds.

(d) A plant flowering after two years produces (on average) 700 seeds.

(e) On average, 25% of seeds result in plants that flower after one year and 40% produce plants that flower after two years. The remaining 35% of seeds fail to produce plants that survive to produce more seeds.

Let $s_n$ be the number of seeds produced by flowers in year $n$. Show that

$$s_{n+2} = 66s_{n+1} + 280s_n, \qquad n = 0, 1, 2, 3, \ldots$$

The botanist tries to establish a colony of the plants in a new area by planting 20 seeds. Deduce that $s_1 = 1320$, and obtain a general expression for $s_n$. Hence show that after many years have passed $s_n \approx 18.9 \times 70^n$.

**5.19** You buy a pair of baby rats which you intend to breed for profit. When one month old, each pair produces two pairs of babies. To maintain a fertile stock you sell pairs after they have bred twice. If $T_n$ denotes the total number of pairs at the start of month $n$, show that

$$T_{n+2} - 2T_{n+1} - 2T_n = 0, \qquad n = 1, 2, 3, \ldots$$

with $T_1 = 1$, $T_2 = 3$. Obtain a general expression for $T_n$.

**5.20** The behaviour of a bee population in a hive can be described by the following rules.

(a) Unfertilized eggs laid by a queen hatch into males, so that male bees do not have a father.

(b) Fertilized eggs develop into females, so female bees have both a father and a mother.

Hence a male's only sexual function is in the production of females.

The ancestry of a single male bee can be traced through the diagram in Figure 5.9, where $M$ denotes male and $F$ female. Reading downwards, this shows that the male bee has a mother; at the next level, a grandmother and a

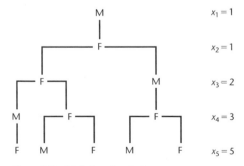

*Figure 5.9* Male bee family tree

grandfather; then one great-grandfather and two great-grandmothers (that is, *three* great-grandparents); and *five* great-great-grandparents.

Let $x_n$ be the number of bees in the $n$th row of the ancestry diagram. Show that $x_n = F_n$, the $n$th Fibonacci number.

**5.21**  In the seating plan for a banquet all the tables are round and seat the same number of people. It has been decided that because more men than women are attending, no two women may sit next to each other. If the tables seat only *two* people, there are *three* different seating arrangements, shown in Figure 5.10, where $M$, $W$ denote man or woman respectively. When there are *three* people to a table there are *four* different arrangements; and with *four* people there are *seven* different arrangements, as shown in Figures 5.11 and 5.12 respectively.

Let $L_n$ be the number of different arrangements when there are $n$ persons per table. By referring to Example 5.16 show that $L_5 = F_6 + F_4$. Verify this by drawing the diagram with five per table.

Show that in general $L_{n+2} = F_{n+3} + F_{n+1}$, and hence deduce that

$$L_{n+2} = L_{n+1} + L_n, \qquad n = 1, 2, 3, \dots$$

This is the same recurrence formula as for Fibonacci numbers in (5.40), but by beginning with $L_1 = 1$, $L_2 = 3$ you get the sequence

| 1 | 3 | 4 | 7 | 11 | 18 | 29 | ... |
|---|---|---|---|----|----|----|-----|
| $L_1$ | $L_2$ | $L_3$ | $L_4$ | $L_5$ | $L_6$ | $L_7$ | |

These numbers are called **Lucas numbers**, and like Fibonacci numbers they arise in many situations. Prove that

$$L_n = \left(\frac{1 + \sqrt{5}}{2}\right)^n + \left(\frac{1 - \sqrt{5}}{2}\right)^n, \qquad n = 1, 2, 3, \dots$$

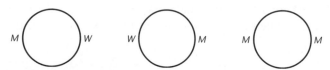

*Figure 5.10*  Two per table

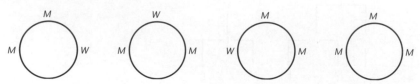

*Figure 5.11*  Three per table

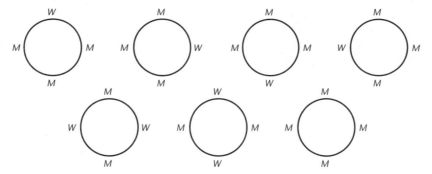

*Figure 5.12* Four per table

**5.22** Prove that every sixth Fibonacci number is divisible by 8. Use this together with the results in Example 5.19 and Exercise 5.45 to guess a general result.

**5.23** For the Lucas numbers defined in Problem 5.21 prove that for $n = 1, 2, 3, \ldots$

(a) $L_n + L_{n+2} = 5F_{n+1}$

(b) $L_1 + L_2 + \ldots + L_n = L_{n+2} - 3$

(c) $L_1^2 + L_2^2 + \ldots + L_n^2 = L_n L_{n+1} - 2$

(d) $F_{2n} = F_n L_n$.

**5.24** Based on census data and electoral results, it is estimated that on average 5% of the British population moves each year to the south-eastern corner of England, attracted by better employment prospects. However, because of a deteriorating quality of life, 10% of the population of the South East leaves that region each year to live elsewhere in Britain. Let $x_n$ denote the proportion of the population living in the South East in year $n$, and $y_n$ the proportion living outside. Assume that the total population remains constant. Obtain a pair of difference equations expressing $x_{n+1}, y_{n+1}$ in terms of $x_n$ and $y_n$.

If initially one quarter of the population lives inside the South East, compute the proportion living there after five years.

Assuming these trends are continued, what will be the proportion living in the South East in the long term?

**5.25** A well-known puzzle is illustrated in Figure 5.13. The $8 \times 8$ square in Figure 5.13(a) when cut up into the four shapes I, II, III, IV can apparently be reassembled into a $5 \times 13$ rectangle in Figure 5.13(b). However, the area of the square is 64 units, one unit less than the area of the rectangle!

To explain this anomaly, draw a large-scale diagram and confirm that $ACDB$ is not a straight line but a very thin rhombus, shown in exaggerated form in Figure 5.13(c). Use the identity in Exercise 5.46(d) to obtain a generalization in terms of a square of side $F_{n+1}$ and a rectangle with sides $F_n$ and $F_{n+2}$ (the above example has $n = 5$).

What happens when $n$ is even? Try drawing the case $n = 4$.

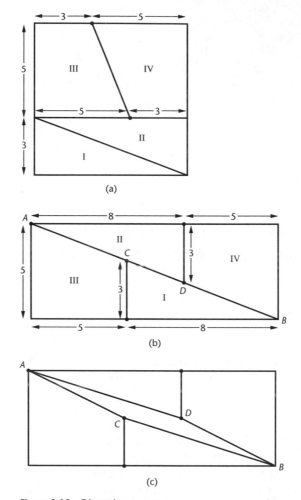

*Figure 5.13* Dissecting a square

**5.26** Suppose that an insect population behaves like that in Example 5.22, with the following alterations:

(1) one-half is replaced by one-sixteenth

(2) two-thirds is replaced by one-quarter

(4) the birth rates are 7 instead of 2, and 6 instead of 3.

Show that $x_n$ satisfies the equation

$$x_{n+3} = \tfrac{7}{16}x_{n+1} + \tfrac{3}{32}x_n, \qquad n = 1, 2, 3, \ldots$$

instead of (5.60).

Assume that the solution of this third order linear difference equation has the form

$$x_n = a_1 \lambda_1^n + a_2 \lambda_2^n + a_3 \lambda_3^n$$

where $\lambda_1, \lambda_2, \lambda_3$ are the roots of the cubic equation obtained by substituting $x_n = \lambda^n$ into the difference equation. Hence deduce that eventually a colony of insects obeying these rules becomes extinct.

**5.27**   The **Padovan sequence** is defined by

$$P_1 = 1, \; P_2 = 1, \; P_3 = 2, \; P_{n+2} = P_n + P_{n-1}; \qquad n \geq 2.$$

Determine $P_4, P_5, \ldots, P_{20}$ and compute the ratios of successive Padovan numbers. Show that if $P_{n+1}/P_n \rightarrow p$ as $n \rightarrow \infty$ then $p^3 - p - 1 = 0$, and verify that $p \approx 1.324718$.

Prove also that $P_{n+2} = P_{n+1} + P_{n-3}, n \geq 4$.

Geometrical and other properties of Padovan numbers are described by Stewart (1996).

### Student project

**5.28**   Suppose you are keen on fish and have an aquarium which contains 100 litres of water. An important aspect which you want to control is the prevention of the build-up of a high concentration of salt in the water, because this would be dangerous to freshwater fish. The water you use is not distilled, and so contains a small amount of salt at a concentration of $s$ units per litre.

(a)   Suppose that 1% of the water evaporates each week, so if you take no action the salt concentration would steadily increase. What would be the percentage increase in the concentration of salt after 20 weeks of this neglect?

(b)   You decide to try to improve the water quality by topping up with one litre of fresh water (with salt concentration $s$) at the end of each week. Let $x_n$ denote the total amount of salt in the aquarium at the beginning of the $n$th week, after you have brought the water level back to normal. Obtain an expression for $x_n$. Show that after 20 weeks there is a 20% increase in the salt concentration. This is very little better than the situation in part (a).

(c)   In view of the disappointing effect of your efforts in part (b), you decide on a more active policy of removing an *additional* litre at the end of each week, and then adding *two* litres of fresh water so as to refill the tank. Show that the salt content in the water satisfies the difference equation

$$x_{n+1} = \tfrac{98}{99} x_n + 2s, \qquad n = 0, 1, 2, 3, \ldots$$

with $x_0 = 100s$.

Compute $x_{20}$ using (5.19), and hence determine the percentage change in salt concentration after 20 weeks using this scheme. Investigate what happens to the salt concentration after two further 20-week periods. If this scheme was used for a considerable length of time, what would be the long-term change in salt concentration?

(d) You are still dissatisfied with the result of your efforts, and decide to replace more of the water each week, taking out five litres and adding six litres of fresh water. Obtain the new difference equation, and compute the changes in salt concentration after 20 weeks and in the long term.

(e) If you removed $p$ litres (where $p$ is an integer) at the end of each week, and added $p + 1$ litres of fresh water, what would be the concentration in the long term? What value of $p$ should you use in order to obtain only a 2% increase in the long-term salt concentration?

## Student project

**5.29** You encountered Pascal's triangle in Section 3.2.2. Rewrite Table 3.4 as follows:

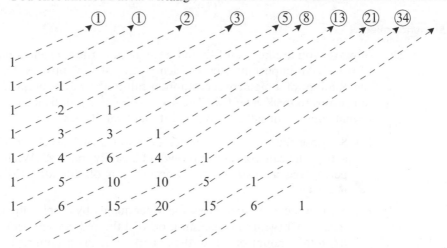

As stated in Property 6 in Section 3.2.2.1, if you add up the numbers along the diagonals marked with dashed lines, the totals shown within the circles are Fibonacci numbers. Verify that this is equivalent to, for example,

$$F_4 = \sum_{i=0}^{1} C(3 - i, i), \quad F_5 = \sum_{i=0}^{2} C(4 - i, i)$$

(recall that $C(k, 0) = 1$).

Prove by the method of induction that in general

$$F_{n+1} = \sum_{i=0}^{p} C(n - i, i)$$

where $p = (1/2)n$ when $n$ is even, and $p = (1/2)(n - 1)$ when $n$ is odd.

## Further reading

Section 5.1    Noble B. (1970). *Numerical Methods: 1, Iteration, Programming and Algebraic Equations*. Edinburgh: Oliver and Boyd

Sections 5.2,    Sandefur J.T. (1990). *Discrete Dynamical Systems*. Oxford: Oxford University Press
5.3.1

Section 5.3.2    Gardner M (1961). 'Phi: The golden ratio.' In *The 2nd Scientific American Book of Mathematical Puzzles and Diversions*. New York: Simon and Schuster, p. 89

Gardner M. (1981). 'Fibonacci and Lucas numbers.' In *Mathematical Circus*. New York: Vintage, p. 152

Garland T.H. (1987). *Fascinating Fibonaccis: Mystery and Magic in Numbers*. Palo Alto CA: Dale Seymour

Hoggatt V.E. Jr. (1969). *Fibonacci and Lucas Numbers*. Boston: Houghton Mifflin

Huntley H.E. (1970). *The Divine Proportion: A Study in Mathematical Beauty*. New York: Dover

March L. and Steadman P. (1974). *The Geometry of the Environment*. London: Methuen, Chapter 9

Rényi A. (1984). *A Diary of Information Theory*. Chichester: Wiley

Schroeder M.R. (1986). *Number Theory in Science and Communication*, 2nd enlarged edn. Berlin: Springer-Verlag, p. 65

Stewart I. (1996). 'Tales of a neglected number.' *Scientific American*, **274**(6), 92–3

Vajda S. (1989). *Fibonacci and Lucas Numbers, and the Golden Section*. Chichester: Ellis Horwood, p. 85

Section 5.4    Barnett S. (1990). *Matrices: Methods and Applications*. Oxford: Oxford University Press

Barnett S. (1995). *Some Modern Applications of Mathematics*. London: Ellis Horwood/Prentice Hall, Chapter 1

# 6 Graphs and Networks

One of the pleasures of holidaying in Switzerland is the extensive and efficient railway system, a small part of which is shown in Figure 6.1 for a south-eastern area of the country. You can use this diagram to plan trips between various stations, but you should be aware that Figure 6.1 is not a map. This would be something like that shown in Figure 6.2. You can see that there are important differences between the two representations:

(1)   The actual rail lines connecting stations are not straight.
(2)   Figure 6.1 is not to scale – that is, the lengths of the lines joining stations in the figure are not proportional to the actual distances, unlike the map in Figure 6.2.
(3)   An alternative version of Figure 6.1 is shown in Figure 6.3, from which it is apparent that the relative positions of the stations need not be geographically accurate.

All three figures are examples of what is called a **graph**, which is a set of points and a collection of lines joining some or all of the points in pairs. It is usual, but

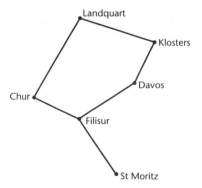

*Figure 6.1*   Swiss rail lines

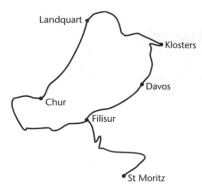

*Figure 6.2*   Swiss rail map

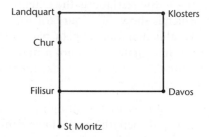

*Figure 6.3*   Alternative form of Figure 6.1

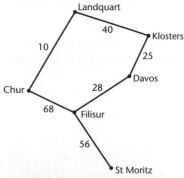

*Figure 6.4*   Swiss rail travel times

not essential, to draw straight lines. This usage of the word 'graph' is of course completely different from what you are familiar with as the graph of a function, perhaps displayed on your graphics calculator.

In everyday language most people would call Figures 6.1 to 6.3 'rail networks', but it's often preferable to reserve **network** as a technical term for a graph in which numerical values, or **weights**, are attached to each line. For example, in Figure 6.4 the times to travel between the stations are shown (in

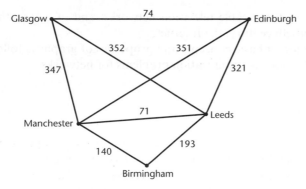

*Figure 6.5* Road network

minutes); Figure 6.5 gives a road network connecting some British cities, with distances (in kilometres) marked on the lines.

In all the above cases a key visual property of a graph is that it reveals which pairs of points are connected to each other. This information can be presented in different ways, for example the distances and connections shown in Figure 6.5 are often laid out in road atlases in a tabular form, like that in Table 6.1.

The intersection between a column and a row in the array gives the distance between the appropriate cities, for example, Edinburgh to Leeds is 321 km. The two empty cells in Table 6.1 indicate that there are no direct road links in Figure 6.5 between Birmingham and Edinburgh or Glasgow (that is, without passing through other cities). The graphical representation in Figure 6.5 is often quicker and easier to use than Table 6.1.

Networks have a variety of applications to transportation problems, such as finding the shortest route by road between two cities; or how to travel between two stations on a complex system like the London Underground or Paris Metro. The international telephone network is the world's largest artificial construct, and routing calls between any two places in the cheapest possible way is a very important problem for the communications industry. Other indispensable features of developed countries include electricity supply grids,

*Table 6.1* Distances in Figure 6.5

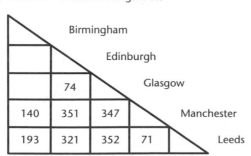

gas pipeline systems, cable television networks, and of course that wonder of the late twentieth century, the Internet.

A description of basic concepts and properties of graphs is followed by some algorithms for solving optimization problems for networks.

## 6.1 Definitions

Because of the many different configurations which graphs can possess, it's necessary to build up quite a large vocabulary of specialized terms. A **graph** $G$ consists of a finite set of points, called **vertices** (singular: 'vertex') or **nodes**, and a set of lines or **edges** joining pairs of vertices. It helps when discussing a graph to label each vertex, often with a lower-case letter as shown in Figure 6.6(a). An edge can then be described by the two vertices at its endpoints, for example in Figure 6.6(a) the edge $(a,b)$ joins vertex $a$ to $b$. The order of the letters does not matter, so the same edge can be denoted by $(b,a)$. An alternative notation is shown in Figure 6.6(b). If the context is clear then in this case vertices can be denoted simply by their suffices – for example, $(2,3)$ stands for the edge $(a_2,a_3)$. Using the graph $G$ in Figure 6.6, the set of vertices $V(G)$ and the set of edges $E(G)$ can be written as either

$$V(G) = \{a,b,c,d\}, \quad E(G) = \{(a,b),(a,d),(b,d),(b,c)\}$$

or

$$V(G) = \{a_1,a_2,a_3,a_4\}, \quad E(G) = \{(1,2),(1,4),(2,4),(2,3)\}$$

Two vertices are called **adjacent** if there is an edge joining them. This edge is said to be **incident** with each of the vertices. The **degree** $d(v)$ of a vertex $v$ is the number of edges incident with it. For example, in Figure 6.6(b) $d(a_1) = d(a_4) = 2$, $d(a_2) = 3$, $d(a_3) = 1$. If there is more than one edge joining two vertices the graph is called a **multigraph**.

If each edge has a direction associated with it (for example, a map of one-way streets) then the edge is called a **directed edge**, and the graph is called a **directed graph**, or **digraph** for short. A graph without directions on its edges can be called **undirected**.

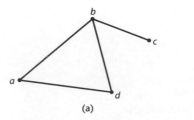
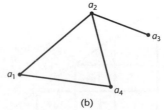

(a)  (b)

*Figure 6.6*  Vertices and edges

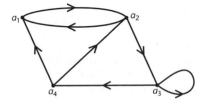

*Figure 6.7* Directed graph

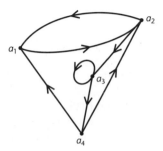

*Figure 6.8* Equivalent version of Figure 6.7

**Example 6.1** **A digraph**

In Figure 6.7 the direction on each edge is indicated by an arrow. The order of the vertices is now crucial when describing an edge, so that in this case the set of edges is

$$\{(1,2), (2,1), (2,3), (3,4), (4,1), (4,2), (3,3)\}$$

Notice that there are two edges joining vertices $a_1$ and $a_2$, and that there is a **loop** (3,3), which is an edge whose initial and final vertices are the same. The **indegree** of a vertex $v$ is the number of edges directed towards $v$, and the **outdegree** is the number of edges directed away from $v$ (loops are ignored). For example, in Figure 6.7 the indegree of $a_4$ is 1 and the outdegree is 2.

The actual physical layout of the vertices and edges of a graph is usually unimportant. For example, the digraph in Figure 6.8 is the same as that in Figure 6.7, since it has the same set of vertices and edges. Two graphs are called **isomorphic** if their sets of edges and vertices are identical (apart from possible relabelling).

Some important definitions for undirected graphs are listed in Table 6.2.

The **length** of a walk (or path) is the number of edges it contains.

According to the definitions in the table, during an open walk from an initial vertex to a final vertex, you proceed from one vertex to an adjacent one. You can visit any vertex more than once, and you can retrace your steps by going

*Table 6.2* Some definitions for undirected graphs

| Name | Description |
| --- | --- |
| walk (open) | A sequence of edges joining two vertices, with repetition of edges and vertices allowed. |
| closed walk | A walk which starts and finishes at the same vertex. |
| trail | A walk in which no edge is repeated. |
| circuit | A trail which starts and finishes at the same vertex. |
| path | A walk in which no edge or vertex is repeated. |
| cycle | A path which starts and finishes at the same vertex. |

from a vertex $x$ to an adjacent vertex $y$, and then back from $y$ to $x$. If your walk is a trail then you do not repeat any edge, but you may pass through a vertex more than once. On a path, however, you do not revisit any vertex, so you cannot repeat any edge. If your walk is a cycle, then you end up at the vertex you started from, without repeating any edges or vertices.

As an informal example, suppose you go on a driving holiday in France and wish to explore a number of towns. If you visit each town exactly once, and return to your starting point at the end of the holiday, then your tour will be a cycle. However, if you visit some towns more than once (even if only passing through) but don't repeat any road between towns, then your route will be a trail, and a circuit if you end up at your starting point.

**Example 6.2**   **Illustrations of the definitions**

Walks can be described either by listing the edges, or just the vertices.

(a)  In Figure 6.6(a) an open walk $a \longrightarrow b \longrightarrow c \longrightarrow b \longrightarrow d$ from $a$ to $d$ consists of the four edges $(a,b)$, $(b,c)$, $(c,b)$, $(b,d)$ and has length 4. Notice that to describe a walk it is necessary to distinguish between $(b,c)$, which indicates going from $b$ to $c$, and $(c,b)$ which goes in the opposite direction. This walk is not a trail since the edge joining $b$ and $c$ is repeated.

(b)  In Figure 6.9, the open walk $a_1 \longrightarrow a_2 \longrightarrow a_6 \longrightarrow a_3 \longrightarrow a_5 \longrightarrow a_6 \longrightarrow a_2$ from $a_1$ to $a_2$ consists of the six edges $(a_1, a_2)$, $(a_2, a_6)$, $(a_6, a_3)$, $(a_3, a_5)$, $(a_5, a_6)$, $(a_6, a_2)$ and has length 6. The edge $(a_2, a_6)$ is repeated, and the vertices $a_2$ and $a_6$ are repeated, so the walk is neither a trail nor a path.

(c)  The open walk in Figure 6.9 $a_2 \longrightarrow a_3 \longrightarrow a_6 \longrightarrow a_5 \longrightarrow a_3 \longrightarrow a_4$ from $a_2$ to $a_4$ has length 5. It is a trail since no edge is repeated, but it is not a path since the vertex $a_3$ is repeated.

(d)  The closed walk in Figure 6.9 of length 3 from $a_1$ to $a_1$ consisting of the edges $(1,2)$, $(2,6)$, $(6,1)$ is a cycle.

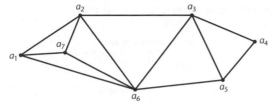

*Figure 6.9* Example of graph

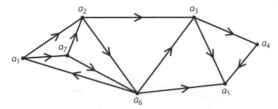

*Figure 6.10* Directed graph

(e) The circuit in Figure 6.9 from $a_2$ to $a_2$ consisting of the edges (2,6), (6,3), (3,4), (4,5), (5,3), (3,2) is not a cycle since the vertex $a_3$ is repeated.

(f) The closed walk $1 \longrightarrow 7 \longrightarrow 6 \longrightarrow 7 \longrightarrow 1$ in Figure 6.9 has length 4. It is neither a circuit nor a cycle, since the edges (1,7) and (7,6) are repeated, and the vertices $a_6$ and $a_7$ are repeated.

(g) The edges (1,7), (7,6), (6,1) in Figure 6.9 form a cycle which can start and finish at any of the vertices $a_1$, $a_6$ or $a_7$.

For a digraph, each of the terms in the table of definitions is preceded by the adjective 'directed'. For example, a **directed trail** is a sequence of directed edges joining two vertices, each edge having an arrow in the same direction, and no edge being repeated.

| Example 6.3 | **Directed graph** |
|---|---|

The directed graph in Figure 6.10 consists of the same edges and vertices as in Figure 6.9, but directions have now been added.

(a) In Figure 6.10 the edges (1,7), (7,2), (2,3), (3,4), (4,5) constitute a directed path of length 5 from $a_1$ to $a_5$.

(b) The directed path $a_1 \longrightarrow a_7 \longrightarrow a_6 \longrightarrow a_1$ is a directed cycle of length 3.

(c) There is no directed walk from $a_3$ to $a_6$.

Be warned that there is no general agreement in the literature over terminology. For example, some books call directed edges 'arcs' and undirected edges 'links'. This diversity can often be confusing, as is illustrated by Table 6.3, which unfortunately is by no means complete.

Table 6.3 Alternative definitions in use for undirected graphs

| This book | Alternatives in use |
| --- | --- |
| open walk | chain |
| closed walk | closed chain |
| path | simple chain, simple path |
| circuit | closed trail |
| cycle | circuit, closed path |
| directed open walk | path |
| directed closed walk | closed path |
| directed path | simple path |
| directed cycle | cycle |

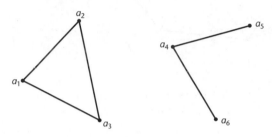

Figure 6.11   Disconnected graph

An undirected graph is called **connected** if there is a path between every pair of vertices. In a **disconnected** graph there are at least two vertices which are not linked by a path.

### Example 6.4   Disconnected graph

The graph in Figure 6.9 is connected. However, the graph $G$ in Figure 6.11 is disconnected, since there is no path from $a_1$ to $a_4$, for example. This graph consists of two pieces, called the **components** of $G$, formed by the two sets of vertices $\{1,2,3\}$ and $\{4,5,6\}$. In general an undirected graph is disconnected if and only if its vertices can be split up into two (or more) sets $V_1$ and $V_2$ such that if $a \in V_1$ and $b \in V_2$ there is no edge $(a,b)$.

Until the opening of the Channel Tunnel the British and European rail networks were disconnected. It is now possible to travel by rail between any two mainline stations in Britain and continental Europe.

If certain vertices and edges of a graph $G$ are deleted then this produces a **subgraph** of $G$. If some edges are deleted but all the vertices are retained then this gives a **spanning subgraph**. For example, the graph in Figure 6.12(a) is a subgraph of that in Figure 6.6(b), whereas the graph in Figure 6.12(b) is a spanning subgraph.

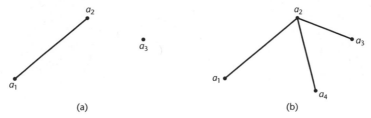

(a)　　　　　　　　　　　　　　(b)

*Figure 6.12*　Subgraphs of Figure 6.6(b)

It is often useful to describe a graph in non-pictorial form, for example to facilitate storage in a computer memory. For an undirected graph this can be done in two different ways. The first method is to define an **adjacency matrix** $A$, which is an array of numbers having one row and one column for each vertex. The entry (or **element**) in row $i$, column $j$ is 1 if vertex $i$ and vertex $j$ are adjacent, and 0 otherwise. Notice that $A$ has the same number of rows as columns, so is a **square** matrix.

**Example 6.5**　**Adjacency matrices**

(a)　The adjacency matrix for the graph in Figure 6.6(b) is as follows:

|  | $a_1$ | $a_2$ | $a_3$ | $a_4$ |
|---|---|---|---|---|
| $a_1$ | 0 | 1 | 0 | 1 |
| row 2 ⟶　$a_2$ | 1 | 0 | 1 | 1 |
| $a_3$ | 0 | 1 | 0 | 0 |
| $a_4$ | 1 | 1 | 0 | 0 |

For example, since $a_2$ is adjacent to $a_1$, $a_3$ and $a_4$ then row 2 has 1's in columns 1, 3 and 4. It is standard notation for matrices to enclose the elements within brackets, so that

$$A = \begin{bmatrix} 0 & 1 & 0 & 1 \\ 1 & 0 & 1 & 1 \\ 0 & 1 & 0 & 0 \\ 1 & 1 & 0 & 0 \end{bmatrix} \tag{6.1}$$

Notice, for example, that since $a_1$, $a_3$ and $a_4$ are adjacent to $a_2$ there are 1's in column 2, rows 1, 3 and 4. This symmetry holds in general for an undirected graph: the element in row $i$, column $j$ of $A$ is the same as the element in row $j$, column $i$. The matrix $A$ is said to be **symmetric**. Only the entries along and below the **principal diagonal** (which runs from the top left corner to the bottom right corner) need to be recorded in order to describe the graph. For example, the graph in Figure 6.6(b) can be described by the lower triangular part of $A$ in (6.1):

$$\begin{bmatrix} 0 & & & \\ 1 & 0 & & \\ 0 & 1 & 0 & \\ 1 & 1 & 0 & 0 \end{bmatrix}$$

The principal diagonal is indicated by the dashed line.

(b) The adjacency matrix for the graph in Figure 6.9 is

$$\begin{array}{ccccccc} a_1 & a_2 & a_3 & a_4 & a_5 & a_6 & a_7 \end{array}$$

$$\text{row 5} \longrightarrow \begin{bmatrix} 0 & 1 & 0 & 0 & 0 & 1 & 1 \\ 1 & 0 & 1 & 0 & 0 & 1 & 1 \\ 0 & 1 & 0 & 1 & 1 & 1 & 0 \\ 0 & 0 & 1 & 0 & 1 & 0 & 0 \\ 0 & 0 & 1 & 1 & 0 & 1 & 0 \\ 1 & 1 & 1 & 0 & 1 & 0 & 1 \\ 1 & 1 & 0 & 0 & 0 & 1 & 0 \end{bmatrix}$$

For example, the fifth row shows that there are edges $(a_5, a_3)$, $(a_5, a_4)$, $(a_5, a_6)$. The degree of vertex $a_5$ is 3, because of the three 1's in the fifth row.

A second way of characterizing a graph is to define an **incidence matrix** $B$ which has one row for each vertex, and one column for each edge, the edges being listed in some suitable order. If the $k$th edge is $(i,j)$ then the $k$th column of $B$ has 1 in its $i$th and $j$th rows, and zeros in the remaining positions in the column. If $(i,i)$ is an edge then the corresponding column is entirely zero.

**Example 6.6**    **Incidence matrix**

In Figure 6.6(b), list the edges as

$$\begin{array}{cccc} 1 & 2 & 3 & 4 \\ (1,2) & (2,3) & (2,4) & (1,4) \end{array}$$

Then the incidence matrix is

$$\begin{array}{c} \text{edges} \\ \begin{array}{cccc} 1 & 2 & 3 & 4 \end{array} \end{array}$$

$$B = \begin{bmatrix} 1 & 0 & 0 & 1 \\ 1 & 1 & 1 & 0 \\ 0 & 1 & 0 & 0 \\ 0 & 0 & 1 & 1 \end{bmatrix} \begin{array}{c} 1 \\ 2 \\ 3 \\ 4 \end{array} \text{ vertices}$$

$$\uparrow$$
$$\text{column 3}$$

For example, the third column of $B$ has 1's in the second and fourth rows, showing that the third edge in the list is (2,4). Again, the number of 1's in row $i$ of $B$ is equal to the degree of vertex $i$.

Notice that $B$ will only be a square matrix (that is, have the same number of rows as columns) if the number of edges of the graph is equal to the number of vertices.

For a directed graph, the adjacency matrix has 1 in row $i$, column $j$ if there is a directed edge $(i, j)$ from vertex $i$ to vertex $j$. Alternatively, if $(i,j)$ is labelled as the $k$th edge then column $k$ of the incidence matrix has 1 in row $i$ and $-1$ in row $j$.

**Example 6.8**    **Matrices for a digraph**

For the directed graph in Figure 6.7 the adjacency matrix is

$$A = \begin{array}{c} \\ \\ \\ \\ \end{array} \begin{array}{cccc} 1 & 2 & 3 & 4 \\ \end{array}$$
$$A = \begin{bmatrix} 0 & 1 & 0 & 0 \\ 1 & 0 & 1 & 0 \\ 0 & 0 & 1 & 1 \\ 1 & 1 & 0 & 0 \end{bmatrix} \begin{array}{c} 1 \\ 2 \\ 3 \\ 4 \end{array}$$

*Notice*: (i) $A$ is not symmetric; (ii) the edges (1,2) and (2,1) correspond to 1's in row 1, column 2 and row 2, column 1; (iii) the loop from $a_3$ to $a_3$ corresponds to 1 in row 3, column 3 of $A$.

To determine the incidence matrix, first list the directed edges in any convenient order:

| 1 | 2 | 3 | 4 | 5 | 6 | 7 |
|------|------|------|------|------|------|------|
| (1,2) | (2,1) | (2,3) | (3,3) | (3,4) | (4,2) | (4,1) |

The incidence matrix is

$$\text{edges}$$
$$B = \begin{array}{c} \\ \\ \\ \text{row 4} \rightarrow \end{array} \begin{array}{cccccccc} 1 & 2 & 3 & 4 & 5 & 6 & 7 \\ \end{array}$$
$$B = \begin{bmatrix} 1 & -1 & 0 & 0 & 0 & 0 & -1 \\ -1 & 1 & 1 & 0 & 0 & -1 & 0 \\ 0 & 0 & -1 & 0 & 1 & 0 & 0 \\ 0 & 0 & 0 & 0 & -1 & 1 & 1 \end{bmatrix} \begin{array}{c} 1 \\ 2 \\ 3 \\ 4 \end{array} \text{ vertices}$$
$$\phantom{B = } \uparrow \qquad \uparrow$$
$$\text{column 4} \quad \text{column 6}$$

For example, the fourth column of $B$ shows there is a loop from $a_3$ to $a_3$, and the sixth column of $B$ shows there is an edge from $a_4$ to $a_2$. The outdegree of vertex $i$ is equal to the number of 1's in row $i$, and the indegree is equal to the number of $-1$'s. For example, row 4 of $B$ shows that $a_4$ has outdegree 2 and indegree 1.

Readers familiar with matrix multiplication are asked to develop applications of the adjacency matrix in Problem 6.30.

**Exercise** **6.1** To encourage people to enjoy the countryside, your local water company has built three small car parks $c_1$, $c_2$, $c_3$ near a reservoir and linked together by footpaths as shown in Figure 6.13. Each footpath is exactly 1 km long. List all the possible walks of length 2 km:

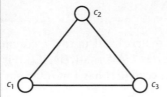

*Figure 6.13   Country park walks*

(a) starting and finishing at the same car park

(b) starting at one car park and finishing at another.

**Exercise** **6.2** Determine the degree of each vertex for the graph in Figure 6.9.

**Exercise** **6.3** Construct a graph with the following vertex and edge sets:

(a) $V(G_1) = \{a,b,c,d,e\}$,   $E(G_1) = \{(a,b),(a,e),(b,c),(b,d),(c,d),(c,e),(d,e)\}$

(b) $V(G_2) = \{a_1,a_2,a_3,a_4,a_5\}$,   $E(G_2) = \{(1,2),(1,3),(1,4),(1,5),(3,5)\}$.

**Exercise** **6.4** Part of the central section of the London Underground system is shown in Figure 6.14;

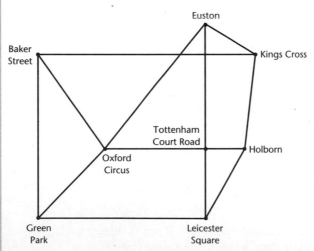

*Figure 6.14   Part of the London Underground system*

(a) list all the walks of length 4 starting and finishing at Oxford Circus

(b) list all the trails of length 5 from Baker Street to Green Park

(c) list all the different paths from Kings Cross to Green Park.

**Exercise**   **6.5**   Consider the Swiss rail system shown in Figures 6.1 to 6.3. Describe the following:

(a)  paths from Landquart to Davos

(b)  a circuit of length 7 starting and finishing at Landquart

(c)  a cycle starting and finishing at Landquart.

**Exercise**   **6.6**   Consider the road graph in Figure 6.5 (that is, ignore the marked distances). Give examples of the following:

(a)  paths of lengths 2, 3 from Leeds to Manchester

(b)  trails of lengths 4, 5, 6 from Leeds to Manchester

(c)  a circuit of length 6 from Manchester

(d)  the longest cycle from Edinburgh.

**Exercise**   **6.7**   For the graph in Figure 6.9, give examples of the following:

(a) three different walks of length 3 from $a_2$ to $a_3$

(b) a trail of length 5 from $a_2$ to $a_6$

(c) a cycle of length 6 from $a_5$

(d) a circuit of length 7 from $a_1$.

**Exercise**   **6.8**   For the digraph in Figure 6.10, give examples of the following:

(a)  three different directed paths of length 4 from $a_1$ to $a_5$

(b)  two directed cycles of length 3 from $a_1$.

**Exercise**   **6.9**   A **friendship graph** can be used to describe which people in a group are friendly with each other. Each vertex represents a person, and two people are friends if their vertices are adjacent. Represent the following table involving five people, where ✓ denotes a friendship, by an appropriate graph.

|          | Betty | Charles | Diana | Ernie |
|----------|-------|---------|-------|-------|
| Andrew   | ✓     | ✓       | ✓     | ✓     |
| Betty    | –     | –       | ✓     | ✓     |

**Exercise**   **6.10**   Determine which of the graphs in Figure 6.15 is a subgraph of the graph in Figure 6.9.

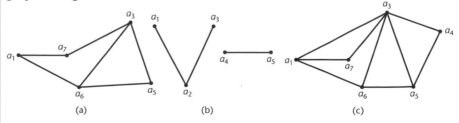

*Figure 6.15*   Graphs for Exercise 6.10

**Exercise** | **6.11** Determine whether the pairs of graphs in Figure 6.16 are isomorphic.

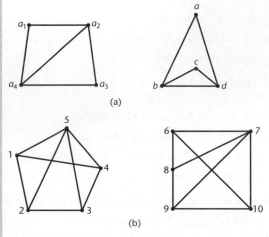

(a)

(b)

*Figure 6.16* Graphs for Exercise 6.11

**Exercise** | **6.12** You are planning a driving holiday in the American Southwest. A package you have purchased from a tour operator includes vouchers which can be used for overnight stays at motels in towns shown in Figure 6.17. The times in hours and minutes to cover each part of the road network are given in the figure. Determine the shortest driving time between each pair of towns, and represent this in tabular form.

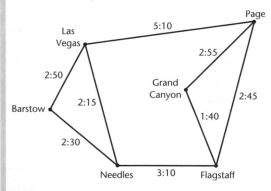

*Figure 6.17* US road network

**Exercise** | **6.13** In the Yorkshire Dales National Park, some footpaths connecting villages in Wharfedale with those in Littondale are represented in Figure 6.18. Suppose you and a friend are on a walking holiday, each with your own car, so that one-way walks are possible.

(a) You decide to do an open trail of length 2 from one village to another. Which pair of villages will give you the greatest number of different trails to choose from?

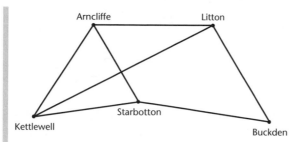

*Figure 6.18* Yorkshire Dales footpaths

(b) Later in the week, being fitter, you decide to do an open walk of length 3. Having got to know the area, you wish to start in Litton and finish in Kettlewell where there is a good choice of cafes and pubs. List all the possible walks you can choose from.

(c) After your friend goes home you decide to do a cycle of length 3. You rule out Starbotton as a place to begin because it is difficult to park there. Which would be the best village to start your cycle so as to give the greatest choice of different cycles? List all the possible cycles you could choose from in this case.

**Exercise**   **6.14** Determine the adjacency matrix and the incidence matrix for the graph in Figure 6.15(c).

**Exercise**   **6.15** Determine the adjacency matrix and the incidence matrix for the digraph in Figure 6.10.

**Exercise**   **6.16** Draw a graph corresponding to each of the following adjacency matrices $A$:

(a) $A$ has first row $[0\ 1\ 1\ 1\ 1]$, the other four rows consisting entirely of zeros

(b) lower triangle of $A = \begin{bmatrix} 1 & & & \\ 1 & 0 & & \\ 0 & 1 & 1 & \\ 0 & 1 & 0 & 1 \end{bmatrix}$, $A$ symmetric

(c) $A = \begin{bmatrix} 0 & 1 & 0 & 0 \\ 0 & 0 & 1 & 0 \\ 0 & 0 & 0 & 1 \\ 1 & 1 & 0 & 0 \end{bmatrix}$

**Exercise**   **6.17** Determine the incidence matrix for each of the graphs in the preceding exercise.

## 6.2 Classes of graphs

### 6.2.1 Bipartite graphs

**Example 6.9** **Assignment of teachers**

Suppose a school has four teachers (labelled $T_1$ to $T_4$) who can teach the subjects mathematics, physics, chemistry and biology according to the following table.

| Teacher | Can teach |
|---------|-----------|
| $T_1$ | mathematics and physics |
| $T_2$ | physics and chemistry |
| $T_3$ | mathematics and chemistry |
| $T_4$ | chemistry and biology |

It is useful to express this information in the form of a graph, as in Figure 6.19. Each teacher is represented by a vertex, which is connected by an edge to the subjects which can be taught. For example, $T_1$ is adjacent to the vertices representing mathematics and physics. The graph in Figure 6.19 is called **bipartite**, since there are two subsets of vertices (the teachers and the subjects) which have no edges joining vertices in the same subset (there are no edges joining teachers, and no edges joining subjects). An assignment problem is to allocate each teacher to an appropriate subject. In terms of the graph this means you must find a set of edges which join just one teacher with each subject – as shown, for example in Figure 6.20. This is an illustration of a **matching**, in which no two edges share a common endpoint. Here **every** teacher is 'matched' with a subject, so the matching is **complete**.

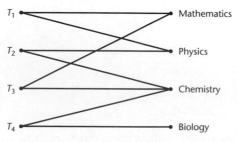

*Figure 6.19* Teacher's capabilities

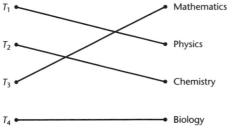

*Figure 6.20* A matching

In general, a graph is **bipartite** if its vertices can be divided into two sets $A$ and $B$ such that no vertices in the same set are adjacent. In other words, the only edges are between vertices in $A$ and vertices in $B$. A **complete** bipartite graph $K_{m,n}$ is one in which every one of the $m$ vertices in the set $A$ is adjacent to every one of the $n$ vertices in the set $B$. It can be shown that a graph is bipartite if and only if it does not contain an **odd cycle**, which is a cycle containing an odd number of vertices.

**Example 6.10**

**Bipartite graphs**

(a) The complete bipartite graphs $K_{2,2}$, $K_{2,3}$ and $K_{1,4}$ are shown in Figure 6.21.

(b) A simple procedure for labelling vertices which tests whether a graph is bipartite or not is as follows. Label any vertex $a_1$, and all vertices adjacent to $a_1$ by $b_1, b_2, b_3, \ldots$. Then label all vertices adjacent to these '$b$' vertices as '$a$' vertices. Continue in this way, labelling vertices adjacent to each newly labelled vertex with the opposite letter. If this procedure can be completed with no adjacent vertices having the same letter then the graph is bipartite; otherwise it is not bipartite. As an example, consider the graph in Figure 6.22. The left-most

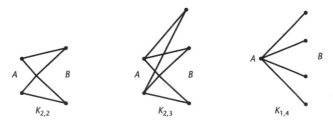

*Figure 6.21* Complete bipartite graphs

*Figure 6.22* Graph to be tested

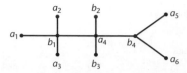

*Figure 6.23* Bipartite labelling

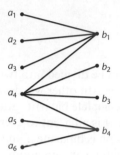

*Figure 6.24* Bipartite form of Figure 6.23

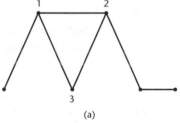 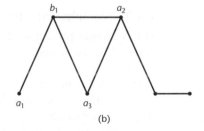

(a)     (b)

*Figure 6.25* Non-bipartite graph

vertex is labelled $a_1$, and labels are attached to vertices as shown in Figure 6.23, in the sequence $b_1$; $a_2, a_3, a_4$; $b_2, b_3, b_4$; $a_5, a_6$. The procedure has been successfully completed, so the graph is bipartite. An isomorphic version in the usual bipartite form can be drawn as in Figure 6.24.

(c) The graph in Figure 6.25(a) is not bipartite because it contains the odd cycle $1 \rightarrow 2 \rightarrow 3 \rightarrow 1$. Attempting the labelling procedure described in part (b) results in Figure 6.25(b). This shows that at the third stage of labelling there are two adjacent vertices $a_2$ and $a_3$, confirming that the graph is not bipartite.

**Exercise** | **6.18** Give another complete matching for the bipartite graph in Figure 6.19.

**Exercise**   **6.19**   Draw the complete bipartite graphs $K_{3,4}$ and $K_{5,2}$.

**Exercise**   **6.20**   Determine which of the graphs in Figure 6.26 are bipartite. Use the labelling procedure in part (b) of Example 6.10 to redraw the bipartite graphs in the usual form.

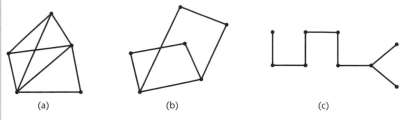

(a)                          (b)                          (c)

*Figure 6.26*   Graphs for Exercise 6.20

**Exercise**   **6.21**   Determine the number of edges in

(a) a complete bipartite graph $K_{m,n}$

(b) a **complete graph** $K_n$, which is defined as a graph with $n$ vertices and an edge joining every pair of vertices.

**Exercise**   **6.22**   Four teachers are available to take school parties on educational trips from Britain to France, Germany, Italy and Spain. The languages spoken by the teachers are as shown in the table:

| Teacher | Can speak |
|---|---|
| 1 | French, German |
| 2 | French, Spanish |
| 3 | Italian, German |
| 4 | Italian, Spanish |

Represent this information as a bipartite graph. It is required that each party is to be accompanied by a teacher who can speak the local language. Determine a complete matching.

**Exercise**   **6.23**   A car dealership chain has branches in five towns selling the makes shown in the table.

| Town | Branches sell |
|---|---|
| 1 | Ford, Nissan |
| 2 | Nissan, Vauxhall |
| 3 | Toyota |
| 4 | Honda, Vauxhall |
| 5 | Toyota |

To achieve an even spread of sales, the group wants to ensure that if one car of each make is purchased this can be done by using dealerships in five different towns. Show that this is not possible with the existing branches. Suggest where a new dealership should be opened so as to satisfy the requirement.

**Exercise**  **6.24**  A **maximum matching** for a bipartite graph is a matching which uses the largest possible number of edges. Show that there is no complete matching for the graph in Figure 6.27, and determine a maximum matching.

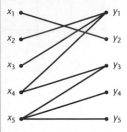

*Figure 6.27*  Graph for Exercise 6.24

**Exercise**  **6.25**  Figure 6.28 represents a cube in three dimensions. Show that the edges form a bipartite graph, and redraw it in the usual form.

*Figure 6.28*  Cube

### 6.2.2  Trees

A 'family tree' is a graph which represents the relationships between generations – for example, Figure 6.29 would be your family tree if your parents were Gwen and Harry, and your grandparents were Anne, Bill, Charles, Diana. There must be a path from any one family member (represented by a vertex) to any other – recall from Section 6.1 that this requires that the graph is **connected**. Further examples of family trees were given in Chapter 5 for Fibonacci's rabbits (Figure 5.3) and for bees (Figure 5.9). In general, a **tree** is defined to be a connected, loop-free, undirected graph which does not contain a cycle as a subgraph.

**Example 6.11**  **United Kingdom postcodes**

Trees provide useful representations of sorting procedures. In the United Kingdom mail is sorted according to 120 postcode areas, each denoted by two

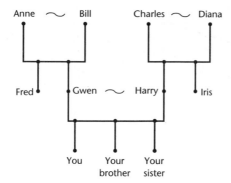

*Figure 6.29*  Family tree

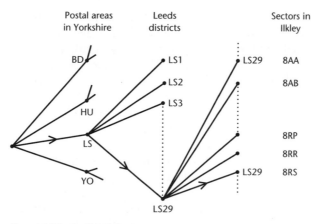

*Figure 6.30*  Sorting letters

letters – for example, BD stands for Bradford, HU for Hull, LS for Leeds and YO for York. This is followed by one or two digits denoting the postcode district, for example LS1 is part of central Leeds, LS24 is Tadcaster and LS29 is Ilkley. The remaining part of the postcode consists of one digit (0 to 9) which denotes the sector within the district, and two more letters which pinpoint the address to around 15 letterboxes. The graph in Figure 6.30 is a tree which shows how a letter addressed to a destination with postcode LS29 8RS can be sorted in stages.

Notice that since a tree contains no cycles then it certainly contains no odd cycles, and so by a result in the previous section it follows that every tree is a bipartite graph. The graph in Figures 6.22 and 6.23 is an example of a tree whose bipartite form is shown in Figure 6.24. It also follows directly from the

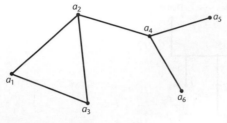

*Figure 6.31* Bridge

*Figure 6.32* Tree with one edge

definition of a tree that every pair of vertices is linked by exactly one path (see Problem 6.9). Hence the removal of any edge produces a graph which is disconnected, and so is not a tree. In general, for any graph whether connected or not, if removal of an edge increases the number of components of the graph, then this edge is called a **bridge**. For example, the edge $(a_2, a_4)$ in Figure 6.31 is a bridge, since removing this edge from the graph produces the disconnected graph with two components in Figure 6.11 (of course, the number of vertices is not altered). A consequence of this definition is that every edge of a tree is a bridge. A vertex having degree one in a tree is called an **end**. It can be shown that every tree (excluding the trivial tree consisting of a single vertex) has at least two ends (see Problem 6.11).

An important property of a tree is that if it has $n$ vertices then there are $n - 1$ edges. This can be proved by the method of induction as follows (again refer to the Interlude for details of the method). The result is certainly true for $n = 2$, as is obvious for the tree with one edge in Figure 6.32, so Step 1 of the induction procedure is established. For Step 2, assume the result is true for $n(>2)$. It is then required to show that the result is true for $n + 1$, that is to show that a tree $T$ with $n + 1$ vertices has $n$ edges. Let $v$ be an end of $T$, and denote by $T_1$ the graph obtained by removing from $T$ the vertex $v$ and the single edge $e$ incident to $v$. Recall that by definition $T$ is connected, that is, there is a path joining every pair of vertices in $T$. However, since $v$ is an end only paths beginning or ending at $v$ would use the edge $e$, so it follows that $T_1$ is also connected. Furthermore, since $T$ contains no cycles neither does $T_1$. Therefore $T_1$ is by definition also a tree, and has $n$ vertices. By the induction assumption $T_1$ therefore has $n - 1$ edges, so adding back the edge $e$ shows that $T$ has $n$ edges. This completes Step 2 of the induction procedure, and it follows from Step 3 that the result is true for *all* values of $n$.

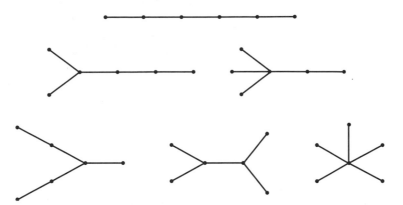

*Figure 6.33* Trees with six vertices

**Example 6.12**  **Trees with six vertices**

The trees shown in Figure 6.33 constitute the complete set of trees having six vertices, in the sense that any other tree with six vertices is isomorphic to one of the examples shown. Notice that each tree in Figure 6.33 (i) has five edges, (ii) has a path between any two vertices, (iii) becomes disconnected if any one edge is removed.

Trees are very useful in counting all the possible ways in which a finite sequence of events can occur (see Chapter 3); in searching problems where a particular item has to be located; and in sorting problems, as illustrated by the postcode in Example 6.11. To develop these applications it's necessary to introduce some more definitions. A **rooted tree** is a tree having a designated vertex $r$ called the **root** of the tree. Since there is a unique path *from r to* every other vertex $v$ of the tree, the selection of a root can be regarded as giving a direction to the edges of the tree. For example, the graph in Figure 6.30 is a rooted tree, with the left-most vertex being the root. The length of the path from $r$ to $v$ (that is, the number of edges used in going from $r$ to $v$) is called the **level** of the vertex, and the **height** of the rooted tree is the greatest level of any vertex. It is customary to draw a tree 'upside down' with the root at the top, as shown in Figure 6.34(a), but the alternative in Figure 6.34(b) is allowable. Any vertex of degree one (apart from the root) of a rooted tree is called a **leaf** – for example there are four leaves of the rooted tree in Figure 6.34(a). The other vertices (including the root) are called **internal**. Sometimes the edges are called **branches**. This arboreal terminology is used because if the root is drawn at the bottom, the graph may look something like a real tree. However, in addition, the language of the 'family tree' is also adopted. Thus, if a vertex $v$ has level $k$, then a vertex

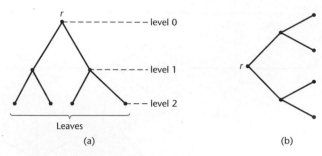

Figure 6.34  Rooted tree of height 2

adjacent to $v$ at level $k - 1$ is called a **parent** of $v$, and a vertex adjacent to $v$ at level $k + 1$ is called a **child** of $v$. In particular, the root is the only vertex having no parent, and leaves have a parent but no children. Two vertices with a common parent are called **siblings**. If $v_1$ and $v_2$ are two vertices and $v_1$ has the smaller level number then $v_1$ is an **ancestor** of $v_2$, and $v_2$ is a **descendant** of $v_1$. The subgraph consisting of all the descendants of any vertex $v$ is the **subtree** having $v$ as its root.

**Example 6.13**  **Illustration of the definitions**

In Figure 6.35 the root is $r$, and the levels of the vertices are as shown in the table

| Level | Vertices |
|-------|----------|
| 1 | $v_1, v_9$ |
| 2 | $v_2, v_3, v_{10}, v_{11}$ |
| 3 | $v_4, v_5, v_7, v_8, v_{12}, v_{13}$ |
| 4 | $v_6, v_{14}, v_{15}$ |

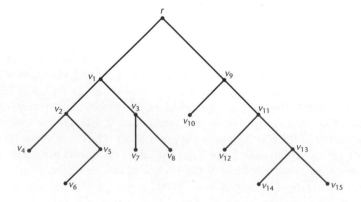

Figure 6.35  Rooted tree of height 4

Since the greatest level is 4, the rooted tree has height 4. The vertices $v_4$, $v_6$, $v_7$, $v_8$, $v_{10}$, $v_{12}$, $v_{14}$ and $v_{15}$ are leaves. The vertex $v_1$ is the parent of the siblings $v_2$ and $v_3$; and $v_2$ and $v_3$ are children of $v_1$. The vertex $v_{11}$ is an ancestor of $v_{12}$, $v_{13}$, $v_{14}$ and $v_{15}$; alternatively, $v_{14}$ and $v_{15}$ are descendants of $v_9$. The subtree obtained by regarding $v_1$ as a root consists of the vertices $v_1$ to $v_8$ inclusive.

A special kind of tree which has many applications is where each vertex has at most two children. Specifically, a **binary rooted tree** is a rooted tree whose root has degree at most 2, and the degree of all other vertices is at most 3. If every internal vertex has exactly two children then the binary tree is said to be **regular**. For example, the graph in Figure 6.34 is a regular binary rooted tree, whereas the graph in Figure 6.35 is a binary rooted tree but is not regular because $v_5$ has only one child. Notice that even for a regular rooted tree, not all the leaves need be at the greatest level. If a binary tree has three or more vertices, then deletion of the root and its incident edges produces two smaller binary trees, called (when the root is drawn at the top) **left** and **right subtrees**. For example, in Figure 6.35 if the root $r$ and its two incident edges are removed then the left and right subtrees have roots $v_1$ and $v_9$ respectively.

**Example 6.14**  **Applications of binary rooted trees**

(a) Suppose two players Anne and Barry compete in a chess match. The rules are that the winner is the player who first wins two consecutive games, or who wins a total of three games. The complete set of possible outcomes can be nicely represented by the binary rooted tree in Figure 6.36. Each edge represents a game, and $A$ is attached to a vertex if the game is won by Anne, and $B$ denotes a game won by Barry. There are ten leaves, representing the ten different ways in which the match can end up. A directed path from the start to a leaf gives the sequence of games which results in the match being won by the player attached

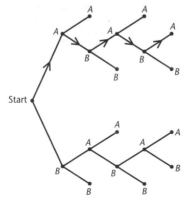

*Figure 6.36*  Outcomes of chess match

to the leaf. For example, Anne wins when the path indicated by arrows in Figure 6.36 is followed, producing the sequence $ABABA$. Notice that a regular binary tree occurs because each game has *two* possible outcomes.

(b) Now consider a tennis tournament involving several players. A player is eliminated after losing a match, and the winner of the final match is the overall champion. The situation when there are 16 players is represented by the regular binary tree in Figure 6.37. Each *circular* vertex represents a match between two players. Because the tournament ends with the final match, the edges can be thought of as directed *towards* the root. The first round (at level three in the tree) consists of eight matches between 16 players. The eight winners proceed to the second round (at level two). The four winners of these matches go into the semi-finals (at level 1) and the final match determines the overall winner. The total number of matches played is $8 + 4 + 2 + 1 = 15$. Including the final there are four rounds and the tree has height 3. Similarly, if the tournament is between 32 players then there will be an extra round of 16 matches at level four in the tree, and the total number of matches will be $16 + 15 = 31$. It is not difficult to prove that if there are $2^N$ players, where $N$ is a positive integer, then there will be $N$ rounds in the tournament and a total of $2^N - 1$ matches (see Problem 6.13). In fact there is an easy and direct way of counting the total number of matches, whatever the number $m$ of players: this relies on the flash of insight that every player loses exactly one match and is then out of the competition, *except* the overall winner who does not lose any match. Hence the total number of matches is $m - 1$.

If the number of players is not equal to a power of 2 then the tournament can still be represented by a regular binary tree with some modifications. For example, when there are nine competitors the tournament is described by the binary tree in Figure 6.38(a). As before, each circular vertex represents a match between two players, but the *square* vertex at level three represents a 'match' involving only one player, that is, this player receives a bye into the second round. The three leaves at level two in the tree represent matches between six other players who also do not compete until the second round. Again, the total number of matches in the tournament is equal to the total number of circular vertices in Figure 6.38(a), namely eight, which agrees with the earlier argument since $8 = 9 - 1$. The initial number of players is equal to twice the number of

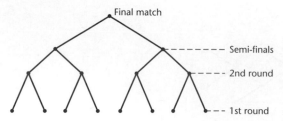

*Figure 6.37*   Sixteen person tennis tournament

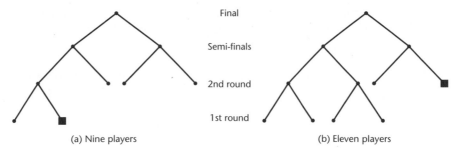

Final

Semi-finals

2nd round

1st round

(a) Nine players                    (b) Eleven players

*Figure 6.38*   More tennis tournaments

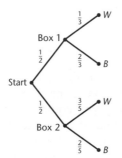

*Figure 6.39*   Choosing a ball

circular leaves plus the player corresponding to the square leaf (in Figure 6.38(a) there are four circular leaves and $2 \times 4 + 1 = 9$). Similarly, the binary tree for 11 players and 10 matches is shown in Figure 6.38(b). There are five circular leaves, ten circular vertices and one square leaf, and $2 \times 5 + 1 = 11$. In general a square leaf is needed in the tree only if the number of competitors is odd.

(c) Suppose there are two boxes, the first containing one white ball and two black balls, and the second containing three white balls and two black balls. You select a box at random, and then take one ball out of it (also at random). All the possible outcomes can be represented by the regular binary tree in Figure 6.39. The numbers on the edges represent probabilities. Beginning at the root, since there are two boxes the probability of selecting either box is 1/2, and these values are attached to the first two edges. If box 1 containing three balls is selected, the probability of selecting a white ball (*W*) is 1/3, and 2/3 for a black ball (*B*). Similarly, box 2 contains five balls and the probabilities of selecting a white or black ball are 3/5 and 2/5 respectively. From the multiplication rule for probability (see Section 3.5.1) the probability of selecting box 1 *and* a white ball is $(1/2) \times (1/3) = (1/6)$. Similarly, the probability of selecting box 2 and a white ball is $(1/2) \times (3/5) = (3/10)$. Hence by the addition rule for probability (again see

Section 3.5.1) the overall probability of obtaining a white ball is $(1/6) + (3/10) = (7/15)$. Similarly, the probability of obtaining a black ball is

$$\tfrac{1}{2} \times \tfrac{2}{3} + \tfrac{1}{2} \times \tfrac{2}{5} = \tfrac{8}{15}$$

Of course the sum of the probabilities is 1, since you are certain to get either a white or black ball. A tree like that in Figure 6.39 where values (or **weights**) are attached to the edges, is called a **weighted tree**.

More generally, a rooted tree is called **ternary** if every vertex has at most three children, and $m$-**ary** if every vertex has at most $m$ children. The qualifier **regular** is applied if every internal vertex has exactly three, or $m$, children respectively.

**Example 6.15**    **Binary search tree**

Suppose I have chosen one number $i$ from the set of eight numbers $\{0,1,2,\dots,7\}$. You have to determine $i$ by asking me questions to which I answer either 'yes' ($Y$) or 'no' ($N$). Obviously you could ask the sequence of questions: 'Is $i$ equal to 1? Is $i$ equal to 2?' and so on, up to 'Is $i$ equal to 7?' However, this would require as many as eight questions, if I have chosen $i = 7$.

The most efficient way of searching for $i$ can be represented by the rooted tree in Figure 6.40. For example, '$i > 5$?' represents the question 'is $i$ greater than 5?' Since at each stage there are two possible answers to each question, the tree is binary. The value of $i$ is obtained at level three, showing that only three questions are required (the height of the search tree is three). For example, if I selected $i = 5$, then starting at the root the answers will be 'yes', 'no' and 'yes', and the path to $i = 5$ is indicated by the arrows in Figure 6.40.

You may well recognize the problem as the same as the one posed in Exercise 4.2. There each of the numbers 0 to 7 was represented by a 3-bit binary number from $(000)_2$ to $(111)_2$, and the questions asked were whether a digit is 1 or 0.

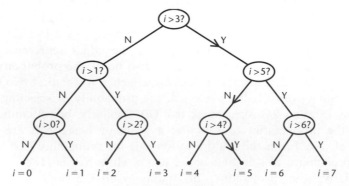

*Figure 6.40*   Binary search tree for decimal numbers

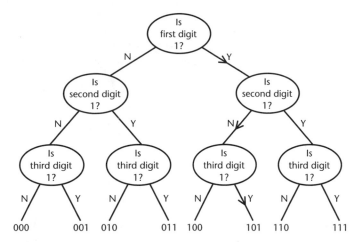

*Figure 6.41* Binary search tree for binary numbers

The search tree in this case is shown in Figure 6.41. The correspondence between the two trees is apparent – for example, the arrowed routes to $i = 5 = (101)_2$ are identical.

Binary search trees are useful in computer science. In general, at each stage half the remaining elements are eliminated from the search until you isolate the element you are looking for.

**Example 6.16** **Binary decision tree: weighing problem**

Suppose you have four coins which are visually identical. Three of the coins are perfect but the fourth is counterfeit, and weighs more. You wish to identify the fake coin using a pair of scales to compare sets of coins. Label the coins 1,2,3,4 and put coins 1,2 in the left pan and coins 3,4 in the right, that is, compare the weight of the set {1,2} against that of the set {3,4}, as indicated in Figure 6.42.

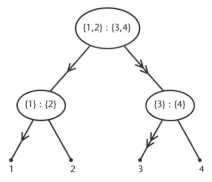

*Figure 6.42* Binary decision tree

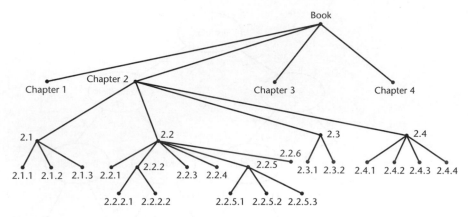

*Figure 6.43*  Ordered tree for Chapter 2

If the left pan tips downwards then follow the left path in Figure 6.42 indicated by the arrow. Next, put coin 1 in the left pan and coin 2 in the right. Again, if the left pan tips down then follow the arrowed left edge in the tree to vertex 1, showing coin 1 is counterfeit; similarly, if the right pan tips down then coin 2 is counterfeit. In the same way, the directed path if coin 3 is counterfeit is indicated by double arrows in Figure 6.42. At each successive level the number of coins being tested is halved, so in this example the counterfeit coin is found at level two, after two weighings.

**Example 6.17**   **Ordered rooted tree**

The section numbers for Chapter 2 of this book can be represented by the rooted tree shown in Figure 6.43. This is an example of an **ordered tree** since the edges leaving each internal vertex are labelled in order from left to right. In general for an ordered tree, a vertex at level $n$ has a label $a_1 . a_2 . a_3 . \ldots . a_n$, where the $a_i$ are positive integers. If a vertex $v$ at level $n$ has a label $a$, then the children of $v$ have labels $a.1, a.2, a.3, \ldots$, going from left to right. For example, in Figure 6.43 the children of the level three vertex 2.2.5 are 2.2.5.1, 2.2.5.2, 2.2.5.3. This ordering of the chapters and sections (as listed on the Contents pages of this book) is called **lexicographic**, since it resembles the alphabetical ordering in a dictionary (or 'lexicon').

A related application of rooted trees to coding is given in Problem 6.20.

**Exercise**   **6.26**   For the UK postcode system described in Example 6.11, what is the maximum possible number of different postcodes in any given postal area?

**Exercise** **6.27** If each of the *m* components of a graph *G* is a tree, then *G* is called a **forest**. Show that if *G* has *n* vertices then it has $n - m$ edges.

Hence prove that if any graph has *n* vertices, $n - 1$ edges and contains no cycles then it is a tree.

**Exercise** **6.28** Give further examples of parents, children, siblings, ancestors and descendants in Figure 6.35.

**Exercise** **6.29** List the remaining nine outcomes of the chess match in Example 6.14(a).

**Exercise** **6.30** Draw:

(a) all binary trees with five or fewer vertices

(b) all regular binary trees with two, three or four leaves.

**Exercise** **6.31** For the knockout tennis tournament in part (b) of Example 6.14, draw the binary trees when there are (a) 10, (b) 13, (c) 15, (d) 19, (e) 22 competitors.

**Exercise** **6.32** Two players compete against each other in a tennis tournament. The overall winner is the player who either wins three games in a row, or wins a total of four games. Draw a rooted tree to represent all the possible outcomes.

**Exercise** **6.33** Three players compete in a succession of three-handed card games. The overall winner is the first player to win a total of two games. Represent all the possible outcomes by a ternary rooted tree.

**Exercise** **6.34** Three boxes contain balls as shown in the table.

|       | White balls | Black balls |
|-------|-------------|-------------|
| Box 1 | 1           | 2           |
| Box 2 | 2           | 3           |
| Box 3 | 3           | 1           |

You select a box at random, and take one ball out of it. Draw a weighted tree showing all possible outcomes. Is it more likely you will get a white ball or a black ball?

**Exercise** **6.35** A frog hops along a straight line, which can be regarded as the *x*-axis. It begins at $x = 0$ and each jump has unit length either to the right or to the left. The frog stops either when it has made a total of four jumps, or when it reaches $x = 2$ or $x = -3$. Draw a rooted tree which represents all the journeys the frog can make.

**Exercise** **6.36** Draw a binary search tree to determine a selected number from the set $\{1,2,3,\ldots,15,16\}$.

**Exercise** **6.37** A laboratory contains 25 computer terminals which must be connected to a network using extension leads. The network input and the leads each have four connectors. Draw a 4-ary rooted tree which shows the minimum number of leads to be used so that all terminals are linked to the network.

**Exercise**    **6.38**   In a binary search for a number which has been selected from the set $\{1,2,3,4,\ldots,500\}$, what is the maximum number of questions which must be asked (that is, the height of the binary search tree)?

**Exercise**    **6.39**   Extend Example 6.16 by drawing a binary decision tree which identifies a single heavier counterfeit coin in a set of eight otherwise identical coins. How many weighings are required?

**Exercise**    **6.40**   For the problem in the previous exercise, suppose you begin by weighing coins $1, 2, 3$ against coins $6, 7, 8$. There are now three possibilities: the balance tips to the left, to the right, or stays level. Draw a ternary rooted tree which shows how to identify the counterfeit coin in only two weighings.

**Exercise**    **6.41**   Draw an ordered rooted tree which represents the contents list of Chapters 1 and 3 of this book.

**Exercise**    **6.42**   Label the vertices of the tree in Figure 6.35 using lexicographic ordering.

### 6.2.3   Planar graphs

Another application of graphs is to the analysis and design of electrical networks. Electronic devices use networks which are printed or etched onto circuit boards or silicon chips. Since these are plane (that is, flat) surfaces it is important to avoid crossing wires if possible. A simple electrical network is shown in Figure 6.44(a), where the 'hump' in the centre indicates that one wire crosses over another (the meanings of the other symbols are irrelevant here). The corresponding graph is shown in Figure 6.44(b), and this can be redrawn as in Figure 6.45(a) without any crossing edges; the equivalent electrical network is in Figure 6.45(b). The graph in Figure 6.44(b) is an example of a **planar graph**. This is defined as a graph which can be redrawn on a plane with the same set of vertices and edges but with no crossing edges; in other words, edges intersect only at vertices.

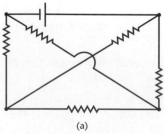
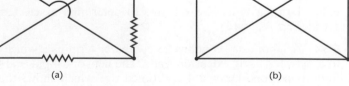

(a)                        (b)

*Figure 6.44*   Electrical network

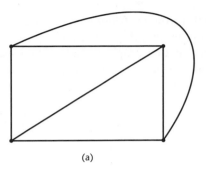 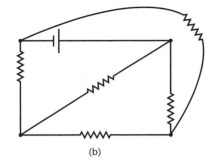

(a)                                              (b)

*Figure 6.45*  Plane version

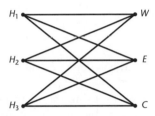

*Figure 6.46*  Three houses and three services

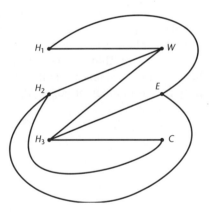

*Figure 6.47*  Unsuccessful connections

<table>
<tr><td>Example 6.18</td></tr>
</table>

**Example 6.18**  **Three houses and services**

This is an old puzzle: can the three houses $H_1$, $H_2$, $H_3$ in Figure 6.46 be connected with terminals of three services water ($W$), electricity ($E$) and cable television ($C$) in such a way that no pipes or cables cross? However hard you try you are doomed to end up with a situation like that shown in Figure 6.47, where one of the houses lacks one of the services (here $H_1$ is not connected to $C$).

You may recognize the graph in Figure 6.46 as the complete bipartite graph $K_{3,3}$. The puzzle has no solution because $K_{3,3}$ is *not* a planar graph. Before this can be proved some properties of planar graphs must be established.

If a planar graph is redrawn (if necessary) on a plane so that no two edges intersect except at vertices it is then called a **plane graph**. An example is given in Figure 6.45(a). A plane graph divides the plane on which it is drawn into regions called **faces**. A face is the region enclosed by a cycle which does not contain smaller cycles. For example, the faces $f_1$, $f_2$, $f_3$ for the graph $G_1$ in Figure 6.45(a) are shown in Figure 6.48. The region exterior to the graph is called the **infinite face**, labelled $f_4$ in Figure 6.48. Notice in particular that since a tree by definition does not contain any cycles, then any tree only produces the infinite face.

A formula due to **Euler** is the following: if a **connected** plane graph has $v$ vertices, $e$ edges and $f$ faces (including the infinite face) then

$$v - e + f = 2 \tag{6.2}$$

For example, in Figure 6.48 you can see that $v = 4$, $e = 6$ and $f = 4$, confirming that (6.2) is satisfied in this case.

Euler's formula (6.2) can now be proved by using the method of induction (described in the Interlude) where the variable is the number of edges. The result (6.2) is certainly true for $e = 1$, since in this case Figure 6.49 shows that $v = 2$ and $f = 1$ (the only face is the infinite face), so that $v - e + f = 2 - 1 + 1 = 2$. This establishes Step 1 of the induction procedure. For Step 2, assume that the formula (6.2) is true for $e$: it is required to prove it is true for $e + 1$. Let $G$ be a connected plane graph with $e + 1$ edges, $v$ vertices and $f$ faces, so it is necessary to prove that $v - (e + 1) + f = 2$, that is, $v - e + f = 3$. If $G$ is a tree, then $f = 1$, and from a result in Section 6.2.2 the number of vertices is one more than the number of edges, that is, $e + 2$, so in this case

$$v - e + f = e + 2 - e + 1 = 3$$

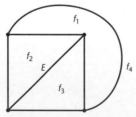

*Figure 6.48*   Faces of a plane graph $G_1$

$f_1$

*Figure 6.49*   Graph with one edge

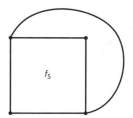

*Figure 6.50*  The graph $G_1 - E$

as required. If $G$ is not a tree, then since it is connected it contains cycles. Let $E$ be any edge lying on a cycle of $G$, and consider the graph $G - E$ which is defined as $G$ with the edge $E$ deleted (but leaving in place the endpoints of $E$). For example, if $G_1$ is the graph in Figure 6.48 and $E$ is the diagonal shown, then $G_1 - E$ is the graph in Figure 6.50; the regions $f_2$ and $f_3$ in Figure 6.48 reduce to the single region $f_5$ in Figure 6.50, so $G_1 - E$ has one fewer face than $G_1$. The same thing happens in general, that is, $G - E$ has $f - 1$ faces. Because $E$ is on a cycle, $G - E$ is still connected. However, $G - E$ has $v$ vertices but only $e$ edges (since one has been removed) so by the induction assumption applied to $G - E$ it follows that

$$v - e + (f - 1) = 2, \qquad \text{that is, } v - e + f = 3$$

again as required. This completes Step 2 of the induction procedure, so by Step 3 it follows that (6.2) is true for *all* values of $e$.

**Example 6.19**  **$K_{3,3}$ is not planar**

Euler's formula can now be used to prove that $K_{3,3}$ is not planar, thus showing that the puzzle in Example 6.18 cannot be solved. The proof is by contradiction: suppose $K_{3,3}$ in Figure 6.46 *is* planar, then it can be redrawn as a plane graph $P$ with six vertices and nine edges. Hence from (6.2) the number of faces of $P$ is

$$f = 2 - v + e = 2 - 6 + 9 = 5$$

Because $K_{3,3}$ is bipartite so is $P$. Therefore, because there are no edges joining vertices within each of the two vertex subsets of $P$, it follows that no face of $P$ can have three edges. Hence each of the five faces of $P$ is enclosed by at least *four* edges. Therefore, if $N$ is defined as the total sum of the number of edges on the boundary of each face of $P$, it follows that $N \geqslant 4 \times 5 = 20$. However, none of the nine edges of $P$ can be the boundary of more than *two* faces, so counting the edges this way shows that $N \leqslant 9 \times 2 = 18$. This is the required contradiction. The only conclusion which can be drawn, therefore, is that the assumption that $K_{3,3}$ is planar is *wrong*. This proves that $K_{3,3}$ is *not* planar.

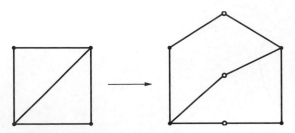

*Figure 6.51* Subdivision of a graph

An obvious question which presents itself is how to determine whether or not a given graph is planar. The solution to this problem requires two further ideas. The first is that of the **complete graph** $K_n$, defined in Exercise 6.21(b) as a graph with $n$ vertices and an edge joining every possible pair of vertices. For example, $K_4$ is the graph in Figure 6.44(b) and is planar, as shown by Figure 6.45(a). However, $K_5$ is *not* planar (see Exercise 6.46).

The second idea is as follows. Let $G$ be a loop-free undirected graph, and remove an edge $(v,u)$ and replace it by two new edges $(v,w)$ and $(w,u)$ where $w$ is not a vertex of $G$. The new graph obtained by adding one or more new vertices, each of degree two, in this way is called a **subdivision** of $G$. An example of a subdivision is shown in Figure 6.51, where the new vertices are indicated by open circles. Notice that if $G$ is planar, so is every subdivision of $G$; and if $G$ is non-planar, so is every subdivision.

It's now possible to state the following theorem on planarity (proved by a Polish mathematician **Kuratowski** in 1930):

A graph $G$ is non-planar if and only if $G$ contains $K_{3,3}$ or $K_5$, or a subdivision of $K_{3,3}$ or $K_5$, as a subgraph.

The proof is too complicated for this book, but it's interesting that the theorem reveals the fundamental role of the archetypal non-planar graphs $K_{3,3}$ and $K_5$. Unfortunately, Kuratowski's theorem does not provide the basis for a simple test for non-planarity.

**Exercise**　**6.43**　Redesign the electrical network shown in Figure 6.52 so that it has no crossing edges.

**Exercise**　**6.44**　Redraw the graph in Figure 6.53 as a plane graph.

**Exercise**　**6.45**　For each of the graphs in Figure 6.54, verify that Euler's formula (6.2) holds.

**Exercise**　**6.46**　Draw the complete graph $K_5$ and attempt to redraw it as a plane graph.

In fact it can be shown that for a loop-free connected planar graph with $e$ edges and $v$ vertices then $3v - 6 \geqslant e$. Use this to prove that $K_5$ is not planar.

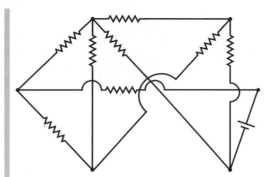

*Figure 6.52* Electrical network for Exercise 6.43

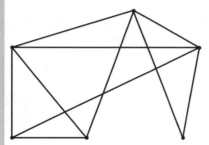

*Figure 6.53* Graph for Exercise 6.44

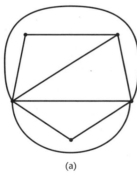

(a)

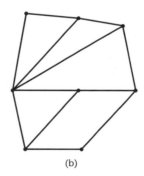

(b)

*Figure 6.54* Graphs for Exercise 6.45

**Exercise**

**6.47** The converse of the result stated in the preceding exercise does not hold – that is, if $3v - 6 \geqslant e$ it does not necessarily follow that the graph is planar. Illustrate this with $K_{3,3}$.

**Exercise**

**6.48** By considering an appropriate subgraph of $K_n$, prove that all complete graphs $K_n$ are non-planar for $n > 5$.

**Exercise**

**6.49** As in the preceding exercise, prove that all complete bipartite graphs $K_{m,n}$ are non-planar for $m \geqslant 3$ and $n \geqslant 3$.

**Exercise**

**6.50**  The 'platonic solids' are three-dimensional objects all of whose faces are identical. For example, a regular tetrahedron has four equilateral triangular faces, and a cube has six square faces. Redraw each of these examples as a plane graph, and verify that Euler's formula (6.2) holds. (This accounts for the use of the term 'faces' for planar graphs.)

### 6.2.4   Other special graphs

#### 6.2.4.1   Digraphs

Recall that a digraph $D$ is a graph in which each edge has a direction assigned to it. When these directions are removed the resulting graph is called the **underlying graph** of $D$; this will be a multigraph if some of the vertices of $D$ are joined by more than one edge, as illustrated in Figure 6.55. A digraph is **connected** if its underlying graph is connected, and **strongly connected** if there is a **directed** path joining every pair of vertices of $D$.

**Example 6.20**   **Connected digraphs**

The digraph in Figure 6.55(a) is connected because there is a path joining every pair of vertices in the underlying graph. However, it is not strongly connected because there is no directed path from $v_3$ to $v_5$, for example.

The digraph in Figure 6.56 is strongly connected because the vertices are all linked by the directed cycle $v_1 \longrightarrow v_2 \longrightarrow v_5 \longrightarrow v_3 \longrightarrow v_4 \longrightarrow v_1$. A directed path between any two vertices is part of this cycle.

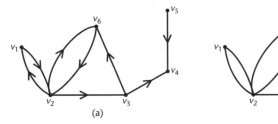

(a)                                  (b)

*Figure 6.55*   Digraph and underlying graph

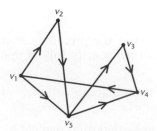

*Figure 6.56*   Strongly connected digraph

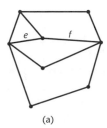 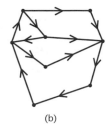

(a)                    (b)

*Figure 6.57*  Orientable graph

In planning a system of one-way streets for a city centre it is important to be able to travel between any two intersections.

This leads to the definition of an **orientable graph** which is a connected graph whose edges can be assigned directions so that the resulting digraph is strongly connected. It can be shown that a connected graph is orientable if and only if it does not contain a bridge.

---

**Example 6.21**  **Orientable graphs**

The connected graph in Figure 6.55(b) is not orientable since it contains the bridge $(v_3, v_4)$: if this edge is removed the resulting graph has two components, and so is not connected. The connected graph in Figure 6.57(a) does not contain a bridge, and so is orientable. The assignment of directions to the edges shown in Figure 6.57(b) produces a strongly connected digraph.

---

### 6.2.4.2  Eulerian graphs

A simple puzzle is to try to trace out all the edges of a graph just once each without taking your pencil off the paper. Specifically, let $G$ be a connected graph or multigraph. Recall from Section 6.1 that a trail from vertex $a$ to vertex $b$ is a set of edges joining $a$ to $b$ in which no edge is repeated, although vertices may be repeated. If a trail from $a$ to $b$ covers *every* edge of $G$ exactly once then it is called an **eulerian trail** (after Euler). If there is such a trail starting and finishing at $a$ then it is called an **eulerian circuit**, and the graph containing the circuit is itself called **eulerian**.

**Euler's theorem** states:

$G$ possesses an eulerian circuit if and only if every vertex of $G$ has even degree. If $G$ contains exactly two vertices with odd degree then there is an eulerian trail joining these two vertices.

If a graph is eulerian then the following procedure determines an eulerian circuit:

(1)  Begin at any vertex.
(2)  Proceed from one vertex to another, and after an edge is used delete it from the graph.
(3)  At every step in (z), never use a bridge unless there is no alternative.

**Example 6.22** **Eulerian graphs**

The graph in Figure 6.9 is not eulerian because there are four vertices having odd degrees ($d(a_1) = 3$, $d(a_5) = 3$, $d(a_6) = 5$ and $d(a_7) = 3$).

The graph in Figure 6.58(a) is a multigraph since there are two edges joining $v_5$ and $v_6$, and it is eulerian because the degree of every vertex is even. To construct an eulerian circuit, begin at $v_1$ and proceed $v_1 \longrightarrow v_2 \longrightarrow v_3$. At this stage the graph is as shown in Figure 6.58(b), since the edges $(v_1, v_2)$ and $(v_2, v_3)$ have been deleted. The edge $(v_3, v_1)$ is a bridge (because deleting it produces a disconnected graph, with $v_1$ an isolated vertex) so use instead the edge $(v_3, v_4)$. Continuing with the algorithm produces the eulerian circuit

$$v_1 \longrightarrow v_2 \longrightarrow v_3 \longrightarrow v_4 \longrightarrow v_5 \longrightarrow v_6 \longrightarrow v_7 \longrightarrow v_2 \longrightarrow v_6 \longrightarrow v_5 \longrightarrow v_3 \longrightarrow v_1$$

The graph in Figure 6.59 is not eulerian because $v_1$ and $v_5$ have odd degrees. However, the other vertices have even degrees, and an eulerian trail is

$$v_1 \longrightarrow v_5 \longrightarrow v_4 \longrightarrow v_3 \longrightarrow v_2 \longrightarrow v_4 \longrightarrow v_1 \longrightarrow v_2 \longrightarrow v_5$$

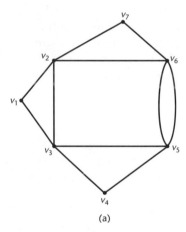

(a)

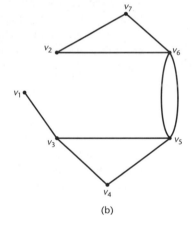

(b)

*Figure 6.58* Eulerian graph

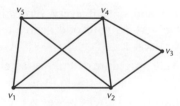

*Figure 6.59* Eulerian trail

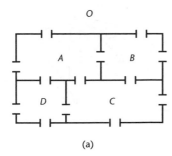

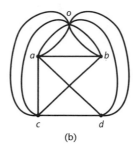

(a)                                    (b)

*Figure 6.60*   Floorplan and graph

**Example 6.23** **Application to a floor plan**

The floor plan of a one-storey building is shown in Figure 6.60(a). There are four rooms *A*, *B*, *C*, *D* and doorways are indicated between rooms, and to the outside *O*. To convert this into the graph shown in Figure 6.60(b), associate a vertex with each room, using the corresponding lower-case letter; the vertex *o* represents the outside. There is an edge joining two vertices if there is a door between the two corresponding regions. For example, there are two edges (*o,c*) since room *C* has two doors to the outside. The vertices *a* and *c* in Figure 6.60(b) have odd degrees, the other vertices have even degrees. Hence there is an eulerian trail from vertex *a* to vertex *c*. Translated back to the floor plan, this means that it is possible to start in room *A* and walk through every doorway exactly once, ending up in room *C*.

For a directed graph, Euler's theorem is modified as follows:

A connected digraph *D* contains a **directed eulerian circuit** if and only if for every vertex *v* the number of edges directed towards *v* is equal to the number of edges directed away from *v* (in other words, the indegree of *v* is equal to the outdegree of *v*). If this holds for all vertices except two, and one vertex has an excess of one indegree, and one vertex has an excess of one outdegree, then there is a directed eulerian path between these two vertices.

**Example 6.24** **Delivery of mail**

Suppose Figure 6.61(a) represents part of a plan for a town, the edges representing streets. A postperson has to deliver mail on foot on both sides of each street. To avoid repeatedly crossing over from one side of a street to another, the deliverer wishes to cover all the addresses by walking along each side of each street exactly once. Each edge in Figure 6.61(a) is replaced by two directed edges, representing the two sides of each street, as shown in Figure 6.61(b). The postperson's problem is to find a directed eulerian circuit. You can

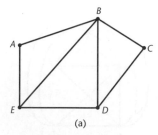 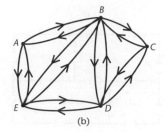

(a)    (b)

*Figure 6.61*  Mail delivery

easily check that for each vertex the inwards and outwards degrees are equal, so such a circuit does indeed exist. Notice, however, that the undirected graph in Figure 6.61(a) does not have an eulerian circuit, although there is an eulerian path (see Exercise 6.59).

If the digraph is not eulerian then some of the sides of the streets will have to be walked along more than once. The postperson's objective is then to select a directed closed walk of the shortest possible length. This problem is often called the 'Chinese postperson' (or 'postman' in older literature) problem, not because of any special features of the Chinese mail delivery system, but because it was first studied by a Chinese mathematician in 1962. Similar problems arise when refuse collection trucks pick up garbage bags from both sides of local streets, or when street-sweeping vehicles clean both sides of streets.

### 6.2.4.3 Hamiltonian graphs

The Irish mathematician Hamilton invented a game called 'Around the World' in 1859 in which each of the 20 faces of a solid dodecahedron was labelled with the name of a major city. A regular dodecahedron has 12 faces each of which is a regular pentagon, and its name arises from the Greek word 'dodeka', meaning twelve. The object of the game is to find a round trip in which each city is visited exactly once. For this reason, if a graph or multigraph has a path going through every vertex exactly once, then the path is called **hamiltonian**; it is a **hamiltonian cycle** if it starts and finishes at the same vertex. Contrast this with an eulerian trail or circuit which passes along every edge exactly once. Unfortunately, there is no useful link between eulerian and hamiltonian graphs, and indeed in the latter case there is no known necessary and sufficient condition for a graph to be hamiltonian (that is, to contain a hamiltonian cycle). At best, there are a number of conditions which are only *sufficient* for a graph to be hamiltonian. For example, if an undirected graph has no loops and $n(\geqslant 3)$ vertices, and if the degree of all vertices is greater than or equal to $(1/2)n$ then the graph has a hamiltonian cycle. However, there may be a hamiltonian cycle even if this condition is not satisfied, as the following example illustrates.

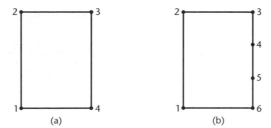

*Figure 6.62* Hamiltonian circuits

**Example 6.25** **Hamiltonian cycles**

In Figure 6.62(a) there are four vertices, each having degree two, and $(1/2)n = 2$. The condition $d(v) \geqslant (1/2)n$ is therefore satisfied for each vertex $v$, and it is obvious that $1 \longrightarrow 2 \longrightarrow 3 \longrightarrow 4 \longrightarrow 1$ is a hamiltonian cycle.

In Figure 6.62(b) there is a hamiltonian cycle $1 \longrightarrow 2 \longrightarrow 3 \longrightarrow 4 \longrightarrow 5 \longrightarrow 6 \longrightarrow 1$. However $n = 6$ and all the vertices have degree $2 < (1/2)n$.

An example of a directed cycle which is actually hamiltonian was given in Example 6.20 for the digraph in Figure 6.56.

**Example 6.26** **Seating plan**

Suppose five people are to be seated around a circular table. How many different seating plans are there if on each occasion no one has the same two neighbours? To answer this, consider the complete graph $K_5$ with five vertices labelled 1 to 5, one for each person, and edges joining all possible pairs of vertices. A seating arrangement corresponds to a hamiltonian cycle in $K_5$. For example, the cycle $1 \longrightarrow 2 \longrightarrow 3 \longrightarrow 4 \longrightarrow 5 \longrightarrow 1$ produces the seating plan shown in Figure 6.63(a). For two arrangements to be different in the sense described above, this means that the two corresponding hamiltonian cycles must have no edges in common. You should verify that the only other cycle having this property is $1 \longrightarrow 3 \longrightarrow 5 \longrightarrow 2 \longrightarrow 4 \longrightarrow 1$,

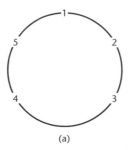

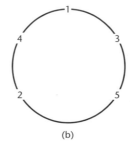

*Figure 6.63* Seating plans

corresponding to the seating plan in Figure 6.63(b), so there are only two different seating plans.

---

A much-studied application of hamiltonian cycles is the so-called 'travelling salesperson' (or 'salesman' in older literature) problem. In this, a sales representative has to make a round trip which involves visiting a number of towns once each. The travelling times, or alternatively the distances, between each pair of towns are known, and the objective is to make the tour in the shortest possible time, or by travelling the shortest possible distance. There is no really efficient algorithm known which completely solves this problem.

**Exercise**　**6.51**　For the graph in Figure 6.57(a), assign directions to the edges so as to produce a digraph which is not strongly connected.

**Exercise**　**6.52**　Suppose that the edge $e$ in Figure 6.57(a) is removed. Assign directions to the remaining edges so as to produce a strongly connected digraph. If the edge $f$ is also removed, does the graph remain orientable?

**Exercise**　**6.53**　Determine the eulerian path in Example 6.23.

**Exercise**　**6.54**　Mark eight points arbitrarily on a piece of paper. Without taking your pencil off the paper, join up the points with as many lines as you can without tracing any line twice. Verify that the condition for an eulerian path is satisfied.
　　　　Repeat this for a digraph.

**Exercise**　**6.55**　The graph in Figure 6.64 represents roads linking seven villages in a country area. These roads are to be kept free from ice in winter. Is it possible for a truck spreading salt and grit to travel along each road exactly once, starting and finishing at the depot? If so, determine a suitable route.

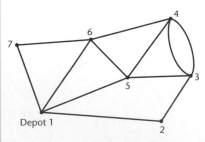

*Figure 6.64*　Country roads

**Exercise**　**6.56**　For a building whose floor plan is shown in Figure 6.65, is it possible to take a route which passes through each doorway exactly once? If so, determine how this can be done.

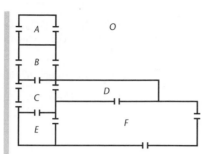

*Figure 6.65*   Floorplan

**Exercise**

**6.57** In the eighteenth-century East Prussian town of Königsberg, seven bridges connected the banks of the River Pregel and two islands, as shown schematically in Figure 6.66. The story goes that the townspeople consulted Euler in 1736 as to whether it was possible to make a round trip which crossed each bridge exactly once. By drawing an appropriate graph, determine Euler's answer.

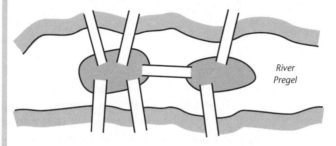

*Figure 6.66*   Bridges of Königsberg

**Exercise**

**6.58** In the preceding exercise, suppose two additional bridges are built, one directly connecting the two river banks and another joining the two islands. Show that an eulerian circuit is now possible, and describe one which starts and finishes on the river bank.

**Exercise**

**6.59** Find a directed eulerian circuit for the mail delivery problem in Example 6.24. Find also an eulerian path for the undirected graph in Figure 6.61(a).

**Exercise**

**6.60** An undirected graph is called **simple** if it is not a multigraph and contains no loops. Let *EC* stand for eulerian circuit and *HC* for hamiltonian cycle. Draw simple connected graphs that contain

(a) *EC* and *HC*

(b) *EC* but not *HC*

(c) *HC* but not *EC*

(d) neither *EC* nor *HC*.

**Exercise** | **6.61** A graph has vertices $v_1, v_2, \ldots, v_6$ and edges $(v_1, v_3)$, $(v_1, v_4)$, $(v_1, v_6)$, $(v_2, v_3)$, $(v_2, v_5)$, $(v_2, v_6)$, $(v_4, v_5)$ and $(v_5, v_6)$. Draw the graph and find a hamiltonian cycle.

**Exercise** | **6.62** For what values of $n$ is the complete graph $K_n$ (a) eulerian? (b) hamiltonian?

**Exercise** | **6.63** Analyze the problem in Example 6.26 when there are seven people to seat around the table.

## 6.3 Optimization in networks

Recall that a **network** is a graph or digraph, each of whose edges has a numerical value, called a **weight**, attached to it. These values may represent distances between vertices, times to travel between vertices, or some other kind of 'costs'. The objective is to find some path, or other subgraph, which is in a certain sense optimal. Four important types of network optimization problems are dealt with in this section.

### 6.3.1 Shortest path

**Example 6.27** **Road network**

A road network linking five major British cities was given in Figure 6.5, the distances between pairs of cities being marked in kilometres. Suppose you want to drive from Birmingham to Glasgow by the shortest route. The simple-minded way of finding this route is to list all the possible paths, and choose the shortest, as shown in the table.

| Route | Distance (km) |
|---|---|
| Birmingham–Manchester–Glasgow | $140 + 347 = 487$ |
| Birmingham–Leeds–Glasgow | $193 + 352 = 545$ |
| Birmingham–Leeds–Edinburgh–Glasgow | $193 + 321 + 74 = 588$ |
| Birmingham–Manchester–Edinburgh–Glasgow | $140 + 351 + 74 = 565$ |

The shortest path between Birmingham and Glasgow is therefore the first in the list, namely via Manchester, giving a total distance of 487 km. Notice that the term **shortest path** for a network means the shortest length of the path as a sum of the distances between the vertices along the path. This is not to be confused with the definition in Section 6.1 for a graph, where the length of a path is the number of edges constituting the path. In other applications distances may be replaced, for example, by times, so the 'shortest' path would then be the 'quickest' path (an example was given in Exercise 6.12).

Clearly in large and complicated networks, with very many possible paths between pairs of vertices, the method of Example 6.27 would be very inefficient. An algorithm which overcomes these difficulties was developed by E.W. Dijkstra in 1959. It is an example of what are quaintly called **greedy algorithms**, perhaps by comparison with a glutton who always chooses the biggest portions at each course of a meal. The idea of a greedy algorithm is that at each step the best possible choice is made, irrespective of past choices or of those choices yet to be made. Dijkstra's method will be described for an undirected network, where it is required to find the shortest path between some starting vertex $s$ and final vertex $f$. It relies on the observation that if the vertices $s, v_1, v_2, \ldots, v_n, f$ describe a shortest path from $s$ to $f$, then $s, v_1, v_2, \ldots, v_n$ must be a shortest path from $s$ to the vertex $v_n$ – for if not, then it would be possible to find a shorter path from $s$ to $f$ by using such a shorter path from $s$ to $v_n$. Notice that there may be several shortest paths from $s$ to $f$, all having the same length.

At each stage of the algorithm a **label** $L(i)$ is assigned to every vertex $v_i$ in the network. The value of $L(i)$ is equal to the distance to $v_i$ from $s$ using the shortest path found so far. Each label has the status at each stage of the algorithm of either permanent $(P)$ or temporary $(T)$. If a label is permanent then the problem of finding the length of the shortest path to that particular vertex is solved. If a label is temporary then it is uncertain whether the current path from $s$ to this vertex is the shortest possible or not. The algorithm begins with $L(s) = 0$ having permanent status, and all other vertices having temporary labels denoted by '$\infty$', which represents a number which is much larger than any of the weights on the edges of the network. At each stage when applying the algorithm the number of vertices with temporary labels is reduced by *one*. To do this, update each temporary label by comparing its current value with the length of the path from $s$ via the most recent permanently-labelled vertex, followed by an edge from that vertex. The vertex with the *smallest* new temporary label is then made permanent (see Step 2 below). This procedure is repeated, with one new label being made permanent at each iteration. The algorithm stops when the final vertex $f$ receives a permanent label, and $L(f)$ is then the required shortest distance.

A formal description of **Dijkstra's algorithm** follows.

## Dijkstra's algorithm

**Step 1:** Set $L(s) = 0$, status $P$; set $L(i) = \infty$, status $T$, for all other vertices. Denote by $p$ the last vertex $v_p$ at any stage to be given a permanent label, and set $p = s$.

**Step 2:** For each vertex $v_i$ with a temporary label, compute its new label according to

$$\underbrace{L(i)}_{\text{new label}} = \min\,[\,\underbrace{L(i),\, L(p)}_{\text{old labels}} + d(p, i)\,] \qquad (6.3)$$

In (6.3) $d(p, i)$ is the weight attached to the edge $(v_p, v_i)$, the labels inside the square bracket are the values at the *previous* iteration, and 'min' means take the smaller of the two values inside the square bracket. Let vertex $v_m$ hold the *smallest* new temporary label. Set $p = m$ and make $L(p)$ permanent.

**Step 3:** If the vertex $f$ has a temporary label, repeat Step 2 using the updated table of labels. Otherwise, $L(f)$ is the required shortest distance.

**Step 4:** For each permanently labelled vertex $v_k$ other than the vertex $s$, define a vertex $u(v_k)$ according to:

$$u(v_k) = v_i, \quad \text{where } L(k) = L(i) + d(i,k), \quad i \neq k \tag{6.4}$$

If $v_i$ is not unique this means the shortest path is not unique. A shortest path is

$$s \longrightarrow w_r \longrightarrow \dots \longrightarrow w_4 \longrightarrow w_3 \longrightarrow w_2 \longrightarrow w_1 \longrightarrow f$$

where the vertices are obtained in reverse order from

$$w_1 = u(f), \ w_2 = u(w_1), \ w_3 = u(w_2), \ w_4 = u(w_3), \dots, u(w_r) = s \tag{6.5}$$

Despite the apparent complexity of the formal description the algorithm is quite easy to apply in practice.

---

**Example 6.28**   **Application of Dijkstra's algorithm**

It is required to determine the shortest path from vertex $s$ to vertex $f$ in Figure 6.67.

*Step 1:*   Assign the initial values of the labels as follows:

| Vertex | $s$ | $v_1$ | $v_2$ | $v_3$ | $v_4$ | $f$ |
|---|---|---|---|---|---|---|
| Label $L(i)$ | 0 | $\infty$ | $\infty$ | $\infty$ | $\infty$ | $\infty$ |
| Status | P | T | T | T | T | T |

Set $p = s$.

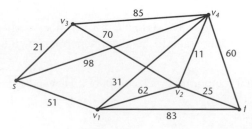

*Figure 6.67*   Network

*Step 2*: Compute the new values of the temporary labels using (6.3). For $i = 1$ this gives

$$L(1) = \min[L(1), L(s) + d(s,1)]$$
$$= \min[\infty, \quad 0 + 51] = 51$$

from table
in Step 1       from Figure 6.67

Similarly with $i = 2, 3, 4, f$ in (6.3) the new values of the labels are computed as follows:

$$L(2) = \min[L(2), L(s) + d(s,2)]$$
$$= \min[\infty, 0 + \infty] = \infty$$

Notice that in this case $d(s,2) = \infty$, since there is no edge joining vertex $s$ to vertex $v_2$. This rule applies in general: $d(a,b) = \infty$ if there is no edge $(a,b)$.

$$L(3) = \min[L(3), L(s) + d(s,3)] \quad L(4) = \min[L(4), L(s) + d(s,4)]$$
$$= \min[\infty, 0 + 21] = 21 \qquad\qquad = \min[\infty, 0 + 98] = 98$$
$$L(f) = \min[L(f), L(s) + d(s,f)]$$
$$= \min[\infty, 0 + \infty] = \infty$$

The smallest of these new temporary labels is $L(3) = 21$, so $m = 3$. This is the 'greedy' aspect of the algorithm. Referring to Figure 6.67 you can see that the vertex 'nearest' to $s$, namely $v_3$, is selected.

Now set $p = 3$ and make $L(3)$ permanent, so the updated table of labels is:

| Vertex | s | $v_1$ | $v_2$ | $v_3$ | $v_4$ | f |
|---|---|---|---|---|---|---|
| Label $L(i)$ | 0 | 51 | $\infty$ | 21 | 98 | $\infty$ |
| Status | P | T | T | P | T | T |

Since vertex $f$ has a temporary label, repeat Step 2 to recalculate the temporary labels as follows, by setting $i = 1, 2, 4, f$ in (6.3):

$$L(1) = \min[L(1), L(3) + d(3,1)]$$
$$= \min[51, \quad 21 + \infty] = 51$$

from updated
table       from Figure 6.67

$$L(2) = \min[L(2), L(3) + d(3,2)] \quad L(4) = \min[L(4), L(3) + d(3,4)]$$
$$= \min[\infty, 21 + 70] = 91 \qquad\qquad = \min[98, 21 + 85] = 98$$
$$L(f) = \min[L(f), L(3) + d(3,f)]$$
$$= \min[\infty, 21 + \infty] = \infty$$

The smallest of these new temporary labels is $L(1) = 51$, so $m = 1$. Set $p = 1$ and make $L(1)$ permanent. Again, notice how the 'greedy' aspect of the algorithm works: in Figure 6.67 the vertex $v_1$ is chosen because it gives the shortest path from $s$ going either directly, or via the previously selected vertex $v_3$. The updated table of labels is now:

| Vertex | $s$ | $v_1$ | $v_2$ | $v_3$ | $v_4$ | $f$ |
|---|---|---|---|---|---|---|
| Label $L(i)$ | 0 | 51 | 91 | 21 | 98 | $\infty$ |
| Status | P | P | T | P | T | T |

Since $L(f)$ is still temporary, Step 2 is repeated with $p = 1$ and $i = 2, 4, f$, giving:

$$L(2) = \min[L(2), L(1) + d(1,2)] \quad L(4) = \min[L(4), L(1) + d(1,4)]$$
$$= \min[91, 51 + 62] = 91 \qquad\qquad = \min[98, 51 + 31] = 82$$
$$L(f) = \min[L(f), L(1) + d(1,f)]$$
$$= \min[\infty, 51 + 83] = 134$$

The smallest of these new temporary labels is $L(4) = 82$, so $m = 4$. Set $p = 4$ and make $L(4)$ permanent. The updated table of labels is:

| Vertex | $s$ | $v_1$ | $v_2$ | $v_3$ | $v_4$ | $f$ |
|---|---|---|---|---|---|---|
| Label $L(i)$ | 0 | 51 | 91 | 21 | 82 | 134 |
| Status | P | P | T | P | P | T |

Another repetition of Step 2 produces:

$$L(2) = \min[L(2), L(4) + d(4,2)]$$
$$= \min[91, 82 + 11] = 91$$
$$L(f) = \min[L(f), L(4) + d(4,f)]$$
$$= \min[134, 82 + 60] = 134$$

The smallest of these values is $L(2) = 91$, so $p = 2$ and the next table of labels is:

| Vertex | $s$ | $v_1$ | $v_2$ | $v_3$ | $v_4$ | $f$ |
|---|---|---|---|---|---|---|
| Label $L(i)$ | 0 | 51 | 91 | 21 | 82 | 134 |
| Status | P | P | P | P | P | T |

One more repetition of Step 2 gives

$$L(f) = \min[L(f), L(2) + d(2,f)]$$
$$= \min[139, 91 + 25] = 116$$

and this is the length of the shortest path from $s$ to $f$. The final table of labels is:

| Vertex | $s$ | $v_1$ | $v_2$ | $v_3$ | $v_4$ | $f$ |
|---|---|---|---|---|---|---|
| Label $L(i)$ | 0 | 51 | 91 | 21 | 82 | 116 |
| Status | $P$ | $P$ | $P$ | $P$ | $P$ | $P$ |

The actual shortest path is now found from:

Step 4    Set $k = 1$ in (6.4) to get

$$L(1) = L(i) + d(i,1)$$

Using the *final* table of labels, it is therefore required to find the value of $i \neq 1$ such that

$$51 = L(i) + d(i,1) \tag{6.6}$$

From Figure 6.67 the weights $d(i,1)$ are

$$d(s,1) = 51, \quad d(2,1) = 62, \quad d(4,1) = 31, \quad d(f,1) = 83$$

It is easy to see that (6.6) is satisfied when $i = s$, so in (6.4) $u(v_1) = s$.

Repeat the solution of (6.4) for the other values of $k$ as follows:

$$(k = 2); \quad L(2) = 91 = L(i) + d(i,2)$$

This is satisfied by $i = 3$ since $L(3) = 21$, $d(3,2) = 70$, so $u(v_2) = v_3$. Similarly:

$$(k = 3) \quad u(v_3) = s, \quad \text{since } L(3) = 21 = L(s) + d(s,3)$$

$$(k = 4) \quad u(v_4) = v_1, \quad \text{since } L(4) = 82 = L(1) + d(1,4)$$
$$\underset{i=1}{\llcorner \quad \lrcorner}$$

$$(k = f) \quad u(f) = v_2, \quad \text{since } L(f) = 116 = L(2) + d(2,f)$$
$$\underset{i=2}{\llcorner \quad \lrcorner}$$

From (6.5) the vertices on the shortest path are given, in reverse order, by

$$w_1 = u(f) = v_2$$
$$w_2 = u(w_1) = u(v_2) = v_3$$
$$u(w_2) = u(v_3) = s$$

Hence the shortest path is $s \longrightarrow w_2 \longrightarrow w_1 \longrightarrow f$, that is, $s \longrightarrow v_3 \longrightarrow v_2 \longrightarrow f$.

Notice that the final table of permanent labels gives the lengths of the shortest paths from $s$ to each of the other vertices.

Dijkstra's algorithm can be applied to find the shortest directed path between two vertices of a directed network by setting $d(i,j) = \infty$ if there is no directed arc from vertex $v_i$ to vertex $v_j$.

**Exercise** **6.64** Use Step 4 of Dijkstra's algorithm to find the shortest path between $s$ and $v_4$ for the network in Figure 6.67.

**Exercise** **6.65** Solve the problem in Example 6.27 using Dijkstra's algorithm.

**Exercise** **6.66** For the US road network in Figure 6.17, find the route which gives the shortest time of travel between Barstow and Page.

**Exercise** **6.67** For the graph in Figure 6.9, suppose weights are attached to the edges according to the following table:

|       | $a_2$ | $a_3$ | $a_4$ | $a_5$ | $a_6$ | $a_7$ |
|-------|-------|-------|-------|-------|-------|-------|
| $a_1$ | 19    | –     | –     | –     | 4     | 7     |
| $a_2$ | –     | 17    | –     | –     | 21    | 12    |
| $a_3$ |       | –     | 5     | 3     | 18    | –     |
| $a_4$ |       |       | –     | 6     | –     | –     |
| $a_5$ |       |       |       | –     | 10    | –     |
| $a_6$ |       |       |       |       | –     | 11    |

so for example $(a_4,a_5)$ has weight 6. By listing routes from $a_1$ to $a_4$, determine:

(a) the shortest path    (b) the longest trail    (c) the longest path,

between these two vertices.

**Exercise** **6.68** The edges and attached weights for an undirected network having seven vertices are given below (a null entry means there is no edge joining the appropriate pair of vertices). Find the shortest path from $v_1$ to $v_7$ using Dijkstra's method.

|       | $v_2$ | $v_3$ | $v_4$ | $v_5$ | $v_6$ | $v_7$ |
|-------|-------|-------|-------|-------|-------|-------|
| $v_1$ | 17    | 15    | 19    | –     | –     | –     |
| $v_2$ | –     | –     | –     | 18    | 22    | –     |
| $v_3$ |       | –     | 8     | 21    | –     | –     |
| $v_4$ |       |       | –     | 5     | –     | –     |
| $v_5$ |       |       |       | –     | 11    | 22    |
| $v_6$ |       |       |       |       | –     | 6     |

**Exercise** **6.69** What is the effect on the solution to the preceding exercise if $d(5,7)$ changes from 22 to 17?

**Exercise** **6.70** Apply Dijkstra's algorithm to the network in Figure 6.68 to find the shortest directed path

(a) from $v_1$ to $v_6$   (b) from $v_6$ to $v_1$.

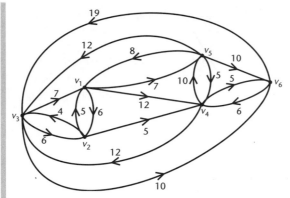

*Figure 6.68*  Directed network for Exercise 6.70

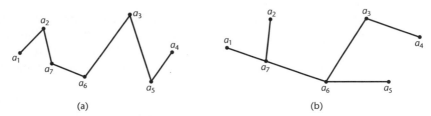

(a)                                            (b)

*Figure 6.69*  Spanning trees for Figure 6.9

## 6.3.2    Minimal spanning tree

A tree was defined in Section 6.2.2 as a connected graph containing no cycles. In particular, this means that there is a path between every pair of vertices. If $G$ is a loop-free, connected, undirected graph, then any spanning subgraph of $G$ which is a tree is called a **spanning tree** of $G$. Recall that a spanning subgraph contains all the vertices of $G$, so you can think of a spanning tree as a sort of skeletal framework which is just enough to ensure that all the vertices of $G$ are joined together.

**Example 6.29**    **Minimal spanning trees**

For the graph in Figure 6.6(b), the spanning subgraph in Figure 6.12(b) is a spanning tree. Recall that a spanning subgraph is obtained from a graph by deleting edges but not vertices.

There will in general be more than one spanning tree for a given graph. Two spanning trees of the graph in Figure 6.9 are given in Figure 6.69. Recall that the number of edges in a tree is one less than the number of vertices.

If the spanning tree is a subgraph of an undirected network then, as before, the **length** of the tree is the sum of the weights on its edges. In this case, any

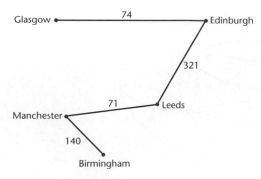

*Figure 6.70* Minimal spanning tree for Figure 6.5

spanning tree which has the smallest possible length is called a **minimal spanning tree**. Because a minimal spanning tree is a connected network it is sometimes known as a **minimal connector**.

As an example of a practical application, suppose a communications company wishes to connect together all the cities shown in Figure 6.5 by laying cables along the main roads in the network. The least length of cable would be required if the minimal spanning tree in Figure 6.70 is used, having a length of 606 km.

Similar problems arise, for example, when designing a network to connect computers on a college campus, or when siting warehouses to be linked by a road network.

It would be very inefficient to determine a minimal spanning tree by computing the lengths of all possible spanning trees, of which there will in general be a large number. A systematic way of finding a minimal spanning tree was produced by R.C. Prim in 1957. The method begins with a tree having one edge. Another edge is added which is the shortest of the remaining edges having exactly *one* vertex in the tree (edges having no vertices in the tree are rejected because adding them would produce a disconnected graph; and edges having two vertices in the tree would produce a cycle). The procedure is continued, adding one edge at each iteration, until a spanning tree is obtained. The method of construction ensures it will be minimal. Like Dijkstra's method, Prim's algorithm is of the greedy type. A formal description is quite straightforward.

## Prim's Algorithm

For a network with $n$ vertices:

**Step 1:** Select any vertex. Choose an edge from those edges incident to this vertex having the shortest length. Call this graph $T$.

**Step 2:** Select an edge $(i,j)$ with the smallest length from amongst all edges $(i,k)$ with $i$ in $T$ and $k$ not in $T$ (if there is a tie, the tree under construction is not unique). Add this edge to $T$.

**Step 3:** If $T$ has $n - 1$ edges then it is a minimal spanning tree. Otherwise, repeat Step 2.

**Example 6.30** **Application of Prim's algorithm**

Return to the network with six vertices in Figure 6.67, and determine a minimal spanning tree.

*Step 1:* Begin at vertex $v_2$, and choose the edge $(v_2,v_4)$ which is the shortest edge incident to $v_2$, so that $T = \{(v_2,v_4)\}$.

*Step 2:* From Figure 6.67 you can see that the edges having exactly one vertex in $T$ are

$$(v_2,v_1), (v_2,v_3), (v_2,f), (v_4,v_1), (v_4,s), (v_4,v_3), (v_4,f)$$

The shortest of these is $(v_2,f)$ so this edge is added to $T$, producing the updated graph

$$T = \{(v_2,v_4),(v_2,f)\}$$

This has two edges, so is not a minimal spanning tree. Three more repetitions of Step 2 are needed to produce a minimal spanning tree with five edges, as shown in the table.

| Edges with one vertex in T | Shortest edge | Updated T |
|---|---|---|
| $(v_2,v_1)$, $(v_2,v_3)$, $(v_4,v_1)$, $(v_4,s)$, $(v_4,v_3)$ | $(v_4,v_1)$ | $(v_2,v_4)$, $(v_2,f)$, $(v_4,v_1)$ |
| $(v_1,s)$, $(v_2,v_3)$, $(v_4,v_3)$, $(v_4,s)$ | $(v_1,s)$ | $(v_2,v_4)$, $(v_2,f)$, $(v_4,v_1)$, $(v_1,s)$ |
| $(v_3,s)$, $(v_3,v_2)$, $(v_4,v_3)$ | $(v_3,s)$ | $(v_2,v_4)$, $(v_2,f)$, $(v_4,v_1)$, $(v_1,s)$, $(v_3,s)$ |

The minimal spanning tree is the final updated $T$ in the table, shown in Figure 6.71. The length of this tree is $21 + 51 + 31 + 11 + 25 = 139$.

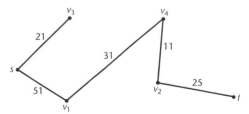

*Figure 6.71* Minimal spanning tree for Figure 6.67

**Exercise**  **6.71**  Repeat Example 6.30, starting at vertex $s$.

**Exercise**  **6.72**  Find the length of each spanning tree for the network in Figure 6.72. Hence determine a minimal spanning tree.

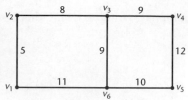

*Figure 6.72*  Network for Exercise 6.72

**Exercise**  **6.73**  Use Prim's algorithm starting at vertex $v_1$ to find a minimal spanning tree for the network specified by the following table of edges and weights:

|       | $v_2$ | $v_3$ | $v_4$ | $v_5$ | $v_6$ |
|-------|-------|-------|-------|-------|-------|
| $v_1$ | 16    | –     | –     | 12    | 11    |
| $v_2$ | –     | 20    | –     | –     | 8     |
| $v_3$ |       | –     | 19    | –     | 15    |
| $v_4$ |       |       | –     | 9     | 9     |
| $v_5$ |       |       |       | –     | 13    |

**Exercise**  **6.74**  Use Prim's algorithm starting at vertex $v_7$ to find a minimal spanning tree for the network specified by the following table of edges and weights.

|       | $v_2$ | $v_3$ | $v_4$ | $v_5$ | $v_6$ | $v_7$ |
|-------|-------|-------|-------|-------|-------|-------|
| $v_1$ | 25    | 38    | –     | –     | 25    | 27    |
| $v_2$ | –     | 42    | 31    | –     | –     | 25    |
| $v_3$ |       | –     | 25    | 29    | –     | –     |
| $v_4$ |       |       | –     | 23    | –     | 27    |
| $v_5$ |       |       |       | –     | 27    | 15    |
| $v_6$ |       |       |       |       | –     | 17    |

Repeat by starting at $v_1$, to obtain a different minimal spanning tree.

**Exercise**  **6.75**  A communications company decides to connect five towns $A, B, C, D, E$ with the latest type of fibre optics cable. Because of differing terrains, the estimated costs (in millions of pounds) of laying cables between each pair of towns are set out in the following table. Determine the cheapest way of linking the towns.

|     | $B$   | $C$   | $D$   | $E$   |
|-----|-------|-------|-------|-------|
| $A$ | 1.00  | 1.25  | 1.20  | 1.10  |
| $B$ | –     | 0.75  | 0.65  | 1.30  |
| $C$ |       | –     | 0.70  | 0.55  |
| $D$ |       |       | –     | 0.50  |

**Exercise** **6.76** In the preceding exercise it is decided for operational reasons that there must be a direct connection between towns *A* and *D*. Determine the cheapest way of linking all the towns in this case.

**Exercise** **6.77** To promote its interplanetary holidays using British Spaceways, a travel company wishes to translate its brochure, written in English, into Jovian, Martian and Venusian. The costs of translations between each pair of languages are shown in the table below, in millions of Earth Dollars. Determine the cheapest way of carrying out the translations, using an appropriate directed network.

| From \ To | English | Jovian | Martian | Venusian |
|---|---|---|---|---|
| English | – | 2.1 | 1.5 | 2.8 |
| Jovian | 2.1 | – | 3.1 | 4.7 |
| Martian | 1.5 | 3.1 | – | 2.9 |
| Venusian | 2.8 | 4.7 | 2.9 | – |

## 6.3.3 Critical path

Every complex project involves many interrelated activities. When planning such a project it is important to estimate the duration of each activity, and to define precedence relations – that is, to determine which activities cannot be started until others have finished. If the activities and their relationships are represented by a digraph, then a **critical path** from the beginning to the end identifies those **critical activities** whose delay would postpone the completion of the entire project.

**Example 6.31** **Construction project**

Suppose a prefabricated building is to be erected on a 'green-field' site. Some of the activities involved in this project are shown in Table 6.4 below. This is a greatly simplified version of a real-life situation where there will be hundreds or thousands of activities. The 'immediate predecessors' in Table 6.4 are those activities which must be completed before the appropriate current activity can commence – for example, before the building can be erected (activity *H*) the foundations must be laid (activity *G*) and the building kit must be delivered (activity *F*).

An alternative way of representing the activities is by the **activity network** in Figure 6.73. Each activity is represented by a directed edge in the network, the associated weight being the duration of the activity. For example, activity *H* is represented by the directed edge (3,4) from vertex 3 (called a **start event**) to vertex 4 (called a **finish event**). The start event for *H* is the finish event for

*Table 6.4*   Some activities in a construction project

| Activity | Description | Duration (days) | Immediate predecessors |
|----------|-------------|-----------------|------------------------|
| A | Place order for foundations | 4 | – |
| B | Place order for fencing | 30 | – |
| C | Place order for building | 4 | – |
| D | Await delivery of fences | 28 | B |
| E | Await start for foundations | 30 | A |
| F | Await delivery of building kit | 50 | C |
| G | Lay foundations | 10 | E |
| H | Erect building | 7 | F, G |
| I | Invite estimates and place contracts for interior work | 20 | C |
| J | Erectors leave site | 3 | H |
| K | Erect fences | 5 | B |

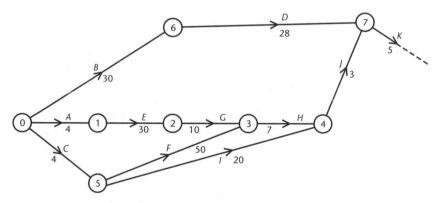

*Figure 6.73*   Construction project

activities *F* and *G*, as indicated in Table 6.4. Suppose the stage of the project described by Table 6.4 and Figure 6.73 is regarded as completed when the shell of the building has been erected (that is, the finish event 4 is reached). By looking at Figure 6.73 it is easy to see that the *longest* directed path from the overall start event 0 to vertex 4 is 0⟶5⟶3⟶4, corresponding to the sequence of activities *C*, *F*, *H*, and has the total length of 61 days. This is the *shortest* time in which the project can be completed, up to vertex 4. The path 0⟶5⟶3⟶4 is the **critical path** for this stage of the project. If any activities along this path take longer than expected then the overall completion time of 61 days is increased. However, this does not apply to activities not on the critical path. For example, if the time taken to lay the foundations (activity *G*) were to increase from ten days to 12 days, this would have no effect on the overall completion time; laying the foundations is not a critical activity.

There are a few points to note when constructing an activity network with $n + 1$ vertices.

(1) An activity *Y* is represented by a directed edge $(i, j)$ and its duration is $d_{ij}$. *Y* is said to **depend on** activities $X_1, X_2, X_3, \ldots$ if it cannot commence until $X_1, X_2, X_3, \ldots$ have been completed. The start vertex *i* for *Y* is the finish vertex for $X_1, X_2, X_3, \ldots$ as shown in Figure 6.74.

(2) There is a unique initial vertex labelled 0, which is the start vertex for all those activities which do not depend upon any others.

(3) There is a unique final vertex labelled *n*, which is the finish vertex for all those activities upon which no other activities depend.

(4) Every vertex (except the final vertex) has at least one activity starting from it; and every vertex (except the initial vertex) has at least one activity finishing at it.

(5) The network cannot contain any directed cycles. For example, the cycle in Figure 6.75 would imply that '*C* depends on *B* which depends on *A* which depends on *C* ...' which reveals a logical error in the interrelationships between activities *A*, *B* and *C*.

(6) By convention, there is at most one edge between each pair of vertices. For example, the situation shown in Figure 6.76(a) must be replaced by that in

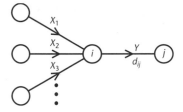

*Figure 6.74* Activity dependencies

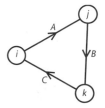

*Figure 6.75* Directed cycle

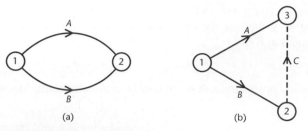

Figure 6.76   Dummy activity

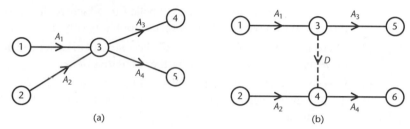

(a)

(b)

Figure 6.77   Use of dummy activity

Figure 6.76(b), where $C$ is a **dummy activity** which has zero duration, and uses no resources. It is customary to denote a dummy by a dashed edge in the network.

(7)  Dummy activities are useful in other circumstances. For example, suppose part of a table of activities contains the following:

| Activity | $A_3$ | $A_4$ |
|---|---|---|
| Depends on | $A_1$ | $A_1, A_2$ |

The network shown in Figure 6.77(a) does not represent these dependencies, since it implies that $A_3$ depends upon $A_1$ and $A_2$. The correct representation is given in Figure 6.77(b), where a dummy activity $D = (3,4)$ has been introduced. Since $D$ depends on $A_1$, and $A_4$ depends upon $A_2$ and $D$, then $A_4$ depends on $A_1$ and $A_2$, as required.

In simple problems it is usually not difficult to construct an activity network from a table of activities and dependencies. For large, complex projects formal procedures for constructing networks are needed, but details lie outside the scope of this book.

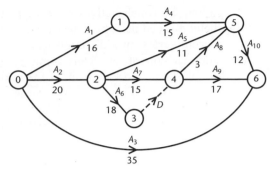

*Figure 6.78*  Activity network

**Example 6.32**  **Simple activity network**

Consider the following table of dependencies:

| Activity | $A_1$ | $A_2$ | $A_3$ | $A_4$ | $A_5$ | $A_6$ | $A_7$ | $A_8$ | $A_9$ | $A_{10}$ |
|---|---|---|---|---|---|---|---|---|---|---|
| Depends on | – | – | – | $A_1$ | $A_2$ | $A_2$ | $A_2$ | $A_6, A_7$ | $A_6, A_7$ | $A_4, A_5, A_8$ |
| Duration (days) | 16 | 20 | 35 | 15 | 11 | 18 | 15 | 3 | 17 | 12 |

Activities $A_3$, $A_9$ and $A_{10}$ do not have any other activities depending on them, and so their finish vertex is the final vertex, labelled 6 in the activity network shown in Figure 6.78. Notice the dummy activity $D = (3, 4)$ which has zero duration and is needed to distinguish between $A_6$ and $A_7$, in view of point (6) above.

In such a small network with only ten activities it is quite easy to identify the critical path. This is the *longest* directed path from the initial vertex 0 to the final vertex 6, namely $0 \longrightarrow 2 \longrightarrow 3 \longrightarrow 4 \longrightarrow 6$ with a total length of $20 + 18 + 0 + 17 = 55$ days. An alternative way of describing the critical path is by its sequence of edges, that is, the activities $A_2$, $A_6$, $D$, $A_9$. You can easily check in this example that any other directed path, for example $0 \longrightarrow 1 \longrightarrow 5 \longrightarrow 6$, has a shorter duration.

For a large network with many activities, a systematic procedure for determining a critical path is required. This is done by finding two special times associated with each vertex. The first is the **earliest time** $e_i$ when a vertex $i$ can be reached with all the incoming activities for that vertex completed. This earliest time is the largest of all the times to travel from an adjacent vertex along a directed edge. In view of this definition, $e_n$ is the shortest time in which the whole project can be completed. The second special time $l_i$ is the **latest time** at which vertex $i$ can be left without extending the length of the critical path – in other words $l_i$ is the latest time at which activities starting at vertex $i$ can commence without falling behind schedule. This latest time is the smallest of the times taken from an adjacent vertex, in a direction *opposite* to that of the directed edge. The computations consist of determining $e_1, e_2, e_3, \ldots, e_n$ by making a *forward* sweep through the network starting at the initial vertex, and then making a *backward* sweep starting at the final vertex to determine $l_{n-1}, l_{n-2}, \ldots, l_2, l_1$. A vertex where the two times $e_i$ and $l_i$ are equal is a **critical vertex** (or **event**), and the critical path goes through these critical vertices. These rules can be stated formally as follows.

## Critical path algorithm

Let the network have $n + 1$ vertices, with the initial and final vertices labelled $0$ and $n$ respectively.

**Step 1:** Set $e_0 = 0$, and compute

$$e_i = \max_j (e_j \mid d_{ji}), \qquad i - 1, 2, \ldots, n$$

where the maximum is taken over all vertices $j$ for which there is a directed edge $(j, i)$.

**Step 2:** Set $l_n = e_n$, $l_0 = 0$, and compute

$$l_i = \min_j (l_j - d_{ij}), \qquad i = n - 1, n - 2, \ldots, 2, 1$$

where the minimum is taken over all vertices $j$ for which there is a directed edge $(i, j)$.

**Step 3:** A critical path from the initial vertex to the final vertex has length $e_n$, and consists of the critical vertices for which $e_i = l_i$.

**Example 6.33** **Critical path for Figure 6.78**

Return to the activity network in Figure 6.78. Step 1 of the algorithm gives:

$$e_0 = 0, \quad e_1 = \max(e_0 + d_{01}) = \max(0 + 16) = 16$$
$$e_2 = \max(e_0 + d_{02}) = \max(0 + 20) = 20$$
$$e_3 = \max(e_2 + d_{23}) = \max(20 + 18) = 38$$
$$e_4 = \max(e_2 + d_{24}, e_3 + d_{34})$$
$$= \max(20 + 15, 38 + 0) = 38$$
$$e_5 = \max(e_1 + d_{15}, e_2 + d_{25}, e_4 + d_{45})$$
$$= \max(16 + 15, 20 + 11, 38 + 3) = 41$$
$$e_6 = \max(e_0 + d_{06}, e_4 + d_{46}, e_5 + d_{56})$$
$$= \max(0 + 35, 38 + 17, 41 + 12) = 55$$

Step 2 of the algorithm gives:

$$l_6 = e_6 = 55, \quad l_0 = 0$$
$$l_5 = \min(l_6 - d_{56}) = \min(55 - 12) = 43$$
$$l_4 = \min(l_5 - d_{45}, l_6 - d_{46})$$
$$\quad = \min(43 - 3, 55 - 17) = 38$$
$$l_3 = l_4 - 0 = 38, \quad \text{since } d_{34} = 0$$
$$l_2 = \min(l_5 - d_{25}, l_4 - d_{24}, l_3 - d_{23})$$
$$\quad = \min(43 - 11, 38 - 15, 38 - 18) = 20$$
$$l_1 = \min(l_5 - d_{15}) = \min(43 - 15) = 28$$

Since $e_2 = l_2$, $e_3 = l_3$, and $e_4 = l_4$, the critical path is $0 \longrightarrow 2 \longrightarrow 3 \longrightarrow 4 \longrightarrow 6$ and has length $e_6 = 55$, which agrees with what was found in Example 6.32.

Notice that for any critical activity represented by an edge $(i, j)$ on the critical path then $d_{ij} = l_j - e_i$. For example, in Figure 6.78 the critical activities are $A_2$, for which $20 = l_2 - e_0$; $A_6$, for which $18 = l_3 - e_2$; and $A_9$, for which $17 = l_6 - e_4$. For any non-critical activity $(i, j)$ then $d_{ij} < l_j - e_i$ and the quantity $f_{ij} = l_j - e_i - d_{ij}$ is called the **total float** of activity $(i, j)$. The value of $f_{ij}$ is the maximum possible delay which can occur in activity $(i, j)$ without increasing the length of the critical path, providing there are no delays elsewhere. For example, in Figure 6.78 the activity $A_5$ has total float

$$f_{25} = l_5 - e_2 - d_{25}$$
$$\quad = 43 - 20 - 11 = 12$$

This means that the time to carry out activity $A_5$ can be extended by up to 12 days before $A_5$ becomes critical, assuming there are no delays elsewhere. For example, if the duration of $A_5$ extends to 24 days then the new critical path would be $0 \longrightarrow 2 \longrightarrow 5 \longrightarrow 6$, having length 56 days.

There are two other types of slack in the duration of a non-critical activity. The **free float** of an activity $(i, j)$ is $e_j - e_i - d_{ij}$, and is the maximum possible amount by which the activity can be delayed without affecting any subsequent activity in the network. The **independent float** of an activity $(i, j)$ is $e_j - l_i - d_{ij}$, and is the maximum possible amount by which the activity can be delayed without affecting any other activity, either previous or subsequent. Again referring to the network in Figure 6.78, the free float of activity $A_5$ is $e_5 - e_2 - d_{25} = 41 - 20 - 11 = 10$. This means that the duration of $A_5$ can be extended by up to 10 days before the earliest starting time of activity $A_{10}$ is affected (notice that if $d_{25}$ becomes greater than 21 then from the expression for $e_5$ in Example 6.33, $e_5$ becomes greater than 41). Similarly, the independent float of $A_3$ is $e_6 - l_0 - d_{06} = 55 - 0 - 35 = 20$. This means that the duration of $A_3$ can be extended by up to 20 days without having any effect on the timing of any other activities in the network. However, if the duration of $A_3$ extends beyond 55 days then this will increase the length of the critical path.

**Exercise** **6.78** Some further activities for the construction project in Example 6.31 are given in Table 6.5.

*Table 6.5* Further activities in the construction project

| Activity | Description | Duration (days) | Depends on |
|---|---|---|---|
| L | Plumbing | 10 | H |
| M | Connect to water supply | 2 | L |
| N | Joinery | 14 | H |
| P | Electrical wiring | 3 | H |
| Q | Purchase paint | 2 | – |
| R | Painting | 15 | M, N, P, Q |
| S | Connect to electricity supply and test wiring | 6 | R |
| T | Order furniture and await delivery | 30 | – |
| U | Install furniture | 4 | R |

Draw the remainder of the activity network, continuing from Figure 6.73. Determine by inspection a critical path from vertex 4 to the new final vertex. What is the shortest time in which the whole project can be completed?

**Exercise** **6.79** Construct a table of activities and dependencies for the activity network in Figure 6.79.

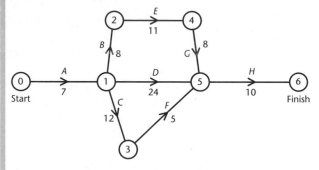

*Figure 6.79* Network for Exercise 6.79

**Exercise** **6.80** Express the following table as an activity network.

| Activity | A | B | C | D | E | F | G | H | I |
|---|---|---|---|---|---|---|---|---|---|
| Depends on | – | A | A | B | C, D | C, D | B | F | E, G, H |
| Duration | 5 | 12 | 7 | 16 | 15 | 10 | 14 | 8 | 6 |

**Exercise** **6.81** Express the following table as an activity digraph, using dummy activities as appropriate.

| Activity | $A_1$ | $A_2$ | $A_3$ | $A_4$ | $A_5$ | $A_6$ | $A_7$ |
|---|---|---|---|---|---|---|---|
| Depends on | – | – | $A_1$ | $A_2$ | $A_1, A_3, A_4$ | $A_3, A_4$ | $A_5, A_6$ |

**Exercise**   **6.82**   Determine by inspection the critical path for the network you found in Exercise 6.80.

**Exercise**   **6.83**   Repeat the preceding exercise using the critical path algorithm.

**Exercise**   **6.84**   For the activity network in Figure 6.79, determine the critical path

(a) by inspection   (b) using the algorithm.

**Exercise**   **6.85**   In the preceding exercise, what can the duration of activity $D$ extend to before it becomes part of a critical path?

**Exercise**   **6.86**   Use the algorithm to find the critical path for the complete construction project described by Figure 6.73 and the network you found in Exercise 6.78.

**Exercise**   **6.87**   For the construction project in the preceding exercise, bad weather delays the start of laying the foundations. How many days' delay can be tolerated before the planned completion date of the whole project is endangered?

**Exercise**   **6.88**   Draw an activity network for the following table, and determine a critical path.

| Activity | $A_1$ | $A_2$ | $A_3$ | $A_4$ | $A_5$ | $A_6$ | $A_7$ | $A_8$ | $A_9$ | $A_{10}$ | $A_{11}$ | $A_{12}$ | $A_{13}$ |
|---|---|---|---|---|---|---|---|---|---|---|---|---|---|
| Depends on | – | – | – | $A_2$ | $A_2$ | $A_5$ | $A_1, A_4$ | $A_1, A_4$ | $A_2$ | $A_5$ | $A_3, A_6$ | $A_8, A_9, A_{10}, A_{11}$ | $A_3, A_6$ |
| Duration | 5 | 10 | 2 | 8 | 8 | 9 | 3 | 4 | 8 | 10 | 4 | 5 | 8 | 5 |

Determine also the values of the floats for activities $A_4$ and $A_7$, and interpret them.

## 6.3.4   Maximal flow

In this section a directed network is regarded as describing a flow of some commodity from a starting point to a final destination. For example, the edges of the network may be pipelines through which a fluid flows, or roads along which traffic travels, or communication links along which messages are sent. In each case the weight attached to an edge represents the maximum capacity of that edge, and the problem is to find the **maximal flow** which can be achieved from the start to the destination.

For networks of this type (sometimes called **transportation** networks) the starting vertex $s$ is called the **source**, and is the only vertex whose indegree (that is, the number of edges directed towards the vertex) is zero. The destination vertex $t$ is called the **sink**, and is the only vertex whose outdegree (that is, the number of edges directed away from the vertex) is zero. The weight $c(e)$ is the maximum capacity (assumed non-negative) of an edge $e$. If $f(e)$ is an actual flow along $e$, then $c(e) \geqslant f(e) \geqslant 0$; and if $f(e) = c(e)$ the edge is said to be **saturated**. For any vertex other than the source and the sink, the amount of

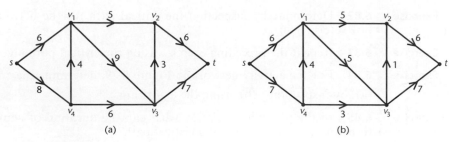

Figure 6.80  Network capacities and flow

material flowing into a vertex must be equal to the amount flowing out. It can also be assumed that the total flow leaving the source is equal to the total flow into the sink.

**Example 6.34**   **Flow in a transportation network**

A transportation network with capacities for each edge is given in Figure 6.80(a). A flow through the network is shown in Figure 6.80(b). You can easily check that the condition on flows for each vertex is satisfied, for example vertex $v_3$ has inflow $5 + 3$ and outflow $1 + 7$. The edges $(s, v_1)$, $(v_1, v_2)$, $(v_4, v_1)$, $(v_2, t)$ and $(v_3, t)$ are saturated, since for each one the flow is equal to the capacity. The flow from the source is $6 + 7 = 13$, and this is equal to the flow into the sink. In this simple example it is easy to see that the flow in Figure 6.80(b) is in fact the greatest possible flow from source to sink.

For a large transportation network it is difficult to find the maximal flow by inspection, as was done in the preceding example. It is necessary to define a **cut** $C$, which is a set of edges such that every directed path from $s$ to $t$ contains at least one edge in $C$. In other words, removal of the edges in $C$ from the network results in a disconnection of $s$ from $t$. The reason behind the name 'cut' is clear from Figure 6.81, where each dashed line can be thought of as cutting up the network in Figure 6.80 into two parts $A$ and $B$. In Figure 6.81(a) the cut $C_1$ consists of the edges $(v_1, v_2)$, $(v_1, v_3)$ and $(v_4, v_3)$; similarly $C_2$ in Figure 6.81(b) consists of $(v_1, v_2)$, $(v_1, v_3)$, $(v_4, v_1)$ and $(s, v_4)$. Each edge in a cut has one vertex in $A$ and one vertex in $B$. The **capacity** of a cut $C$ is the sum of the capacities of those edges in $C$ which are directed from a vertex in $A$ to a vertex in $B$. Thus in Figure 6.81(a) the capacity of $C_1$ is $5 + 9 + 6 = 20$, whereas in Figure 6.81(b) the capacity of $C_2$ is $5 + 9 + 8 = 22$, since the edge $(v_4, v_1)$ is directed from $B$ to $A$. The capacity of a cut is therefore the maximum amount which can flow from $A$ to $B$ across the cut. A key theorem due to the American mathematicians Dantzig, Ford and Fulkerson in 1956 states that the **maximal flow** through a transportation network is equal to the smallest of the capacities of all possible cuts, called the **minimum cut**. Continuing with the network in Figure 6.80, three

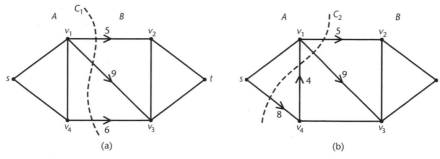

*Figure 6.81* Cuts for Figure 6.80

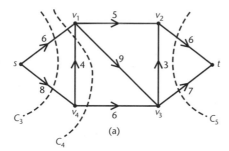

*Figure 6.82* More cuts for Figure 6.80

more cuts $C_3$, $C_4$, $C_5$ are shown in Figure 6.82, with capacities 14, 16 and 13 respectively. You should draw all the other possible cuts for this network, and confirm that $C_5$ is the minimum cut. This agrees with the maximal flow of 13 found in Example 6.34.

### Example 6.35   Maximal flow

To find the minimum cut for the network in Figure 6.83, it is helpful to display the calculations in tabular form. Each cut divides the network into a part $A$ containing the source $s$ and a part $B$ containing the sink $t$. The complete set of cuts is obtained by selecting as the vertices in $A$ all possible combinations of $s$ with the other vertices, excluding $t$. The minimum cut $C$, having the smallest capacity 48, is the sixth in the list and is indicated by the dashed line in Figure 6.83. The maximal flow is therefore 48.

Once the value of the maximal flow has been found, an actual set of flows along the edges can usually be determined by inspection in simple cases.

In general a somewhat complicated algorithm is required to determine the maximal flow and its route through the network, and you are invited to investigate this in Problem 6.33.

| Edges in cut | Vertices in A | Vertices in B | Capacity of cut |
|---|---|---|---|
| $(s,a), (s,b), (s,c)$ | $s$ | $a, b, c, t$ | $16 + 14 + 29 = 59$ |
| $(a,t), (a,b), (s,b), (s,c)$ | $s, a$ | $b, c, t$ | $28 + 7 + 14 + 29 = 78$ |
| $(s,a), (a,b), (b,t), (b,c), (c,b), (s,c)$ | $s, b$ | $a, c, t$ | $16 + 12 + 5 + 29 = 62$ |
| $(s,a), (s,b), (c,b), (b,c), (c,t)$ | $s, c$ | $a, b, t$ | $16 + 14 + 4 + 20 = 54$ |
| $(a,t), (b,t), (c,b), (b,c), (s,c)$ | $s, a, b$ | $c, t$ | $28 + 12 + 5 + 29 = 74$ |
| $(s,a), (a,b), (b,t), (c,t)$ | $s, b, c$ | $a, t$ | $16 + 12 + 20 = 48$ |
| $(a,t), (a,b), (s,b), (c,b), (b,c), (c,t)$ | $s, a, c$ | $b, t$ | $28 + 7 + 14 + 4 + 20 = 73$ |
| $(a,t), (b,t), (c,t)$ | $s, a, b, c$ | $t$ | $28 + 12 + 20 = 60$ |

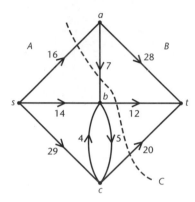

*Figure 6.83* Minimum cut

**Exercise** **6.89** The flow through a transportation network is described by the table below. Each entry is the directed flow between the corresponding vertices – for example, the flow along the directed edge $(v_1, v_3)$ is 5. A null entry indicates there is no directed edge between those vertices. Draw the network, and determine the values of the flows $x_1$, $x_2$ and $x_3$.

| From \ To | $v_1$ | $v_2$ | $v_3$ | $v_4$ | $v_5$ | $t$ |
|---|---|---|---|---|---|---|
| $s$ | 20 | – | – | 10 | 16 | – |
| $v_1$ | – | $x_1$ | 5 | – | – | – |
| $v_2$ | – | – | – | – | – | 42 |
| $v_3$ | – | – | – | 8 | – | 4 |
| $v_4$ | – | $x_2$ | – | – | 0 | – |
| $v_5$ | – | 9 | $x_3$ | – | – | – |

**Exercise**  **6.90**  A transportation network is described by the table of capacities below, where a null capacity indicates there is no directed edge joining the corresponding vertices. Draw the network, and determine by inspection a set of flows along the edges which gives the maximal flow through the network from the source $s$ to the sink $t$.

| From \ To | $v_1$ | $v_2$ | $v_3$ | $t$ |
|---|---|---|---|---|
| $s$ | 60 | 13 | – | – |
| $v_1$ | – | 20 | 14 | 10 |
| $v_2$ | – | – | 35 | – |
| $v_3$ | – | – | – | 51 |

**Exercise**  **6.91**  Draw the diagram for the transportation network whose table of capacities is given below. Determine the value of the maximal flow by finding the minimum cut.

| From \ To | $v_1$ | $v_2$ | $v_3$ | $v_4$ | $t$ |
|---|---|---|---|---|---|
| $s$ | 11 | – | – | 20 | – |
| $v_1$ | – | 15 | – | 8 | – |
| $v_2$ | – | – | 13 | 6 | 13 |
| $v_3$ | – | 12 | – | – | 16 |
| $v_4$ | – | – | 13 | – | – |

**Exercise**  **6.92**  In the preceding exercise, find by inspection an actual maximal flow along the edges.

**Exercise**  **6.93**  Determine by inspection a set of flows along the edges which produces the maximal flow through the network in Figure 6.83.

## Miscellaneous problems

**6.1**  Prove that for any undirected graph, the sum of the degrees of its vertices is equal to twice the number of its edges.

**6.2**  List the indegree and outdegree for each vertex of the digraph in Figure 6.10.
  Prove that for any digraph the sum of the indegrees is equal to the sum of the outdegrees.

**6.3**  If a graph (not a multigraph) has $n$ vertices, show that the maximum possible number of edges is $C(n,2)$. What is this number for a digraph?

**6.4**  A vertex of an undirected graph is called **odd** or **even** according to whether its degree is odd or even. Prove that every graph contains an even number of odd vertices.

**6.5** For an undirected multigraph with $p$ edges connecting vertex $i$ to vertex $j$, the element of the adjacency matrix in row $i$, column $j$ is equal to $p$. Draw the undirected multigraph for which the lower triangle of the adjacency matrix is

$$\begin{bmatrix} 2 & & & \\ 1 & 0 & & \\ 0 & 1 & 1 & \\ 0 & 1 & 2 & 0 \end{bmatrix}$$

**6.6** An undirected graph is called **regular of degree two** if every vertex has degree two.

(a) Draw regular graphs of degree two with three, four and five vertices.

(b) Show that regular graphs of degree *three* must have an even number of vertices (use the result in Problem 6.1). Draw a regular graph of degree three with six vertices.

**6.7** An undirected graph is call **k-regular** if all of its $n$ vertices have the same degree $k$. Prove that it has $(1/2)nk$ edges.

**6.8** Suppose the two sets of vertices in a bipartite graph are $X$ and $Y$. It can be shown that there is a complete matching if for some positive integer $p$ the degrees of all the vertices $x \in X$, $y \in Y$, satisfy $d(x) \geqslant p \geqslant d(y)$. This condition is only a *sufficient* one for a complete matching to exist. It is possible to have a bipartite graph for which there is a complete matching even though this condition is not satisfied.

Use this result to show that if there is a group of six men and six women, and each man likes exactly two women, and each woman is liked by exactly two men, then it is possible for the group to pair off so that every man is with a woman he likes, and every woman is with a man who likes her.

This is an example of the so-called 'marriage problem', perhaps better called the 'partner problem' nowadays.

**6.9** Prove that any two vertices in a tree are linked by exactly one path.

**6.10** Show that if every vertex of an undirected graph $G$ has degree at least two, then $G$ contains a cycle. Hence deduce that every tree has at least one end.

**6.11** Prove that if a tree has $n$ vertices then the sum of the degrees of its vertices is equal to $2n - 2$. Hence prove that every tree with more than one vertex has at least two ends.

**6.12** Prove by induction on $k$ that a binary rooted tree has at most $2^k$ vertices at level $k$.

**6.13** Use the result in the preceding problem to prove that a binary rooted tree of height $n$ has at most $2^{n+1} - 1$ vertices in total. (*Hint*: use the formula for the sum of a geometric series in Section 5.2.2.)

**6.14** Suppose $T$ is a regular $m$-ary rooted tree with $n$ vertices, $p$ internal vertices and $r$ leaves. Show that:

(a) $r = (m - 1)p + 1$   (b) $n = mp + 1$.

**6.15**  Using the notation of the preceding problem, show that

$$p = \frac{n-1}{m} = \frac{r-1}{m-1}$$

Hence determine the number of extension leads needed in Exercise 6.37.

**6.16**  Prove by induction on $n$ that an $m$-ary rooted tree of height $n$ has at most $m^n$ leaves. Hence deduce that the height of the tree is at least $\log_m r$, where $r$ is the number of leaves.

**6.17**  Suppose I play a 'yes/no' game with you like that in Example 6.15. This time I have selected a number $i$ belonging to the set $\{0, 1, 2, 3\}$. Draw a binary search tree which determines $i$ with two questions.

To make things harder for you, I announce that I will lie at most once when giving you my answers. Draw a binary search tree which ensures that you determine $i$ with at most five questions. (*Hint*: first analyze the situation with $i \in \{0, 1\}$ and show that if at most one lie is permitted then three questions are required, instead of just one if I am always truthful.)

**6.18**  Suppose that in the preceding problem I now select $i$ from the set $\{0, 1, 2, 3, \ldots, 7\}$ and again I may lie at most once in answer to your 'yes/no' questions. How many questions will you need to be sure of determining $i$?

**6.19**  Generalize the weighing problem in Example 6.16 and Exercises 6.39 and 6.40 as follows: prove by induction that if a set of $3^m$ coins contains exactly one which is lighter than the others, then the single counterfeit coin can be detected in $m$ weighings. (*Hint*: as in Exercise 6.40, use the three positions of the scales, and first prove true for $m = 1, m = 2$.)

What happens if there are $3^m + 1$ coins, including one counterfeit?

**6.20**  Rooted trees can be used to decode what are called **instantaneous binary codes**, in which codewords have variable lengths. For example, suppose five codewords are:

| Letter | A | B | C | D | E |
|---|---|---|---|---|---|
| Codeword | 0 | 10 | 110 | 1110 | 1111 |

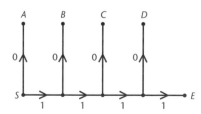

*Figure 6.84*  Decoding tree

These codewords can be represented by the tree in Figure 6.84. Each codeword is reached by a directed path from the start vertex $S$. In order for there to be no ambiguity, no codeword can be an initial part of any other codeword. For example, 100 is not an allowable codeword since 10 denotes B.

Any received string of bits (assumed error-free) can be decoded uniquely and immediately (hence the name). For example, to decode a received message 11001110, start at S in Figure 6.84. The bits 110 take you to C; returning to S, the bit 0 takes you to A; finally, 1110 takes you from S to D so the message is decoded as CAD.

Construct the tree for the code:

| Letter | A | B | C | D | E | K |
|---|---|---|---|---|---|---|
| Codeword | 00 | 01 | 100 | 101 | 1110 | 1111 |

Use this tree to decode the received messages

01001001111, 10111101001111

Determine as many additional codewords as possible having length at most four bits.

**6.21** An instantaneous ternary code has codewords as follows:

| Letter | A | B | C | D | E | I | R | S | T |
|---|---|---|---|---|---|---|---|---|---|
| Codeword | 0 | 10 | 11 | 12 | 20 | 21 | 220 | 221 | 222 |

Represent these codewords by a suitable ternary tree.
Use this tree to decode the received message 1221221112202022220.

**6.22** For a plane graph, the **degree of a face** is defined to be the number of edges which form the boundary of the face. For example, in Figure 6.85 degree($f_2$) = 3 and degree($f_1$) = 5 – notice that the edge $a$ is counted twice for the infinite face $f_1$.

Prove that for a connected plane graph the sum of the degrees of all the faces is equal to twice the number of edges. For example, in Figure 6.85, $3 + 5 = 2 \times 4$.

Verify that this result holds for the graph in Figure 6.48.

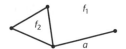

*Figure 6.85* Two faces

**6.23** Let $e$ be an edge of a connected graph $G$ and suppose $e$ does not lie on any cycle of $G$. Show that assuming $e$ is not a bridge produces a contradiction. This proves that if $e$ does not lie on any cycle of $G$ then it must be a bridge. The converse of this result can also be proved, namely that if $e$ is a bridge then it does not lie on any cycle of $G$.

**6.24** Suppose a digraph has no directed eulerian circuit. It can be shown that a directed eulerian trail exists if and only if all except at most two vertices have equal indegree and outdegree, and for at most two vertices the indegree and outdegree differ by one.

Use the result in Problem 6.2 to show that if a digraph has a directed eulerian trail, and exactly one vertex with an excess of one indegree, then there will be exactly one vertex with an excess of one outdegree.

**6.25** Let $G$ be a graph with $n$ vertices. Suppose the vertices can be labelled so that their degrees satisfy $d_1 \leqslant d_2 \leqslant d_3 \leqslant \ldots \leqslant d_n$. Then $G$ is hamiltonian if for each value of $i < (1/2)n$ such that $d_i \leqslant i$ then $d_{n-i} \geqslant n - i$.

Use this result to show that a graph whose vertices have degrees 2,2,2,3,5,7,8,8,9 is hamiltonian.

**6.26** How many different hamiltonian cycles are there in the complete graph $K_n$, for $n \geqslant 3$?

**6.27** Prove that the complete bipartite graph $K_{m,n}$ is hamiltonian if and only if $m = n$.

**6.28** The **$n$-cube** $Q_n$ is a graph with $2^n$ vertices which carry the labels $(x_1, x_2, \ldots, x_n)$ where $x_i = 0$ or $1$. There is an edge between two vertices if and only if they differ in exactly one co-ordinate. For example, $Q_2$ is shown in Figure 6.86. It can be shown that $Q_n$ is hamiltonian for all $n \geqslant 2$.

A **Gray code** is defined as a hamiltonian cycle in $Q_n$. For example, the code 00, 01, 11, 10 is obtained from $Q_2$. Draw $Q_3$ and find a hamiltonian cycle.

Gray codes are used for analogue to digital conversion.

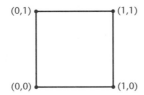

*Figure 6.86* $Q_2$

**6.29** Prove that every connected graph has a spanning tree. (*Hint*: consider a spanning subgraph with the smallest possible number of edges.)

**Student project**

**6.30** (a) Consider the adjacency matrix $A$ in (6.1) for the graph in Figure 6.6. Evaluate $A^2$. Verify that the element in row $i$, column $j$ of $A^2$ is equal to the number of different walks of length 2 from vertex $a_i$ to vertex $a_j$, by writing down all such possible walks. For example, the element in row 1, column 1 of $A^2$ is 2, and the two walks of length 2 from $a_1$ to $a_1$ are $1 \longrightarrow 2 \longrightarrow 1$, $1 \longrightarrow 4 \longrightarrow 1$.

(b) Let $A$ be the adjacency matrix for an undirected graph $G$ with $n$ vertices and at most one edge between any pair of vertices. Prove that the element in row $i$, column $j$ of $A^2$ is equal to the number of different walks of length 2 from vertex $i$ to vertex $j$.

(c) Prove by induction that for the matrix $A$ in part (b), the element in row $i$, column $j$ of $A^p$, where $p$ is a positive integer, is equal to the number of different walks of length $p$ from vertex $i$ to vertex $j$. Check this result for $p = 3$ with $A$ in part (a).

(d) Prove that $G$ is connected if and only if no entry of the matrix

$$A + A^2 + A^3 + \ldots + A^{n-1}$$

is zero. Check this result for the graph in Figure 6.11.

Can you give a result which generalizes this example?

## Student project

**6.31** The following theorem (sometimes called the 'marriage theorem') gives a necessary and sufficient condition for a bipartite graph $G = (V, E)$ with $V$ partitioned into $V = A \cup B$, to have a complete matching from $A$ to $B$. It solves the 'partner problem' illustrated in Problem 6.8.

Let $X$ be any subset of $A$, and let $Y$ be the subset of $B$ consisting of those vertices which are adjacent to at least one vertex in $X$. Then a complete matching exists if and only if for every subset $X$ then $n(X) \leqslant n(Y)$.

(a) Apply this to the graph in Figure 6.27 by listing all the possible sets $X$.

(b) Write out a detailed proof of the theorem with the aid of appropriate textbooks.

(c) Use the literature to describe a detailed algorithm for finding a maximum matching.

## Student project

**6.32** Let $K_n^*$ denote a digraph obtained by assigning directions to the edges of the complete graph $K_n$ with $n$ vertices. That is, for any pair of vertices $a$, $b$, exactly one of the directed edges $(a, b)$ or $(b, a)$ is in $K_n^*$. Such a graph is called a **tournament**, because of the application in (d) below.

(a) Let $v$ be any vertex of $K_n^*$ which has the maximum outdegree of all vertices in $K_n^*$. Prove that the length of a directed path from $v$ to any other vertex is either 1 or 2.

(b) If $u$ is any vertex of $K_n^*$, denote by $K_n^* - u$ the digraph obtained by deleting $u$ and all edges incident with $u$. Prove that $K_n^* - u$ is also a tournament.

(c) Prove by induction that $K_n^*$ contains a directed hamiltonian path.

(d) Consider a 'round robin' tournament involving a number of teams, in which every possible pair of teams plays each other, with no ties allowed. The winner is the team with the most victories (there may be more than one winning team). Draw a digraph to represent the tournament in which there is a directed edge $(T_i, T_j)$ if team $T_i$ beats team $T_j$. Hence deduce that it is possible to draw up a final list of the teams such that each team has beaten the next one on the list. For example, with three teams, if $T_1$ beats $T_2$ and $T_3$, and $T_3$ beats $T_2$ then the list is $T_1, T_3, T_2$.

## Student project

**6.33** Using appropriate textbooks, develop a detailed description of the Ford–Fulkerson algorithm for computing the flows along the edges of a transportation network so as to produce a maximal flow. Use it to solve Exercises 6.91 and 6.92.

## Further reading

| Sections 6.1–6.2 | Chartrand G. (1985). *Introductory Graph Theory*. New York: Dover, Chapters 1, 2 and 9 |
|---|---|
| | Dierker P.F. and Voxman W.L. (1986). *Discrete Mathematics*. San Diego: Harcourt Brace Jovanovich, Chapter 3 |
| | Finkbeiner II D.T. and Lindstrom W.D. (1987). *A Primer of Discrete Mathematics*. New York: W.H. Freeman, Chapter 4 |
| | Grimaldi R.P. (1994). *Discrete and Combinatorial Mathematics, An Applied Introduction*, 3rd edn. Reading MA: Addison-Wesley, Chapters 11 and 12 |
| | Jackson B.W. and Thoro D. (1990). *Applied Combinatorics with Problem Solving*. Reading MA: Addison-Wesley, Chapters 5 and 6 |
| | Molluzzo J.C. and Buckley F. (1986). *A First Course in Discrete Mathematics*. Belmont CA: Wadsworth, Chapter 8 |
| | Ore O. (1962). *Theory of Graphs*. Providence RI: American Mathematical Society |
| | Ore O. (1963). *Graphs and Their Uses*. New York: Random House |
| | Roberts F.S. (1984). *Applied Combinatorics*. Englewood Cliffs NJ: Prentice-Hall, Chapters 3 and 11 |
| | Spode Group (1990). *Decision Mathematics*. Chichester: Ellis Horwood, Chapter 5 |
| | Wilson R.J. and Watkins J.J. (1990). *Graphs, An Introductory Approach*. New York: Wiley |
| Section 6.3 | Chartrand G. (1985). *op. cit.*, Chapter 4 |
| | Dierker P.F. and Voxman W.L. (1986). *op. cit.*, Chapter 4 |
| | Grimaldi R.P. (1994). *op. cit.*, Chapter 13 |
| | Jackson B.W. and Thoro D. (1990). *op. cit.*, Chapters 6 and 10 |
| | Roberts F.S. (1984). *op. cit.*, Chapter 13 |
| | Spode Group (1990). *op. cit.*, Chapter 6 |
| Section 6.3.3 | French S., Hartley R., Thomas L.C. and White D.J. (1986). *Operational Research Techniques*. London: Edward Arnold, Chapter 6 |
| | Molluzzo J.C. and Buckley F. (1986). *op. cit.*, p.454 |
| | Spode Group (1990). *op. cit.*, Chapter 14 |

Section 6.3.4    Finkbeiner II D.T. and Lindstrom W.D. (1987). *op. cit.*, p.306

Ford L.R. Jr. and Fulkerson D.R. (1962). *Flows in Networks*. Princeton NJ: Princeton University Press

Gass S.I. (1975). *Linear Programming*, 4th edn. New York: McGraw Hill, Section 11.4

Hu T.C. (1969). *Integer Programming and Network Flows*. Reading MA: Addison-Wesley, Chapter 8

Spode Group (1990). *op. cit.*, Chapter 10

# Answers to Exercises

## Chapter 1

**1.1**  495

**1.2**  2538, which goes into itself

**1.4**  0.037037037 . . .

**1.5**  (a) $\frac{181}{333}$  (b) $\frac{1183}{1100}$  (c) $\frac{35813}{33300}$

**1.6**  Cyclic arrangement of 153846. 0.076923076923 . . .

**1.8**  5, 7 produce endless loops. 9 ends at 1.

**1.10**  (a) 3, 4  (b) 3, 4, 7  (c) 4, 7  (d) 3, 7, 13

**1.11**  (a) *abcabc*  (b) *ab0ab*

      (c) *aaaaaaaaaaaa*, or *abcabcabcabc*, or *ab0ab0ab0ab*, etc.

**1.12**  (a) 10003  (b) 100002  (c) 1000006

**1.13**  Divisible by 3 and 4

**1.18**  (a) $-1 + 17i$  (b) $-155 - 180i$  (c) $16 + 30i$  (d) $861 + 327i$

      (e) $\frac{24}{53} - \frac{31}{53}i$  (f) $\frac{99}{578} + \frac{5}{578}i$

**1.19**  (a) $-6 \pm 2i$  (b) $-\dfrac{7}{4} \pm \dfrac{i\sqrt{23}}{4}$

**1.22**  40077.29749 km, 0.5 cm

**1.23**  Correct to 6 decimal places; 4 decimal places

**1.24**  3.14151

**1.25**  9

## Interlude

**I.2**  $n(n+1)$

**I.5**  (a) 66  (b) 100  (c) 13  (d) $x_2 + x_5 + x_8$

**I.6**  (a) $\displaystyle\sum_{i=1}^{200}(2i - 1)$  (b) $\displaystyle\sum_{i=10}^{100} 2i$  (c) $\displaystyle\sum_{i=1}^{18} 7i$  (d) $\displaystyle\sum_{i=0}^{8} x_{7+4i}$  (e) $\displaystyle\sum_{i=0}^{7} 7i^2$

**I.7**  (a) $98x$  (b) $4\displaystyle\sum_{j=3}^{20} j - 54$  (c) 12  (d) $6\displaystyle\sum_{j=-5}^{20} j$

**I.9**  $\dfrac{n}{n+1}$

**I.14**  $\left(1 - \dfrac{1}{2^2}\right)\left(1 - \dfrac{1}{3^2}\right)\cdots\left(1 - \dfrac{1}{n^2}\right) = \dfrac{n+1}{2n}, \qquad n \geq 2$

## Chapter 2

**2.1** (a) 3 o'clock  (b) 9 o'clock  (c) 2 o'clock

**2.2** (a) 20 minutes after the hour
(b) 10 minutes before the hour

**2.3** 4:30

**2.4** Thursday; Friday

**2.5** Thursday

**2.6** Monday

**2.7** $110°$; $340°$

**2.8** (a) 0.5  (b) 1  (c) 0  (d) 0.5

**2.9**

| $B$ | $C$ | $a$ | $b$ | $c$ | $d$ | 1 | 2 | 3 |
|------|------|------|------|------|------|------|------|------|
| 1002 | 1003 | 1201 | 1202 | 1203 | 1210 | 0301 | 0302 | 0303 |

**2.10** (a) 45  (b) 386  (c)1143

**2.11** (a) $1\frac{5}{8}$  (b) $13\frac{9}{25}$  (c) $120\frac{120}{121}$

**2.13** (a) 1011011

(b)
| $A$ | $B$ | $C$ | $b$ | $c$ | $d$ | 1 | 2 | 3 |
|------|------|------|------|------|------|------|------|------|
| 101 | 102 | 103 | 142 | 143 | 144 | 061 | 062 | 063 |

**2.14** (a) 47  (b) 242  (c) 2983

**2.15**

| 11 | 12 | 13 | 14 | 15 | 16 | 17 |
|------|------|------|------|------|------|------|
| $(1011)_2$ | $(1100)_2$ | $(1101)_2$ | $(1110)_2$ | $(1111)_2$ | $(10000)_2$ | $(10001)_2$ |

| 18 | 19 | 20 |
|------|------|------|
| $(10010)_2$ | $(10011)_2$ | $(10100)_2$ |

**2.16** (a) $(1010111)_2$  (b) $(1111001)_2$  (c) $(110101000)_2$
(d) $(10011100010000)_2$

**2.17** (a) $(101010)_2, 23 + 19 = 42$
(b) $(1011011)_2, 37 + 54 = 91$
(c) $(110111)_2, 21 + 15 + 19 = 55$
(d) $(101101010)_2, 34 + 17 + 101 + 210 = 362$

**2.18** (a) $(1000)_2, 22 - 14 = 8$
(b) $(10110)_2, 101 - 79 = 22$
(c) $(1000010)_2, 212 - 191 + 45 = 66$

**2.19** (a) $(1101001000)_2, 21 \times 40 = 840$
(b) $(10110010011101)_2, 243 \times 47 = 11421$
(c) $(1101100000100111)_2, 93 \times 35 \times 17 = 55335$

**2.20** (a) $(10000000000000)_2 = 8192$
(b) $(1001001100)_2 = 588$
(c) $(100011100110)_2 = 2278$

**2.21** (a) $(100001)_2 = 33$
(b) $(1101)_2 = 13$, remainder $(11)_2 = 3$
(c) $(1000011)_2 = 67$
(d) $(10000011)_2 = 131$, remainder $(1111)_2 = 15$

**2.22** (a) 3.375  (b) 0.109375  (c) 1.71875

**2.24** (a) yes  (b) no

**2.25**

| 0 | 1 | 2 | 3 | 4 | 5 | 6 | 7 | 8 |
|---|---|---|---|---|---|---|---|---|
| 00000 | 00001 | 00010 | 00011 | 00100 | 00101 | 00110 | 00111 | 01000 |
| 9 | 10 | 11 | 12 | 13 | 14 | 15 | $-1$ | $-2$ |
| 01001 | 01010 | 01011 | 01100 | 01101 | 01110 | 01111 | 11111 | 11110 |
| $-3$ | $-4$ | $-5$ | $-6$ | $-7$ | $-8$ | $-9$ | $-10$ | $-11$ |
| 11101 | 11100 | 11011 | 11010 | 11001 | 11000 | 10111 | 10110 | 10101 |
| $-12$ | $-13$ | $-14$ | $-15$ | $-16$ | | | | |
| 10100 | 10011 | 10010 | 10001 | 10000 | | | | |

**2.28**  16 bits: $-2^{15}$ to $2^{15} - 1$; $128 = (0000000010000000)_2$;
$-64 = (1111111111000000)_2$
32 bits: $-2^{31}$ to $2^{31} - 1$; 128 has eighth bit = 1, all others zero; $-64$ has first 26 bits = 1, all others zero.

**2.32**

| 17 | 18 | 19 | 20 | 21 | 22 | 23 | 24 |
|----|----|----|----|----|----|----|----|
| $(21)_8$ | $(22)_8$ | $(23)_8$ | $(24)_8$ | $(25)_8$ | $(26)_8$ | $(27)_8$ | $(30)_8$ |

etc. up to $49 = (61)_8$

**2.33**  (a) 2954   (b) 31812

**2.34**  (a) $(101110001010)_2$   (b) $(111110001000100)_2$

**2.35**  (a) $(33764)_8$   (b) $(15243)_8$

**2.36**  (a) $(11104)_8$, $4012 + 664 = 4676$
(b) $(2605)_8$, $483 + 190 + 296 + 444 = 1413$
(c) $(215306)_8$, $30485 + 24830 + 17075 = 72390$

**2.37**  (a) $(365)_8$, $483 - 238 = 245$
(b) $(2027)_8$, $1144 - 395 + 298 = 1047$
(c) $(41626)_8$, $28945 - 11643 = 17302$

**2.38**

| + | 1 | 2 | 3 | 4 | 5 | 6 | 7 |
|---|---|---|---|---|---|---|---|
| 4 | $(5)_8$ | $(6)_8$ | $(7)_8$ | $(10)_8$ | $(11)_8$ | $(12)_8$ | $(13)_8$ |
| 5 | | | | $(11)_8$ | $(12)_8$ | $(13)_8$ | $(14)_8$ |
| 6 | | symmetrical | | | | $(14)_8$ | $(15)_8$ |
| 7 | | | | | | | $(16)_8$ |

| × | 1 | 2 | 3 | 4 | 5 | 6 | 7 |
|---|---|---|---|---|---|---|---|
| 4 | $(4)_8$ | $(10)_8$ | $(14)_8$ | $(20)_8$ | $(24)_8$ | $(30)_8$ | $(34)_8$ |
| 5 | | | | $(24)_8$ | $(31)_8$ | $(36)_8$ | $(43)_8$ |
| 6 | | symmetrical | | | | $(44)_8$ | $(52)_8$ |
| 7 | | | | | | | $(61)_8$ |

**2.39**  (a) $(1213)_8$, $31 \times 21 = 651$
(b) $(102356)_8$, $166 \times 205 = 34030$
(c) $(624)_8 = 404$, remainder $(1)_8$

**2.40**  (a) $(10000000)_8 = 2097152$   (b) $(13602)_8 = 6018$

**2.41**  (a) 5.265625   (b) 0.55859375

**2.42**  number must end with $k$ zeros

**2.44**  (a) yes   (b) no

**2.46**  (a) yes   (b) no

**2.48**  (a) 10894   (b) 11616064

**2.49**　(a) $(1E85)_{16}$　(b) $(BF1D)_{16}$

**2.50**　sum $= (DDA2)_{16}$, difference $= (A098)_{16}$

**2.51**　(a) $(1B863)_{16}$　(b) $(ED250)_{16}$

**2.52**　$(3AF00)_{16} = 241408$

**2.53**　(a) $(1CC)_{16}$　(b) $(395)_{16}$　(c) $(5323)_8$　(d) $(54237500)_8$

**2.54**　(a) $(45144)_{16}$　(b) $(6A3634D9)_{16}$　(c) $(17A6)_{16}$, remainder $(3)_{16}$

**2.55**　(a) $\dfrac{4095}{4096}$　(b) $10\dfrac{487}{2048}$

**2.56**　Divisible by 15 provided sum of digits is divisible by 15
(a) yes　(b) no

**2.57**　$(a_n \ldots a_1)_{16}$ divisible by 17 provided $a_1 - a_2 + a_3 - \ldots + (-1)^{n-1}a_n$ is divisible by 17
(a) yes　(b) yes

**2.58**　(a) 3873　(b) $(101220002)_3$　(c) $(110000)_3$　(d) $(211112)_3$
(e) $(1111222222)_3$

**2.59**　(a) 902　(b) 148818　(c) 2983

**2.60**　(a) $(11436)_7$　(b) 15166

**2.61**　56809, 176951

**2.62**　(a) 12 gross and 3　(b) 10 gross, 10 dozen and 8
(c) 24 gross, 11 dozen and 7

**2.63**　$(9B8B)_{12}$

**2.64**　$(5, 21, 45, 19)_{60}$

**2.65**　Even provided there are an even number of 1's; no, yes.

**2.66**　$(815)_9$, $(70802)_3$

**2.69**　(a) Divisible by 2 and 4
(b) Divisible by 2, but not 4
(c) Not divisible by 2

**2.70**　(a) Divisible by 2 if last digit is 0, 2, 4 or 6
Divisible by 4 if last digit is 0 or 4
(b) Divisible by 2 if last digit is 0, 2, 4, 6, 8, $A$, $C$, $E$
Divisible by 4 if last digit is 0, 4, 8, $C$
Divisible by 8 if last digit is 0 or 8

**2.71**　(a) Divisible by 8 but not 6
(b) Divisible by 10 but not 8
(c) Divisible by 12 but not 10

**2.72**　(a) yes　(b) yes　(c) no

**2.73**　(a) no　(b) yes　(c) yes

**2.75**　Divisible by 3 but not 6

**2.76**　$q = 135, r = 16$

**2.77**　(a) 2, 3, 4, g.c.d. $= 4$　(b) 3, 5, 7, 9, 15, 21, 45, 63, g.c.d. $= 63$

**2.78**　Divisible by 3, 9, 27 but not 81; 23679

**2.79**　$a = b$

**2.84**　Divisible by 7

**2.85**　78, 110, 165

**2.87** (a) 27 (b) 11 (c) 1

**2.88** 6

**2.89** (a) $x = -1, y = 12$ (b) $x = 19, y = -376$

**2.90** $x = -1, y = 4, d = 19$

**2.91** $x = 157, y = -4676$

**2.94** $x = 3400 - 39p, y = -2000 + 23p, p =$ arbitrary integer. $x = 7, y = 1$

**2.95** 16 at 28p, 24 at 23p

**2.96** 175 larger tiles, 1050 smaller tiles using strips 10 cm wide

**2.98** $(F_{16}, F_{24}) = 21 = F_8 = F_{(16,24)}$

**2.100** (a), (b) and (d) are valid

**2.101** $m = 7$ (b) $m = 2, 3, 4, 5, 6, 8, 10, 12, 15, 20, 24, 30, 40, 60$ (c) $m = 13$

**2.102** (a) 4 (b) 20 (c) 0 (d) 105

**2.103** Monday

**2.104** (a) 3 (b) 8

**2.107** (a) $26 \equiv 16 (\text{mod } 5)$ (b) $31 \equiv 1 (\text{mod } 6)$ (c) $43 \equiv 6 (\text{mod } 37)$

**2.112** (a) 2 (b) 1 (c) 3

**2.115** (a) yes (b) yes (c) error not detected

**2.116** 9

**2.117** (a) $8 (\text{mod } 11)$ (b) $4 (\text{mod } 8)$ (c) $2 (\text{mod } 6)$

**2.118** (a) no solution (b) no solution (c) $22 (\text{mod } 29)$ (d) no solution
(e) $6 (\text{mod } 13)$

**2.119** (a) $2 (\text{mod } 35)$ (b) $568 (\text{mod } 725)$ (c) $7 (\text{mod } 14)$

**2.120** $5^{-1} = 9, 6^{-1} = 2, 7^{-1} = 8$

**2.121** $c = 4, x \equiv 2 (\text{mod } 3); c = 8, x \equiv 1 (\text{mod } 3)$

**2.122** (a) $54 (\text{mod } 56)$ (b) $11 (\text{mod } 105)$ (c) $472 (\text{mod } 630)$

**2.123** 441

**2.124** 981

**2.127**

$\mathbb{Z}_6$:

| + | 0 | 1 | 2 | 3 | 4 | 5 |
|---|---|---|---|---|---|---|
| 0 | 0 | 1 | 2 | 3 | 4 | 5 |
| 1 | 1 | 2 | 3 | 4 | 5 | 0 |
| 2 | 2 | 3 | 4 | 5 | 0 | 1 |
| 3 | 3 | 4 | 5 | 0 | 1 | 2 |
| 4 | 4 | 5 | 0 | 1 | 2 | 3 |
| 5 | 5 | 0 | 1 | 2 | 3 | 4 |

| × | 0 | 1 | 2 | 3 | 4 | 5 |
|---|---|---|---|---|---|---|
| 0 | 0 | 0 | 0 | 0 | 0 | 0 |
| 1 | 0 | 1 | 2 | 3 | 4 | 5 |
| 2 | 0 | 2 | 4 | 0 | 2 | 4 |
| 3 | 0 | 3 | 0 | 3 | 0 | 3 |
| 4 | 0 | 4 | 2 | 0 | 4 | 2 |
| 5 | 0 | 5 | 4 | 3 | 2 | 1 |

$-1 = 5, -2 = 4, -3 = 3, -4 = 2, -5 = 1$ $1^{-1} = 1, 5^{-1} = 5$
For $\mathbb{Z}_7$: $-1 = 6, -2 = 5, -3 = 4, -4 = 3, -5 = 2, -6 = 1, 1^{-1} = 1,$
$2^{-1} = 4, 3^{-1} = 5, 4^{-1} = 2, 5^{-1} = 3, 6^{-1} = 6$

**2.128** 3

**2.129** (a) 4   (b) 11   (c) 13   (d) does not exist   (e) 17

**2.130** $1^{-1} = 1, 5^{-1} = 5, 7^{-1} = 7, 11^{-1} = 11, 13^{-1} = 13, 17^{-1} = 17, 19^{-1} = 19,$
$23^{-1} = 23$

**2.131** 20

**2.132**

| $a$ | 1 | 3 | 4 | 5 | 9 |
|---|---|---|---|---|---|
| $\sqrt{a}$ | 1 or 10 | 5 or 6 | 2 or 9 | 4 or 7 | 3 or 8 |

**2.133** (a) 1 or 7   (b) 7 or 10

## Chapter 3

**3.1**   36

**3.2**   9000; 2000

**3.3**   (a) 64   (b) 81

**3.4**   450

**3.5**   (a) 3   (b) 506,000

**3.6**   1024; £86

**3.7**   12; 7

**3.8**   (a) 300   (b) 21   (c) 9261   (d) 7980

**3.9**   28,561; 1014

**3.10**  162

**3.11**  40,320

**3.12**  $2.423 \times 10^{18}; 8.321 \times 10^{81}$

**3.13**  (a) 210   (b) 7920   (c) 342   (d) 145   (e) 40585

**3.14**  (a) $P(13, 3)$   (b) $P(30, 4)$

**3.15**  (a) 720   (b) $n^3 + 6n^2 + 11n + 6$   (c) $n^4 + 2n^3 - n^2 - 2n$

**3.16**  (a) 20,160   (b) 4,989,600

**3.18**  (a) 12,600   (b) 12,600

**3.19**  $\dfrac{40!}{(13!)\,(9!)^3}$

**3.20**  $\dfrac{43!}{(8!)^2(9!)^3 2!3!}$

**3.21**  4,037,880

**3.22**  (a) 360   (b) 1296

**3.23**  (a) 180   (b) 648

**3.24**  1680

**3.25**  600

**3.26**  13,200

**3.27**  260; 5660

**3.28**  9360

**3.29**  120

**3.30** (a) 35 (b) 330 (c) 171 (d) 67,525

**3.31** (a) $\dfrac{16!}{11!5!}$ (b) $\dfrac{17!}{12!5!}$

**3.32** 575,757

**3.34** 270,725 (a) 28,561 (b) 13,182 (c) 880

**3.36** 100

**3.37** 8

**3.38** 15

**3.39** 27,405 (a) 9792 (b) 24,345 (c) 23,850

**3.40** (a) 4368 (b) 900 (c) 258 (d) 1236

**3.41** 7560

**3.42** 21; 262

**3.43** 255

**3.44** (a) 127 (b) 4

**3.45** 8

**3.46** 15

**3.47** (a) 792 (b) 31,824 (c) 364

**3.48** (a) 36 (b) 15 (c) 12

**3.49** $3^7$

**3.50** 99,792

**3.51** 1140

**3.52** (a) $a^5 + 5a^4b + 10a^3b^2 + 10a^2b^3 + 5ab^4 + b^5$

(b) $1 + 7x + 21x^2 + 35x^3 + 35x^4 + 21x^5 + 7x^6 + x^7$

(c) $a^6 + 12a^5b + 60a^4b^2 + 160a^3b^3 + 240a^2b^4 + 192ab^5 + 64b^6$

**3.53** (a) 252 (b) 20 (c) 3876 (d) −448

**3.55** 19,487,171

**3.57** 119,707

**3.63** $r > \frac{1}{2}(n + 1)$ or $r < \frac{1}{2}(n + 1)$

**3.66** $2^k - 1$

**3.67** $\frac{1}{6}, \frac{1}{30}, \frac{1}{60}, \frac{1}{30}, \frac{1}{6}$

**3.68** 677

**3.69** 13

**3.70** (a) 3 (b) 14

**3.71** (a) 4 (b) 26

**3.72** 13

**3.74** (a) 366 (b) 101 (c) 104

**3.81** Yes

**3.82** (a) $\{6, 7, 8, 9, 10, 11\}$ (b) $\{-3, -2, -1, 0, 1, 2, 3\}$ (c) Ø

**3.83** (a) yes (b) no (c) yes

**3.84** (a) Ø (b) $U$ (c) $B$ (d) $\{3, 4\}$ (e) $\{3, 4, 5, 6, 7, 8, 9, 10\}$
(f) $\{1, 2, 3, 4\}$

**3.85** No

**3.87** $U =$ set of all people

**3.88** 37

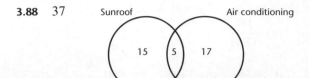

**3.89** 157

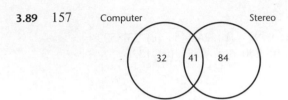

**3.90** 46
**3.91** 75,600
**3.92** 53
**3.93** No, only 110 have at least one A-level pass
**3.94** 8
**3.95** 720
**3.96** 1326; 351
**3.97** 284
**3.100** 9
**3.101** (a) 243   (b) 32   (c) 150
**3.103** $d_n \times 6! = 265 \times 6!$
**3.104** (a) $\frac{1}{16}$   (b) $\frac{1}{4}$   (c) $\frac{11}{16}$
**3.105** (a) $\frac{1}{36}$   (b) $\frac{1}{18}$   (c) $\frac{1}{9}$
**3.106** 2299 to 1
**3.107** (a) $\frac{3}{4}$   (b) $\frac{1}{5}$   (c) $\frac{8}{17}$
**3.108** (a) 11 to 5 against   (b) 11 to 5 on
**3.109** (a) 0.206   (b) 0.059   (c) 0.283
**3.110** (a) 0.888   (b) 0.112   (c) 0.368
**3.111** (a) 0.105   (b) 0.003   (c) 0.895
**3.112** $\frac{1}{4}$
**3.113** 0.0625; 0.176
**3.114** 11
**3.115** (a) 0.932   (b) 0.068   (c) 0.065
**3.118** 38 to 1
**3.119** $\frac{13}{16}$
**3.120** (a) $\frac{80}{243}$   (b) $\frac{192}{243}$. For thin coin, $\frac{5}{16}$ and $\frac{1}{2}$
**3.121** (a) 0.925   (b) 0.23
**3.122** 0.148
**3.123** (a) 0.021   (b) 0.359
**3.125** 75%

**3.126** (a) 0.817 (b) 0.015 (c) 0.016
**3.128** 0.35; 0.23
**3.129** (a) 0.75 (b) 0.5 (c) 0.35 (d) 0.8125
**3.130** 0.4
**3.131** 0.883
**3.132** Yes
**3.133** $\frac{1}{5}$
**3.134** 45% for $A$
**3.136** $P(E_1|E_2) = 0.091$
**3.137** 0.165; yes, because probability before test was 0.01

## Chapter 4

**4.1** Last three digits give elapsed days of year
**4.2** Is first/second/third digit 1?
**4.4** 28 words in total; $2^6$
**4.5** (a) 10001 (b) 11100 (c) 00110 (d) at least 2 errors
**4.6** At least 3 transmission errors
**4.7** Probability $= (0.999)^2$
**4.8** Probability $= (0.999)^4 + 3(0.001)(0.999)^3$
**4.9**

| c | d | 2 | 3 |
|---|---|---|---|
| 11000110 | 11001001 | 01100101 | 01100110 |

**4.13** (a) 010101 (b) error (c) 111100
**4.14** (a) 122101 (b) 010202 (c) error
**4.15** (a) 0120, 2220, 2100, 2121
(b) 1020, 1110, 1122, 2010, 2022, 2112
**4.16** No error, or at least two errors
**4.17** (a) 17514, error (b) 920838
**4.19** 0
**4.20** (a) no (b) no (c) yes
**4.21** (b) 9770963149443
**4.22** $x = 4$
**4.23** Detected in first case, but not in second case when $|x_i - x_{i+1}| = 5$
**4.24** $x_2 + x_4 + x_6 + 3(x_1 + x_3 + x_5 + x_7) + x_8$
**4.25** $x_{12} = 4$
**4.29** (a) $x_i$ becomes $y_i$ with $|x_i - y_i| = 7, i \leq 9$
(b) undetected only when $|x_i - x_{i+1}| = 7, i \leq 8$
**4.30** (a) 41933
**4.31** (a) $x_8 = 1$
**4.32** 6
**4.33** 19101-7260; 3
**4.34** 45701-2979
**4.35** 51593-2067

**4.36**

| Decimal digit | 0 | 1 | 2 | 3 | 4 | 5 | 6 | 7 | 8 | 9 |
|---|---|---|---|---|---|---|---|---|---|---|
| (a) | 0000 | 0001 | 0010 | 0011 | 0100 | 0101 | 0110 | 0111 | 1000 | 1001 |
| (b) | 0000 | 0001 | 0011 | 0100 | 0101 | 0111 | 1000 | 1001 | 1011 | 1100 |
| (c) | 0011 | 0100 | 0101 | 0110 | 0111 | 1000 | 1001 | 1010 | 1011 | 1100 |

(a) = 4-bit binary representation of decimal $d$

(c) = 4-bit binary representation of $d + 3$

**4.37** (a) 2  (b) 3  (c) 3

**4.38** 2, 2, 1, 5

**4.39** (a) 2, 4, 1, 5, 4; 101010

(b) 2, 3, 3, 6, 2; two errors

(c) 2, 1, 5, 4, 4; 010101

**4.40** (a) 010101  (b) 000000

**4.41** 7; 6

**4.42** 2

**4.43** 7

**4.44** (a) 35  (b) 2  (c) detects one

**4.48** (a) yes, $\delta = 1$  (b) no, $\delta = 1$  (c) yes, $\delta = 3$  (d) yes, $\delta = 3$

**4.49** neither contains **0**

**4.51** (a) $N - 1$  (b) $\frac{1}{2}N(N - 1)$

**4.55** (a) $\begin{bmatrix} 1 \\ 1 \end{bmatrix}$  (b) $\begin{bmatrix} 1 \\ 0 \\ 1 \end{bmatrix}$

**4.56** 0000000, 0111001, 1000110, 0100101, 1100011, 0010111, 0001011, 1001101, 0101110, 1010001, 1111111

$\delta = 3$

**4.57** (a) 00000, 00111, 11101, 11010, $k = 2, \delta = 3$

00000, 01110, 11101, 10011, $k = 2, \delta = 3$

(b) 0000000, 1111100, 1110010, 0101001, 0001110

0100111, 1010101, 1011011, $k = 3, \delta = 3$

**4.58** (a) no  (b) correction  (c) detection only

**4.61** (a) 00101  (b) 11110, 11011

**4.62** (a) 26  (b) 6

**4.63** 3 check bits

**4.64** (a) 11  (b) 15

**4.65** 67,108,864

**4.66** (a) $\begin{bmatrix} 1 \\ 1 \\ 0 \end{bmatrix}$  (b) 1011010

**4.67** (a) 11101  (b) more than one error  (c) 11101

**4.68** 11001, 100110; 000000

**4.70** (a) 10011  (b) more than one error

**4.71** 0000000, 1110000, 1001100, 0101010,
1101001, 0111100, 1011010, 0011001,
0100101, 1100110, 1000011, 0010110,
1010101, 0110011, 0001111, 1111111

**4.72** 1101

**4.73** (a) 1110101101  (b) (i) 001001  (ii) more than one error

**4.74** 11101100010

**4.76** Incorrectly decoded as 011001

**4.77** 0000, 0111, 1120, 2021, 0222

**4.78** 1120, 2210

**4.82** (a) 57; $4.32 \times 10^{45}$  (b) 12; $2.59 \times 10^{10}$  (c) 2380; $13^{2376}$

**4.84** (a) 120100, 011010, 201001, 210200, 111201, 100221
(b) 112120; more than one error

**4.85** (a) 625  (b) 311224, 004321  (c) 323140

**4.86** 1, 2, 22, 122

**4.87** (a) yes  (b) yes  (c) no

**4.89** (a) $x = 8$  (b) $x = 7$

**4.90** (b) (i) 0195823273  (ii) 0198583273

**4.91** (b) 3  (c) 0681875356, 0271875356, 0671875256

**4.93** 9780201600445

**4.94** (a) 2104536138  (b) more than one error  (c) no error
(d) more than one error  (e) 3710425859

**4.95** $x_9 = 2, x_{10} = 4$

**4.97** 0000000000, 0505000000

**4.98** (a) 01073894340  (b) error $e$ in $x_4$, $11 - e$ in $x_{10}$, $1 \le e \le 10$

**4.99**

| $a$ | 1 | 2 | 3 | 4 | 5 | 6 | 7 | 8 | 9 | 10 |
|---|---|---|---|---|---|---|---|---|---|---|
| $a^{1/3}$ | 1 | 7 | 9 | 5 | 3 | 8 | 6 | 2 | 4 | 10 |

**4.101** (a) 2706328592  (b) more than two errors  (c) 0800751345

**4.102** $x_7 x_8 x_9 x_{10} = 3966$

**4.105**

| $a$ | 1 | $\alpha$ | $\alpha^2$ | $\alpha^3$ | $\alpha^4$ | $\alpha^5$ | $\alpha^6$ | $\alpha^7$ | $\alpha^8$ | $\alpha^9$ | $\alpha^{10}$ | $\alpha^{11}$ | $\alpha^{12}$ | $\alpha^{13}$ | $\alpha^{14}$ |
|---|---|---|---|---|---|---|---|---|---|---|---|---|---|---|---|
| $\sqrt{a}$ | 1 | $\alpha^8$ | $\alpha$ | $\alpha^9$ | $\alpha^2$ | $\alpha^{10}$ | $\alpha^3$ | $\alpha^{11}$ | $\alpha^4$ | $\alpha^{12}$ | $\alpha^5$ | $\alpha^{13}$ | $\alpha^6$ | $\alpha^{14}$ | $\alpha^7$ |

**4.106**

| 3-bit word | Polynomial | Power of $\alpha$ |
|---|---|---|
| 000 | 0 | 0 |
| 100 | 1 | 1 |
| 010 | $\alpha$ | $\alpha$ |
| 001 | $\alpha^2$ | $\alpha^2$ |
| 110 | $1 + \alpha$ | $\alpha^3$ |
| 011 | $\alpha + \alpha^2$ | $\alpha^4$ |
| 111 | $1 + \alpha + \alpha^2$ | $\alpha^5$ |
| 101 | $1 + \alpha^2$ | $\alpha^6$ |

001, 011

**4.107** 0010, 1110

**4.108** 110110110110110

**4.110** Errors in positions (a) 3, 14 (b) 5, 11

**4.113** (a) 100111001000001 (b) more than two errors
(c) 100111001000001 (d) 110111110011000

**4.117** GPSAR; RING

**4.118** CAHE; TIGER

**4.119**

| Plaintext | E F G ... V W |
|-----------|--------------|
| Ciphertext | R U X ... Q T |

**4.120** LOVE MATHEMATICS

**4.121** FUTUSP; LAST EXERCISE

## Chapter 5

**5.1** 1.81712

**5.2** −4.5895; 1.0895

**5.3** 0.453398

**5.4** 5.5

**5.5** 4.868

**5.6** £1340.10; £1347.35

**5.7** (c)

**5.8** £12000; £8437.87

**5.9** £540.75; $x_{n+1} = 1.04x_n - 300$

**5.10** £1306.68, £999.38

**5.11** (a) $\dfrac{5}{2}(3)^n - \dfrac{7}{2}$ (b) $\dfrac{17}{3}\left(-\dfrac{1}{2}\right)^n - \dfrac{8}{3}$

(c) $4 + 2n$ (d) $4 - \dfrac{1}{2}n + \dfrac{5}{2}n^2$

**5.12** $7500 - 5500(1.04)^n$

**5.14** (a) −21846 (b) −11 (c) 24

**5.16** $x_0 = \dfrac{5}{4}$, $x_{20} = 262145$

**5.18** $\dfrac{5}{3}(3)^{n-4} - \dfrac{7}{2}$

**5.19** £411.54

**5.20** No

**5.21** $x_{15} = 19.99895$; $n \geq 11$

**5.22** (a) $\dfrac{5}{3}(2)^n + \dfrac{7}{3}(5)^n$ (b) $2^{n-1}(8 + 3n)$ (c) $2^{n-1}\left(\dfrac{10}{3} + 3n\right) + \dfrac{7}{3}(5)^n$

**5.23** $x_6 = 5.1859$, an approximate root of the equation $x^3 - 5x^2 - 5 = 0$

**5.24** 69 years

**5.25** 1.62%

**5.26** 40.4 years

**5.27**  1800 years

**5.28**  0.064 cm

**5.29**  0.007 cm; 4.9 m

**5.30**  5.5 cm

**5.31**  7.84 billion, 7.93 billion

**5.32**  £7.02 million

**5.33**  (a) $5(3)^n + 3(-4)^n$  (b) $-(3+i)^n + 5(3-i)^n$  (c) $(4+7n)3^n$

**5.34**  $x_n = 2(3)^n - 2^n + 1$, $x_{10} = 117075$

**5.35**  $-12$

**5.36**  $x_5 = 3.13$; $n \geq 7$

**5.37**  $I_{n+2} = 1.5I_{n+1} - I_n + G$.
Economy collapses by year 9, subsequently rises, only to fall again.

**5.38**  $x_{10} = 2.555$, an approximate root of $x^4 - 3x^3 + 2x^2 - 3x + 2 = 0$

**5.39**  $x_n = 4[(2.5)^n - 1]$; approximately $8\frac{1}{2}$ hours

**5.40**  (a) $2(3)^n - (-4)^n + \frac{1}{9}(5)^n$  (b) $-(3)^{n-1}\left(\frac{1}{7}n + 12\right) - 3(-4)^n$

**5.42**  2731; 14 months

**5.47**  75 cars at city, 40 cars at airport

**5.48**  $a = 0$,  $b = 1$,  $c = -12$,  $d = -7$
$x_n = -7(-4)^n + 9(-3)^n$,  $y_n = 28(-4)^n - 27(-3)^n$

**5.49**  4, $-2$;

$$C^n = \frac{1}{2} \begin{bmatrix} 4^n + (-2)^n & 4^n - (-2)^n \\ 4^n - (-2)^n & 4^n + (-2)^n \end{bmatrix}$$

$$x_n = 2(4)^n + (-2)^n, \quad y_n = 2(4)^n - (-2)^n$$

**5.51**

| $n$ | 1 | 2 | 3 | 4 | |
|-----|-----|------|-----|-----|---|
| $x_n$ | 1.0 | 10.0 | 2.0 | 1.0 | |
| $y_n$ | 1.0 | 0.5 | 5.0 | 1.0 | in thousands |
| $z_n$ | 1.0 | 0.2 | 0.1 | 1.0 | |

$x_{n+3} = x_n, \ y_{n+3} = y_n, \ z_{n+3} = z_n$

## Chapter 6

**6.1**  (a) $1 \longrightarrow 2 \longrightarrow 1$,  $1 \longrightarrow 3 \longrightarrow 1$,  $2 \longrightarrow 1 \longrightarrow 2$,  $2 \longrightarrow 3 \longrightarrow 2$,
$3 \longrightarrow 1 \longrightarrow 3$,  $3 \longrightarrow 2 \longrightarrow 3$
(b) $1 \longrightarrow 2 \longrightarrow 3$,  $1 \longrightarrow 3 \longrightarrow 2$,  $2 \longrightarrow 3 \longrightarrow 1$,  $2 \longrightarrow 1 \longrightarrow 3$,
$3 \longrightarrow 1 \longrightarrow 2$,  $3 \longrightarrow 2 \longrightarrow 1$

**6.2**  $d(a_1) = 3$,  $d(a_2) = 4$,  $d(a_3) = 4$,  $d(a_4) = 2$,  $d(a_5) = 3$,  $d(a_6) = 5$,
$d(a_7) = 3$

**6.3**

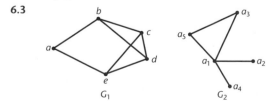

**6.4** (a) 12 walks  (b) 5 trails  (c) 12 paths

**6.5** (a) $L \rightarrow K \rightarrow D$,  $L \rightarrow C \rightarrow F \rightarrow D$

(b) $L \rightarrow C \rightarrow F \rightarrow SM \rightarrow F \rightarrow D \rightarrow K \rightarrow L$

(c) $L \rightarrow C \rightarrow F \rightarrow D \rightarrow K \rightarrow L$

**6.6** (a) $L \rightarrow B \rightarrow M$, $L \rightarrow G \rightarrow M$, $L \rightarrow E \rightarrow M$, $L \rightarrow E \rightarrow G \rightarrow M$

(b) $L \rightarrow E \rightarrow G \rightarrow L \rightarrow M$,  $L \rightarrow E \rightarrow G \rightarrow L \rightarrow B \rightarrow M$,
$L \rightarrow E \rightarrow G \rightarrow L \rightarrow E \rightarrow G \rightarrow M$

(c) $M \rightarrow B \rightarrow L \rightarrow E \rightarrow G \rightarrow L \rightarrow M$

(d) $E \rightarrow L \rightarrow B \rightarrow M \rightarrow G \rightarrow E$

**6.7** (a) $2 \rightarrow 3 \rightarrow 2 \rightarrow 3$,  $2 \rightarrow 3 \rightarrow 4 \rightarrow 3$,  $2 \rightarrow 7 \rightarrow 6 \rightarrow 3$

(b) $2 \rightarrow 3 \rightarrow 4 \rightarrow 5 \rightarrow 3 \rightarrow 6$

(c) $5 \rightarrow 4 \rightarrow 3 \rightarrow 6 \rightarrow 2 \rightarrow 3 \rightarrow 5$

(d) $1 \rightarrow 2 \rightarrow 3 \rightarrow 4 \rightarrow 5 \rightarrow 6 \rightarrow 7 \rightarrow 1$

**6.8** (a) $1 \rightarrow 2 \rightarrow 3 \rightarrow 4 \rightarrow 5$,  $1 \rightarrow 7 \rightarrow 2 \rightarrow 6 \rightarrow 5$,
$1 \rightarrow 2 \rightarrow 6 \rightarrow 3 \rightarrow 5$

(b) $1 \rightarrow 2 \rightarrow 6 \rightarrow 1$,  $1 \rightarrow 7 \rightarrow 6 \rightarrow 1$

**6.9**

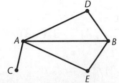

**6.10** (a) no  (b) yes  (c) no

**6.11** (a) yes  (b) yes

**6.12**

|     | LV   | N    | GC   | P    | F    |
|-----|------|------|------|------|------|
| B   | 2:50 | 2:30 | 7:20 | 8:00 | 5:40 |
| LV  | –    | 2:15 | 7:05 | 5:10 | 5:25 |
| N   |      | –    | 4:50 | 5:50 | 3:10 |
| GC  |      |      | –    | 2:55 | 1:40 |
| P   |      |      |      | –    | 2:45 |

**6.13** (a) Litton and Starbotton

(b) $L \rightarrow A \rightarrow S \rightarrow K$,  $L \rightarrow B \rightarrow S \rightarrow K$,  $L \rightarrow A \rightarrow L \rightarrow K$,
$L \rightarrow B \rightarrow L \rightarrow K$,  $L \rightarrow K \rightarrow L \rightarrow K$,  $L \rightarrow K \rightarrow S \rightarrow K$,
$L \rightarrow K \rightarrow A \rightarrow K$

(c) Choice of four cycles from Arncliffe or Kettlewell

**6.14**

$$A = \begin{bmatrix} 0 & 1 & 0 & 0 & 1 & 1 \\ 1 & 0 & 1 & 1 & 1 & 1 \\ 0 & 1 & 0 & 1 & 0 & 0 \\ 0 & 1 & 1 & 0 & 1 & 0 \\ 1 & 1 & 0 & 1 & 0 & 0 \\ 1 & 1 & 0 & 0 & 0 & 0 \end{bmatrix} \begin{matrix} 1 \\ 3 \\ 4 \\ 5 \\ 6 \\ 7 \end{matrix} \text{ vertices}$$

with column headings $1 \quad 3 \quad 4 \quad 5 \quad 6 \quad 7$

edges:  1      2      3      4      5      6      7      8      9

     (1,3)  (1,6)  (1,7)  (3,4)  (3,5)  (3,6)  (3,7)  (4,5)  (5,6)

$$B = \begin{bmatrix} 1 & 1 & 1 & 0 & 0 & 0 & 0 & 0 & 0 \\ 1 & 0 & 0 & 1 & 1 & 1 & 1 & 0 & 0 \\ 0 & 0 & 0 & 1 & 0 & 0 & 0 & 1 & 0 \\ 0 & 0 & 0 & 0 & 1 & 0 & 0 & 1 & 1 \\ 0 & 1 & 0 & 0 & 0 & 1 & 0 & 0 & 1 \\ 0 & 0 & 1 & 0 & 0 & 0 & 1 & 0 & 0 \end{bmatrix} \begin{matrix} 1 \\ 3 \\ 4 \\ 5 \\ 6 \\ 7 \end{matrix}$$

with column labels $1\ 2\ 3\ 4\ 5\ 6\ 7\ 8\ 9$ edges and row labels $1,3,4,5,6,7$ vertices

**6.15**

$$A = \begin{bmatrix} 0 & 1 & 0 & 0 & 0 & 0 & 1 \\ 0 & 0 & 1 & 0 & 0 & 1 & 0 \\ 0 & 0 & 0 & 1 & 1 & 0 & 0 \\ 0 & 0 & 0 & 0 & 1 & 0 & 0 \\ 0 & 0 & 0 & 0 & 0 & 0 & 0 \\ 1 & 0 & 1 & 0 & 1 & 0 & 0 \\ 0 & 1 & 0 & 0 & 0 & 1 & 0 \end{bmatrix} \begin{matrix} 1 \\ 2 \\ 3 \\ 4 \\ 5 \\ 6 \\ 7 \end{matrix}$$

with column labels $1\ 2\ 3\ 4\ 5\ 6\ 7$

edges:  1      2      3      4      5      6      7      8      9      10     11     12

  (1,2) (1,7) (2,3) (2,6) (3,4) (3,5) (4,5) (6,1) (6,3) (6,5) (7,2) (7,6)

$$B = \begin{bmatrix} 1 & 1 & 0 & 0 & 0 & 0 & 0 & -1 & 0 & 0 & 0 & 0 \\ -1 & 0 & 1 & 1 & 0 & 0 & 0 & 0 & 0 & 0 & -1 & 0 \\ 0 & 0 & -1 & 0 & 1 & 1 & 0 & 0 & -1 & 0 & 0 & 0 \\ 0 & 0 & 0 & 0 & -1 & 0 & 1 & 0 & 0 & 0 & 0 & 0 \\ 0 & 0 & 0 & 0 & 0 & -1 & -1 & 0 & 0 & -1 & 0 & 0 \\ 0 & 0 & 0 & -1 & 0 & 0 & 0 & 1 & 1 & 1 & 0 & -1 \\ 0 & -1 & 0 & 0 & 0 & 0 & 0 & 0 & 0 & 0 & 1 & 1 \end{bmatrix} \begin{matrix} 1 \\ 2 \\ 3 \\ 4 \\ 5 \\ 6 \\ 7 \end{matrix}$$

with column labels $1\ 2\ 3\ 4\ 5\ 6\ 7\ 8\ 9\ 10\ 11\ 12$

**6.16**

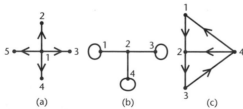

(a)              (b)         (c)

**6.18**   $T_1 \longrightarrow$ mathematics, $T_2 \longrightarrow$ physics, $T_3 \longrightarrow$ chemistry, $T_4 \longrightarrow$ biology

**6.20**   (a) no   (b) yes   (c) yes

**6.21**   (a) $mn$   (b) $C(n, 2)$

**6.22**   $1 \longrightarrow$ French, $2 \longrightarrow$ Spanish, $3 \longrightarrow$ German, $4 \longrightarrow$ Italian

**6.23**   Vauxhall in town 3

**6.24** $x_1 \longrightarrow y_2$, $x_2 \longrightarrow y_1$, $x_4 \longrightarrow y_3$, $x_5 \longrightarrow y_5$

**6.26** 669,240

**6.28** $v_2$ is parent of siblings $v_4$ and $v_5$, $v_6$ is descendant of $v_2$, etc.

**6.29** *AA, ABAA, ABB, ABABB, BB, BAA, BABAA, BABB, BABAB*

**6.31**

| | (a) | (b) | (c) | (d) | (e) |
|---|---|---|---|---|---|
| Number of circular leaves | 5 | 6 | 7 | 9 | 11 |
| Number of square leaves | 0 | 1 | 1 | 1 | 0 |
| Number of matches | 9 | 12 | 14 | 18 | 21 |
| Height of tree | 3 | 3 | 3 | 4 | 4 |

**6.33** 33 leaves, height 4

**6.34**

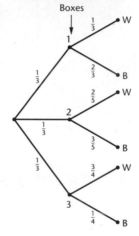

black ball, probability 91/180

**6.35** 12 possible outcomes

**6.37** Tree has 25 leaves, height 3; 7 leads needed

**6.38** 9

**6.39** 3

**6.42**

| $v_1$ | $v_2$ | $v_3$ | $v_4$ | $v_5$ | $v_6$ | $v_7$ | $v_8$ | $v_9$ | $v_{10}$ | $v_{11}$ | $v_{12}$ | $v_{13}$ | $v_{14}$ | $v_{15}$ |
|---|---|---|---|---|---|---|---|---|---|---|---|---|---|---|
| 1 | 1.1 | 1.2 | 1.1.1 | 1.1.2 | 1.1.2.1 | 1.2.1 | 1.2.2 | 2 | 2.1 | 2.2 | 2.2.1 | 2.2.2 | 2.2.2.1 | 2.2.2.2 |

**6.52** no

**6.53** $a \longrightarrow o \longrightarrow d \longrightarrow a \longrightarrow b \longrightarrow o \longrightarrow b \longrightarrow c \longrightarrow o \longrightarrow d \longrightarrow c \longrightarrow a \longrightarrow o \longrightarrow c$

**6.55** $1 \longrightarrow 7 \longrightarrow 6 \longrightarrow 1 \longrightarrow 5 \longrightarrow 6 \longrightarrow 4 \longrightarrow 3 \longrightarrow 4 \longrightarrow 5 \longrightarrow 3 \longrightarrow 2 \longrightarrow 1$

**6.56** $C \longrightarrow O \longrightarrow F \longrightarrow D \longrightarrow C \longrightarrow E \longrightarrow F \longrightarrow O \longrightarrow B \longrightarrow O \longrightarrow A \longrightarrow$
$O \longrightarrow C \longrightarrow B$

**6.57** no

**6.58** $C \longrightarrow A \longrightarrow C \longrightarrow B \longrightarrow A \longrightarrow B \longrightarrow D \longrightarrow A \longrightarrow D \longrightarrow C$

**6.59** $A \longrightarrow B \longrightarrow C \longrightarrow D \longrightarrow E \longrightarrow B \longrightarrow E \longrightarrow D \longrightarrow C \longrightarrow B \longrightarrow A \longrightarrow E \longrightarrow A$
$D \longrightarrow C \longrightarrow B \longrightarrow D \longrightarrow E \longrightarrow A \longrightarrow B \longrightarrow E$

**6.61** $v_1 \longrightarrow v_3 \longrightarrow v_2 \longrightarrow v_6 \longrightarrow v_5 \longrightarrow v_4 \longrightarrow v_1$

**6.62** (a) $n$ odd    (b) all $n$

**6.63** 3 different seating plans

**6.64** $s \longrightarrow v_1 \longrightarrow v_4$

**6.66** Barstow $\longrightarrow$ Las Vegas $\longrightarrow$ Page

**6.67** (a) $a_1 \longrightarrow a_6 \longrightarrow a_5 \longrightarrow a_4$

(b) $a_1 \longrightarrow a_2 \longrightarrow a_7 \longrightarrow a_1 \longrightarrow a_6 \longrightarrow a_2 \longrightarrow a_3 \longrightarrow a_6 \longrightarrow a_5 \longrightarrow a_3 \longrightarrow a_4$

(c) $a_1 \longrightarrow a_2 \longrightarrow a_7 \longrightarrow a_6 \longrightarrow a_3 \longrightarrow a_5 \longrightarrow a_4$

**6.68** $v_1 \longrightarrow v_4 \longrightarrow v_5 \longrightarrow v_6 \longrightarrow v_7$, length 41

**6.69** alternative shortest path is $v_1 \longrightarrow v_4 \longrightarrow v_5 \longrightarrow v_7$

**6.70** (a) $v_1 \longrightarrow v_2 \longrightarrow v_4 \longrightarrow v_6$   (b) $v_6 \longrightarrow v_4 \longrightarrow v_5 \longrightarrow v_1$

**6.72** $v_1 \longrightarrow v_2 \longrightarrow v_3 \longrightarrow v_6 \longrightarrow v_5 \longrightarrow v_4$ or $v_1 \longrightarrow v_2 \longrightarrow v_3 \longrightarrow v_4 \longrightarrow v_5 \longrightarrow v_6$

**6.73** $(1, 5), (1, 6), (2, 6), (3, 6), [(4, 5) \text{ or } (4, 6)]$

**6.74** $(v_1, v_2), (v_2, v_7), (v_6, v_7), (v_7, v_5), (v_5, v_4), (v_4, v_3);$
$(v_1, v_2), (v_1, v_6), (v_6, v_7), (v_7, v_5), (v_5, v_4), (v_4, v_3)$

**6.75** $A \longrightarrow B \longrightarrow D \longrightarrow E \longrightarrow C$

**6.76** $(A, D), (D, E), (E, C), (D, B)$

**6.77** English $\longrightarrow$ Martian $\longrightarrow$ Venusian, Martian $\longrightarrow$ Jovian

**6.78** $N, R, S$; 96 days

**6.79**

| Activity | A | B | C | D | E | F | G | H |
|---|---|---|---|---|---|---|---|---|
| Depends on | – | A | A | A | B | C | E | D, F, G |

**6.80**

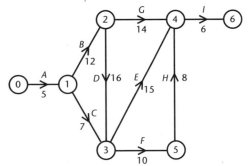

**6.81**

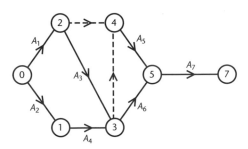

**6.82** $0 \longrightarrow 1 \longrightarrow 2 \longrightarrow 3 \longrightarrow 5 \longrightarrow 4 \longrightarrow 6$

**6.84** $0 \longrightarrow 1 \longrightarrow 2 \longrightarrow 4 \longrightarrow 5 \longrightarrow 6$

**6.86** $C, F, H, N, R, S$

**6.87** 10 days

**6.88** $A_2, A_5, A_6, A_{11}, A_{12}$

|                   | $A_4$ | $A_7$ |
|-------------------|-------|-------|
| Total float       | 1     | 13    |
| Free float        | 0     | 13    |
| Independent float | 0     | 12    |

**6.89** $x_1 = 15, \ x_2 = 18, \ x_3 = 7$

**6.90**

| from \ to | $v_1$ | $v_2$ | $v_3$ | $t$ |
|-----------|-------|-------|-------|-----|
| $s$       | 44    | 13    | –     | –   |
| $v_1$     | –     | 20    | 14    | 10  |
| $v_2$     | –     | –     | 33    | –   |
| $v_3$     | –     | –     | –     | 47  |

**6.91** 24

**6.92** $f(s, v_1) = 11, \ f(s, v_4) = 13, \ f(v_1, v_2) = 11,$
$f(v_2, t) = 11, \ f(v_3, t) = 13, \ f(v_4, v_3) = 13$

**6.93**

| from \ to | $a$ | $b$ | $c$ | $t$ |
|-----------|-----|-----|-----|-----|
| $s$       | 16  | 14  | 18  | –   |
| $a$       | –   | 0   | –   | 16  |
| $b$       | –   | –   | 2   | 12  |
| $c$       | –   | 0   | –   | 20  |

# Answers to Problems

## Chapter 1

**1.2**  4, 9

**1.3**  (a) No unique number corresponding to 6174
       (b) No unique number corresponding to 6174; endless loop

**1.4**  $0.05882352941176470588235\ldots$

**1.5**  $12 \times 483$, $42 \times 138$, $18 \times 297$, $27 \times 198$, $39 \times 186$, $28 \times 157$

**1.6**  3 ends at 1; 5 produces endless loop

**1.8**  First number is divisible by 11, second is not; $a = 1$

**1.9**  987652413

**1.14**  (a) 1, 7, 10, 13, 19, 23, 28, 31, 32, 44, 49, 68, 70, 79, 82, 86, 91, 94, 97, 100
        (b) No fixed end value, cycle endlessly through
$$4 \longrightarrow 16 \longrightarrow 37 \longrightarrow 58 \longrightarrow 89 \longrightarrow 145 \longrightarrow 42 \longrightarrow 20 \longrightarrow 4$$

**1.15**  Divisible by 16, not 32

**1.16**  Divisible by 125, not 625

**1.17**  A0 is $841 \times 1189\,\text{mm}$, A10 is $26 \times 37\,\text{mm}$
        B1 is $707 \times 1000\,\text{mm}$, B10 is $31 \times 44\,\text{mm}$

**1.19**  $\pm 1$, $\dfrac{1}{2} \pm \dfrac{i\sqrt{3}}{2}$, $-\dfrac{1}{2} \pm \dfrac{i\sqrt{3}}{2}$

## Interlude

**I.3**  $\displaystyle\sum_{i=a}^{b} i = n^3 + (n+1)^3$, $a = n^2 + 1$, $b = (n+1)^2$

## Chapter 2

**2.2**  $C$; $G$; $D$

**2.3**

| 1 | 2 | 3 | 4 | 6 | 7 | 8 | 9 |
|---|---|---|---|---|---|---|---|
| 00011 | 00101 | 00110 | 01001 | 01100 | 10001 | 10010 | 10100 |

**2.5**  (a) $13\frac{3}{8}$  (b) $27\frac{26}{64}$

**2.6**  (a) $(1.1011)_2$  (b) $(10.01111)_2$  (c) $(0.011001100110\ldots)_2$

**2.7**  (a) 29, 134
       (b)

| $-4$ | $-3$ | $-2$ | $-1$ | 1 | 2 | 3 | 4 |
|---|---|---|---|---|---|---|---|
| $(1100)_{-2}$ | $(1101)_{-2}$ | $(10)_{-2}$ | $(11)_{-2}$ | $(01)_{-2}$ | $(110)_{-2}$ | $(111)_{-2}$ | $(100)_{-2}$ |

       (c) $(1001)_{-2}$, $(111101)_{-2}$, $(1011111)_{-2}$

**2.11**  Divisible by 5; other factors are 2, 3

**2.12** Divisible by 2, 7 and 9; other factors are 3, 6

**2.13** $(4EC40)_{16}$

**2.15** (a)

| $a$ | 100 | 101 | 110 | 110 | 111 |
|---|---|---|---|---|---|
| $b$ | 000 | 000 | 001 | 000 | 000 |
| $c$ | 100 | 101 | 102 | 110 | 111 |
| | (9) | (10) | (11) | (12) | (13) |

(b)

| $a$ | 0001 | 0010 | | 1000 | | 1110 | | 1111 |
|---|---|---|---|---|---|---|---|---|
| $b$ | 0000 | 0001 | ... | 0110 | ... | 0001 | ... | 0000 |
| $c$ | 0001 | 0002 | | 0120 | | 1102 | | 1111 |
| | (1) | (2) | | (15) | | (38) | | (40) |

**2.17** $12; (a_1, a_2, \ldots, a_n) = (a_1, a_2, \ldots, a_{n-2}, (a_{n-1}, a_n))$

**2.24** 8

**2.26** $a = (a_n \ldots a_1)_{34}$ is divisible by:

3 (or 11) if and only if $a_1 + a_2 + \ldots + a_n \equiv 0 \pmod{3}$ (or $0 \pmod{11}$)

5 (or 7) if and only if $a_1 - a_2 + \ldots + (-1)^{n-1} a_n \equiv 0 \pmod{5}$ (or $0 \pmod{7}$)

$18557 = (16, 1, 27)_{34}$ is divisible by 7 and 11, not 3 or 5.

**2.27** $38 \pmod{143}$

**2.28** $6856 \pmod{8580}$

**2.29** (a) $17 \pmod{18}$   (b) $48 \pmod{50}$

**2.30** $1^{-1} = 1, \quad 3^{-1} = 23, \quad 5^{-1} = 7, \quad 7^{-1} = 5, \quad 9^{-1} = 19, \quad 11^{-1} = 31,$
$13^{-1} = 21, \quad 15^{-1} = 25, \quad 19^{-1} = 9, \quad 21^{-1} = 13, \quad 23^{-1} = 3,$
$25^{-1} = 15, \quad 27^{-1} = 29, \quad 29^{-1} = 27, \quad 31^{-1} = 11, \quad 33^{-1} = 33.$
$p = 3$

**2.32** (b) (i) $(0.1, 0, 1, 0, 0, 0, 1, 0, 0, 0, 1, \ldots)_2$

(ii) $(0.14, 28, 57, 14, 28, 57, \ldots)_{100}$

(iii) $(0.35, 6, 40)_{60}$

(c) $62/117$

(d) $r = x/y$:   $\left.\begin{array}{ll} \text{binary,} & y = 2^k \\ \text{base 60,} & y | 60^k \end{array}\right\} k = \text{positive integer}$

**2.34** (a) $x \equiv 2 \pmod{11}, y \equiv 9 \pmod{11}$

(c) $x \equiv 4 \pmod{7}, y \equiv 5 \pmod{7}, z \equiv 1 \pmod{7}$

## Chapter 3

**3.1** 62

**3.2** 326, 181, 492; 6, 530, 160

**3.3** 6720

**3.5** (a) 308,915,776   (b) 223,149,655   (c) 100,217,250

**3.7** (a) 462   (b) 1585

**3.8** $4^n$

**3.10** (a) $216C(35, 5)$

(b) $25[C(29,4) + 2C(34, 4) + C(19, 4) + C(14, 4)] + 130C(24, 4)$
$$+ 125C(9, 4)$$

**3.11** 826

**3.13** 1820; 4368

**3.14** $C(14, 7)$

**3.16** $3^n$

**3.18** $n2^{n-1}$

**3.20** $\dfrac{r}{2(r+1)(r+2)}; \quad \dfrac{1}{2}$

**3.26** 13

**3.27** $A = \{1, 2, 3, \ldots, 2N\}$, $B =$ subset of $N + 1$ integers

**3.28** (a) 4904 (b) 8297

**3.29** 871

**3.30** $26! - [2(24!) + 23!] + [2(21!) + 22!] - 19!$

**3.31** $d_3 = 2$, $d_4 = 9$, $d_5 = 44$, $d_6 = 265$

**3.32** (a) $265 \times 6!$ (b) $264 \times 6!$ (c) $191 \times 6!$

**3.33** 21; 994, 985, 976, 886, 877 and permutations

**3.34** $3^n - 3(2^n) + 3$

**3.35** (a) $b - a$ to $a$ against (b) $b - a$ to $a$ on

**3.36** (a) 0.03% (b) 0.0034%

**3.37** Probability $\longrightarrow 1/e$

**3.38** 0.052

**3.39** 0.15

**3.41** (a) $\dfrac{1}{2}$ (b) $\dfrac{3}{8}$ (c) $\dfrac{1}{2}$ (d) $\dfrac{5}{8}$

**3.42** $\dfrac{1}{16170}$

**3.44** (a) 0.177 (b) 0.154

**3.45** (a), (c)

|   | $n$ |   |   |   |   |   |   |   |
|---|---|---|---|---|---|---|---|---|
| $m$ | 1 | 2 | 3 | 4 | 5 | 6 | 7 | 8 |
| 1 | 1 |   |   |   |   |   |   |   |
| 2 | 1 | 1 |   |   |   |   |   |   |
| 3 | 1 | 3 | 1 |   |   |   |   |   |
| 4 | 1 | 7 | 6 | 1 |   |   |   |   |
| 5 | 1 | 15 | 25 | 10 | 1 |   |   |   |
| 6 | 1 | 31 | 90 | 65 | 15 | 1 |   |   |
| 7 | 1 | 63 | 301 | 350 | 140 | 21 | 1 |   |
| 8 | 1 | 127 | 966 | 1701 | 1050 | 266 | 28 | 1 |

(d) $\displaystyle\sum_{k=1}^{n} S(m,k)$

**3.48** 60

## Chapter 4

**4.2** $a = 0$: either no errors, or at least two errors

$a \neq 0$: at least one error

$4B8F$

**4.3**   The only errors $x_i \leftrightarrow x_j$ which are detected have $x_j - x_i \not\equiv 0 \pmod{5}$, $i$ and $j$ of opposite parity

**4.4**   $x_4 \leftrightarrow x_5$, birth year can alter by 50
$x_5 \leftrightarrow x_6$, birth year can alter by 45

**4.6**   (a) detected but not correctable
(b) detected and correctable

**4.7**   (b) transpositions undetected if $|x_i - x_{i+1}| = 5$, $i \leq 9$

**4.10**  176

**4.11**  Yes

**4.14**  00110011; transmitted codeword $= 11010010$; two errors

**4.23**  $r_3 - 4$, $r_5 - 2$, $r_9 - 7$

**4.24**  $\alpha^5 = \alpha^2 + 1$, $\alpha^6 = \alpha^3 + \alpha$, $\alpha^7 = \alpha^4 + \alpha^2$, $\alpha^8 = 1 + \alpha^2 + \alpha^3$,
$\alpha^9 = \alpha + \alpha^3 + \alpha^4$, $\alpha^{10} = 1 + \alpha^4$, ..., $\alpha^{18} = 1 + \alpha$

## Chapter 5

**5.2**   1.02987

**5.4**   £589.16; (a) £848.66   (b) 6.737%

**5.5**   733,361

**5.6**   (a) 7.05 New Pounds   (b) no

**5.7**   £575.40; £38,227.63

**5.10**  11 years
$$x_n = 2001.3(1.25)^n - \frac{4}{3}(2)^n$$
$x_{10} = 17,273$; approximately 16 years

**5.13**  (a) $7 - 2^n - 5n$   (b) $3 - 3n - \frac{1}{2}n^2$

**5.14**  $(6 - 2n)3^n + n(n - 1)3^{n-2} - 4^n$

**5.15**

| $n$ | 3 | 4 | 5 | 6 | 7 | 8 |
|---|---|---|---|---|---|---|
| $I_n$ | 3.5G | 5G | 5.75G | 5G | 2.38G | $-1.75G$ |

**5.16**  (a) $I_n \longrightarrow 5G$ as $n \longrightarrow \infty$
(b) Economy oscillates between large surplus and large deficit

**5.17**  $x_0 = -1$, $x_{10} = 184$
$$x_n = -1 + \frac{7}{2}n + \frac{3}{2}n^2$$

**5.18**  $s_n = \frac{1}{37}[700(70)^n + 40(-4)^n]$

**5.19**  $T_n = \frac{\sqrt{3}}{12}[(1 + \sqrt{3})^{n+1} - (1 - \sqrt{3})^{n+1}]$

**5.22**  $F_{kp}$ is divisible by each $F_k$, $p = 1, 2, 3, \ldots$

**5.24**  $0.30$; $\frac{1}{3}$

**5.25**  areas slightly overlap when $n$ even

**5.27**

| $i$ | 4 | 5 | 6 | 7 | 8 | 9 | 10 | 11 | 12 | 13 | 14 | 15 | 16 | 17 | 18 | 19 | 20 |
|---|---|---|---|---|---|---|---|---|---|---|---|---|---|---|---|---|---|
| $P_i$ | 2 | 3 | 4 | 5 | 7 | 9 | 12 | 16 | 21 | 28 | 37 | 49 | 65 | 86 | 114 | 151 | 200 |

**5.28** (a) 22%

(c) 18% after 20 weeks, 33% after 40 weeks, 45% after 60 weeks, 98% long-term

(d) 12%, 18.8%

(e) $p = 25$

## Chapter 6

**6.2**

| | $a_1$ | $a_2$ | $a_3$ | $a_4$ | $a_5$ | $a_6$ | $a_7$ |
|---|---|---|---|---|---|---|---|
| Indegree | 1 | 2 | 2 | 1 | 3 | 2 | 1 |
| Outdegree | 2 | 2 | 2 | 1 | 0 | 3 | 2 |

**6.3** $n(n-1)$

**6.5**

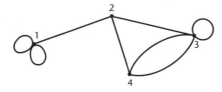

**6.6**

(a)

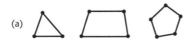

(b)

**6.18** 7

**6.19** $m + 1$ weighings needed

**6.20** BACK, DECK; 110, 1101

**6.21** DISCRETE

**6.26** $\dfrac{1}{2}(n-1)!$

**6.28** $000 \longrightarrow 001 \longrightarrow 011 \longrightarrow 010 \longrightarrow 110 \longrightarrow 111 \longrightarrow 101 \longrightarrow 100 \longrightarrow 000$

**6.30** (d) $G$ is disconnected if $A = \begin{bmatrix} X & 0 \\ 0 & Y \end{bmatrix}$

# Index

96388

LEARNING RESOURCE CENTRE
K COLLEGE
BROOK STREET
TONBRIDGE TN9 2PW

LEARNING RESOURCE CENTRE
K COLLEGE
BROOK STREET
TONBRIDGE TN9 2PW